Genomics in Aquaculture

Genomics in Aquaculture

Edited by

Simon MacKenzie
Institute of Aquaculture
University of Stirling
Stirling, United Kingdom

Sissel Jentoft
Department of Biosciences
Centre for Ecological and Evolutionary Synthesis (CEES)
University of Oslo, Oslo, Norway

Department of Natural Sciences
University of Agder, Kristiansand, Norway

ELSEVIER

AMSTERDAM • BOSTON • HEIDELBERG • LONDON
NEW YORK • OXFORD • PARIS • SAN DIEGO
SAN FRANCISCO • SINGAPORE • SYDNEY • TOKYO
Academic Press is an imprint of Elsevier

Academic Press is an imprint of Elsevier
125 London Wall, London EC2Y 5AS, United Kingdom
525 B Street, Suite 1800, San Diego, CA 92101-4495, United States
50 Hampshire Street, 5th Floor, Cambridge, MA 02139, United States
The Boulevard, Langford Lane, Kidlington, Oxford OX5 1GB, United Kingdom

Library of Congress Cataloging-in-Publication Data
A catalog record for this book is available from the Library of Congress

British Library Cataloguing-in-Publication Data
A catalogue record for this book is available from the British Library

ISBN: 978-0-12-801418-9

Publisher: Nikki Levy
Acquisition Editor: Patricia Osborn
Editorial Project Manager: Jaclyn Truesdell
Production Project Manager: Debbie Clark
Designer: Maria Inês Cruz

Typeset by Thomson Digital

Contents

List of Contributors

José Antonio Álvarez-Dios Department of Genetics, Faculty of Biology (CIBUS), University of Santiago de Compostela; Department of Applied Mathematics, Faculty of Mathematics, University of Santiago de Compostela, Santiago de Compostela, Spain

Peter Aleström Department of Basic Sciences and Aquatic Medicine, Norwegian University of Life Sciences, Oslo, Norway

Michaël Bekaert The Genetics and Reproduction Research Group, Institute of Aquaculture, University of Stirling, Stirling, Scotland, United Kingdom

Camille Berthelot European Molecular Biology Laboratory, European Bioinformatics Institute, Welcome Trust Genome Campus, Cambridge, United Kingdom; CEA-Genomic Institute, French National Sequencing Center, Evry Cedex, France

Julien Bobe INRA, Fish Physiology and Genomics, Rennes, France

Pierre Boudinot Molecular Virology and Immunology, INRA, Université Paris-Saclay, Jouy-en-Josas, France

Carmen Bouza Department of Genetics, Faculty of Veterinary, University of Santiago de Compostela, Lugo, Spain

Manuel Gonzalo Claros Department of Molecular Biology and Biochemistry, Faculty of Sciences, Campus de Teatinos s/n, University of Málaga; Plataforma Andaluza de Bioinformática, University of Málaga, Málaga, Spain

Xavier Cousin IFREMER, Laboratory of Adaptation and Adaptabilities of Animal and Systems (L3AS), Palavas and INRA, Laboratory of Fish Physiology and Genomics (LPGP), Rennes, France

Hugues Roest Crollius Ecole Normale Supérieure (ENS), ENS Biology Institute, IBENS, Paris, France

Jesús Fernández Departament of Animal Breeding, National Institute for Agricultural and Food Research and Technology, Madrid, Spain

Antonio Figueras Marine Research Institute (IIM), Immunology & Genomics group, Spanish National Research Council (CSIC), Vigo, Spain

Antonio Gómez-Tato Department of Genetics, Faculty of Biology (CIBUS), University of Santiago de Compostela; Department of Geometry and Topology, Faculty of Mathematics, University of Santiago de Compostela, Santiago de Compostela, Spain

Carine Genêt Animal Genetics and Integrative Biology, INRA, AgroParisTech, Université Paris-Saclay, Jouy-en-Josas, France

Yann Guiguen INRA, Fish Physiology and Genomics, Rennes, France

Miguel Hermida Department of Genetics, Faculty of Veterinary, University of Santiago de Compostela, Lugo, Spain

Olivier Jaillon CEA-Genomic Institute, French National Sequencing Center, Evry Cedex, France

Kjetill S. Jakobsen Department of Biosciences, Centre for Ecological and Evolutionary Synthesis (CEES), University of Oslo, Oslo, Norway

Sissel Jentoft Department of Biosciences, Centre for Ecological and Evolutionary Synthesis (CEES), University of Oslo, Oslo; Department of Natural Sciences, University of Agder, Kristiansand, Norway

Yanliang Jiang Beijing Key Laboratory of Fishery Biotechnology, Centre for Applied Aquatic Genomics, Chinese Academy of Fishery Sciences, Beijing, China

Xingkun Jin Department of Biosciences, Centre for Ecological and Evolutionary Synthesis (CEES), University of Oslo, Oslo, Norway

Yulin Jin The Fish Molecular Genetics and Biotechnology Laboratory, Aquatic Genomics Unit, School of Fisheries, Aquaculture and Aquatic Sciences, and Program of Cell and Molecular Biosciences, Auburn University, Auburn, AL, United States

Jiongtang Li Beijing Key Laboratory of Fishery Biotechnology, Centre for Applied Aquatic Genomics, Chinese Academy of Fishery Sciences, Beijing, China

Shikai Liu The Fish Molecular Genetics and Biotechnology Laboratory, Aquatic Genomics Unit, School of Fisheries, Aquaculture and Aquatic Sciences, and Program of Cell and Molecular Biosciences, Auburn University, Auburn, AL, United States

Zhanjiang Liu The Fish Molecular Genetics and Biotechnology Laboratory, Aquatic Genomics Unit, School of Fisheries, Aquaculture and Aquatic Sciences, and Program of Cell and Molecular Biosciences, Auburn University, Auburn, AL, United States

Simon MacKenzie Institute of Aquaculture, University of Stirling, Stirling, United Kingdom

Manuel Manchado IFAPA Center El Toruño, IFAPA, Consejeria de Agricultura y Pesca, El Puerto de Santa María (Cádiz), Spain

Lucie Marandel Nutrition, metabolism and aquaculture (NUMEA), INRA, Univ Pau & Pays Adour, Saint Pée sur Nivelle, France

Paulino Martínez Department of Genetics, Faculty of Veterinary, University of Santiago de Compostela, Lugo, Spain

Brendan J. McAndrew The Genetics and Reproduction Research Group, Institute of Aquaculture, University of Stirling, Stirling, Scotland, United Kingdom

Rebeca Moreira Sanmartín Marine Research Institute (IIM), Immunology & Genomics group, Spanish National Research Council (CSIC), Vigo, Spain

Alexander J. Nederbragt Department of Biosciences, Centre for Ecological and Evolutionary Synthesis (CEES), University of Oslo; Department of Informatics, Biomedical Informatics Research Group, University of Oslo, Oslo, Norway

Beatriz Novoa Institute of Marine Research (IIM), Spanish National Research Council (CSIC), Vigo, Spain

Stéphane Panserat Nutrition, metabolism and aquaculture (NUMEA), INRA, Univ Pau & Pays Adour, Saint Pée sur Nivelle, France

Belén G. Pardo Department of Genetics, Faculty of Veterinary, University of Santiago de Compostela, Lugo, Spain

David J. Penman The Genetics and Reproduction Research Group, Institute of Aquaculture, University of Stirling, Stirling, Scotland, United Kingdom

Patricia Pereiro Institute of Marine Research (IIM), Spanish National Research Council (CSIC), Vigo, Spain

Francesc Piferrer Institute of Marine Sciences, Spanish National Research Council (CSIC), Barcelona, Spain

Josep V. Planas Department of Physiology and Immunology, School of Biology, University of Barcelona and Institute of Biomedicine (IBUB), Barcelona, Spain

Edwige Quillet Animal Genetics and Integrative Biology, INRA, AgroParisTech, Université Paris-Saclay, Jouy-en-Josas, France

Laureana Rebordinos Laboratory of Genetics, Faculty of Environmental and Sea Sciences, University of Cadiz, Polígono del Río San Pedro, Puerto Real, Cádiz, Spain

Laia Ribas Institute of Marine Sciences, Spanish National Research Council (CSIC), Barcelona, Spain

Matthew L. Rise Department of Ocean Sciences, Memorial University of Newfoundland, St. John's, NL, Canada

Steven Roberts School of Aquatic and Fishery Sciences University of Washington, Seattle, WA, United Sates

Diego Robledo Department of Genetics, Faculty of Biology (CIBUS), University of Santiago de Compostela, Santiago de Compostela, Spain

Silvia T. Rodríguez-Ramilo Departament of Animal Breeding, National Institute for Agricultural and Food Research and Technology, Madrid, Spain; GenPhySE, National Institute for Agricultural Research, Castanet-Tolosan, France

Juan A. Rubiolo Department of Genetics, Faculty of Veterinary, University of Santiago de Compostela, Lugo, Spain

Bastiaan Star Department of Biosciences, Centre for Ecological and Evolutionary Synthesis (CEES), University of Oslo, Oslo, Norway

Xiaowen Sun Heilongjiang Fishery Research Institute, Chinese Academy of Fisheries Sciences, Harbin, China

Xoana Taboada Department of Genetics, Faculty of Biology (CIBUS), University of Santiago de Compostela, Santiago de Compostela, Spain

Suxu Tan The Fish Molecular Genetics and Biotechnology Laboratory, Aquatic Genomics Unit, School of Fisheries, Aquaculture and Aquatic Sciences, and Program of Cell and Molecular Biosciences, Auburn University, Auburn, AL, United States

Ole Kristian Tørresen Department of Biosciences, Centre for Ecological and Evolutionary Synthesis (CEES), University of Oslo, Oslo, Norway

Ana Viñas *Department of Genetics, Faculty of Veterinary, University of Santiago de Compostela, Lugo, Spain

Jean-Nicolas Volff Institute of Functional Genomics of Lyon (IGFL), Ecole Normale Supérieure de Lyon, CNRS, University of Lyon I, Lyon, France

Stefanie Wehner The Genetics and Reproduction Research Group, Institute of Aquaculture, University of Stirling, Stirling, Scotland, United Kingdom

Hanne C. Winther-Larsen Department of Pharmaceutical Biosciences, School of Pharmacy, University of Oslo, Oslo, Norway

Jian Xu Beijing Key Laboratory of Fishery Biotechnology, Centre for Applied Aquatic Genomics, Chinese Academy of Fishery Sciences, Beijing, China

Peng Xu Beijing Key Laboratory of Fishery Biotechnology, Centre for Applied Aquatic Genomics, Chinese Academy of Fishery Sciences, Beijing; State Key Laboratory of Marine Environmental Science, College of Ocean & Earth Science, Xiamen University, Xiamen, China

Yujia Yang The Fish Molecular Genetics and Biotechnology Laboratory, Aquatic Genomics Unit, School of Fisheries, Aquaculture and Aquatic Sciences, and Program of Cell and Molecular Biosciences, Auburn University, Auburn, AL, United States

Zihao Yuan The Fish Molecular Genetics and Biotechnology Laboratory, Aquatic Genomics Unit, School of Fisheries, Aquaculture and Aquatic Sciences, and Program of Cell and Molecular Biosciences, Auburn University, Auburn, AL, United States

The new era of genome sequencing using high-throughput sequencing technology: generation of the first version of the Atlantic cod genome

1

Ole Kristian Tørresen, Bastiaan Star*, Sissel Jentoft*,**, Kjetill S. Jakobsen*, Alexander J. Nederbragt*,†*

*Department of Biosciences, Centre for Ecological and Evolutionary Synthesis (CEES), University of Oslo, Oslo, Norway; **Department of Natural Sciences, University of Agder, Kristiansand, Norway; †Department of Informatics, Biomedical Informatics Research Group, University of Oslo, Oslo, Norway

Introduction

The first vertebrate genomes to be completed—for example, the human genome (Lander et al., 2001; Venter et al., 2001) and the genome of the teleost, fugu (*Takifugu rupriens*) (Aparicio et al., 2002)—were sequenced using the only available technology at that time, known as Sanger sequencing. This sequencing technology is expensive in terms of cost per base, and due to the size of vertebrate genomes such projects are labor intensive, necessitating large teams, and substantial funding. Still, most of the model organisms were first sequenced using this method. In 2005, the first high-throughput sequencing (HTS) technology, that is, massively parallel pyrosequencing by Roche/454 (Margulies et al., 2005), was introduced, followed by Solexa/Illumina (Bentley et al., 2008) in 2006. These technologies generate millions to billions of bases in a single run, and therefore reduce labor intensity and cost per base by several orders of magnitude. However, the read lengths produced by these technologies when they first became available (25–35 bp for Illumina/Solexa (Illumina hereafter) and 100 bp for Roche/454 (454) were significantly shorter than for Sanger sequencing (600–1000 bp). Due to fast technological improvements, the read lengths from the HTS instruments soon became substantially longer, enabling sequencing and *de novo assembly* (see Glossary) of large genomes using these technologies. As an effect, more groups embarked on these kinds of projects. In 2008, however, such applications had not yet been demonstrated on any vertebrate species, and de novo assembly programs capable of handling the vast amounts of required short read data were not yet available. As a group working on evolutionary and ecological questions in fish, we saw the ongoing technological developments as an opportunity to sequence and assemble a teleost genome using a pure HTS approach, at low cost and with a small research team.

Genomics in Aquaculture. http://dx.doi.org/10.1016/B978-0-12-801418-9.00001-9

In 2008, we were funded by the Research Council of Norway (RCN) to sequence the genome of Atlantic cod (*Gadus morhua* L.) with the 454 sequencing technology. Atlantic cod was chosen because of its impact as an economically important fishery species, its considerable ecological and evolutionary scientific interest and its potential use in aquaculture. Furthermore, Atlantic cod has a relatively moderate genome size (estimated to 830 million base pairs, Mbp)—making it a good target for the first teleost genome sequenced with a pure HTS approach. In this chapter, we will describe the sequencing, assembly, and *annotation* (see Glossary) of the Atlantic cod genome, the specific findings related to the genome and future perspectives.

The first version of the Atlantic cod genome

The sequencing of a wild-caught cod specimen

The Atlantic cod-sequencing project was carried out with DNA isolated from a single specimen. The specimen was wild-caught, male, approximately 8 years old, and originated from the North East Arctic cod (NEAC) population. This population is the most abundant in the world (Ottersen et al., 2014; Kjesbu et al., 2014) and is of most significance for Norwegian fisheries, with a value of the catch close to 4 billion Norwegian Kroner (NOK) per year. The population exhibits seasonal migration from the Barents Sea to its spawning ground along the Norwegian coast, the main one being the Lofoten area, where there has been a traditional seasonal fishery for centuries.

DNA in eukaryotic cells is packaged in chromosomes, containing continuous, linear stretches often of millions of base pairs. In Atlantic cod, there are altogether 23 chromosomes comprising approximately 830 Mbp (Star et al., 2011). HTS technologies available at the time the project started are most efficient in sequencing short stretches of DNA, usually substantially shorter than 1000 bp. An important step in genome sequencing therefore consists of the physical fragmentation of these long chromosomes, resulting in a set of small, randomly broken fragments of DNA. Using standard molecular biology approaches, these small fragments are then placed between specific DNA molecules (adapters), which are recognized by the sequencing machines. This process is called *library preparation*, and forms the start of any sequencing project. The sequencing of a single fragment from these libraries results in a DNA sequence, which is called a *read* (Fig. 1.1). Overall, the procedure of physically breaking the chromosomes and sequencing the resulting pieces is called "shotgun" sequencing, a term which reflects the random fragmentation of the DNA sequences.

At the start of the sequencing project in 2009, two HTS technologies were available; Roche/454 and Illumina. Because of the longer read length of the 454 technology and a wider range of suitable software utilizing those data at that time, we chose to use this technology for the first version of the Atlantic cod genome. Unraveling the genome of a species for the first time requires a large amount of data and often a range of different sequencing libraries. During the first 10 months of 2009, we generated the required amount of data using libraries made with DNA from the single Atlantic cod (NEAC) specimen. A designated team of scientists kept the sequencing machines

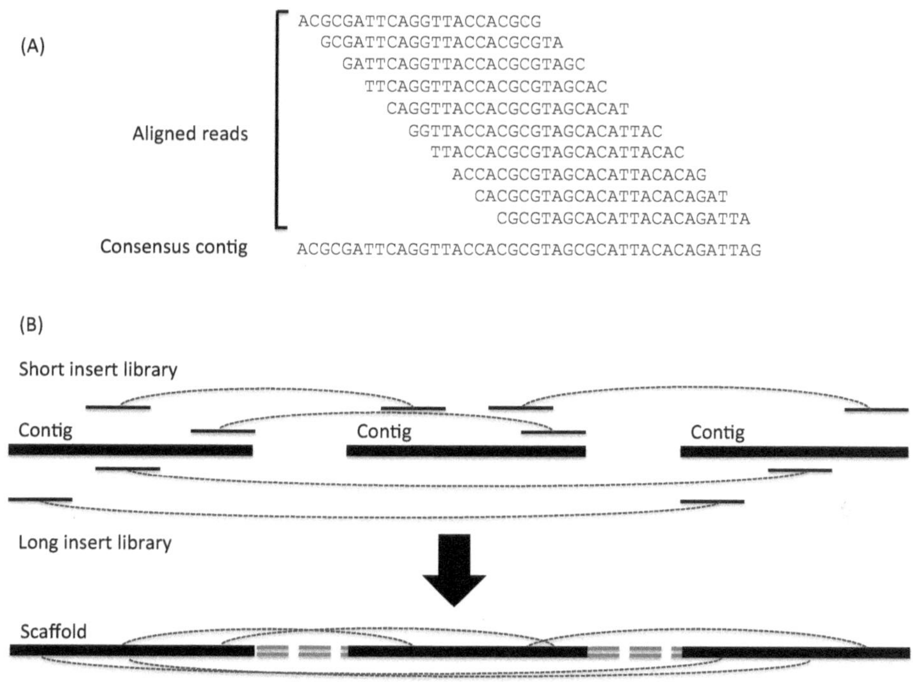

Figure 1.1 The creation of contigs and linking of contigs into scaffolds. (A) The assembler finds reads with similar sequence and creates a contiguous consensus sequence from them, which is called a contig. (B) The mapping of paired read information from libraries of different insert size allows contigs to be ordered and oriented into scaffolds, with gaps between the contigs represented by the *gray dotted lines*.

running as efficiently as possible. The final dataset consisted of approximately 64 million reads and 33 billion base pairs (Gbp). Considering that the Atlantic cod genome is 830 Mbp in size, the size of this dataset means that every nucleotide in the genome is sequenced—on average—about 40 times, providing what is called 40× *coverage* (also called oversampling). Most of the data were generated using the two 454 machines at the Norwegian Sequencing Centre, hosted by the Centre for Ecological and Evolutionary Synthesis (CEES) at the University of Oslo.

The challenges of assembling large, outbred genomes

An important consequence of the random fragmentation of DNA for library preparation is that this process erases all long-range information of where the particular fragments originally belong. Therefore, complex mathematical algorithms have been designed to computationally combine these fragments in a single, coherent structure, which is called an *assembly*. An assembly can be defined as "a hierarchical data structure that maps the sequence data to a putative reconstruction of the target [genome]" (Miller et al., 2010).

There are several algorithms available for assembling sequencing reads, all of which are based on the so-called graph theory in informatics (Compeau et al., 2011; Myers, 2005). The assembly programs Celera Assembler (Miller et al., 2008) and Newbler have been, respectively, specifically adapted to and developed for assembling 454 datasets. A commonly used algorithm specifically designed for reads from the Illumina platform creates what is called a de Bruijn graph (Miller et al., 2010; Pevzner et al., 2001). This algorithm is implemented in various assembly programs such as SOAPdenovo (Li et al., 2010b, Luo et al., 2012), Velvet (Zerbino and Birney, 2008), and ALLPATHS-LG (Gnerre et al., 2011).

All assembly programs use heuristics (rules-of-thumb) to determine outcomes for problems that are not easily solved mathematically or statistically, or are difficult to represent due to assembly-specific constraints. For example, Celera Assembler by default requires two correctly placed pairs of reads as evidence to link together two contigs into a scaffold. In contrast, with SOAPdenovo assembler, five pairs are required to link together contigs. Because such rules introduce arbitrary choices, using the same sequencing data in two different assembly programs will almost certainly lead to alternative assembly outcomes (Earl et al., 2011; Bradnam et al., 2013).

Celera Assembler and Newbler use an algorithm called Overlap-Layout-Consensus (OLC) (Miller et al., 2010), that works by comparing all reads to each other and finding overlaps (similar sequences) between them. From these overlaps, a so-called overlap-graph is created. Contigs are created from the graph based on a consensus of the overlapping reads (Fig. 1.1a). After making a set of contigs, the assembler uses the information from *paired reads* to order and orient the contigs into scaffolds (Fig. 1.1b). Thus, scaffolds consist of ordered and oriented contigs that are separated by gaps. The size of the gap itself is estimated from the paired reads information. While an optimal assembly consists of one continuous sequence per chromosome, preferably without gaps, in reality, most assemblies of large genomes end up with several thousand scaffolds (including gaps). In recent years, assembly programs have substantially improved, shortening the time for the algorithm to complete computation, while simultaneously decreasing the memory requirement. New sequencing technologies (see Chapter 3) that create much longer reads have also enabled more complete genomes that can be assembled in shorter time.

Multiple complexities make the assembly process challenging. First, genome size and the amount of repetitive sequences in the genome are highly positively correlated. A genome needs to be sequenced to a minimum coverage, depending on the sequencing technology and how complete the assembly should be, and therefore larger genomes need more sequencing data than smaller. To analyze the larger amount of sequencing data, more computational resources need to be utilized and/or special approaches need to be used. The Norway spruce genome, for example, is extremely large, comprising about 20 Gbp, and a 2 TB RAM server (2048 GB memory, 500 times that of a common laptop computer) had to be used in the most computationally intensive part of its genome assembly (Nystedt et al., 2013). Such computational resources are expensive and outside the budgets of many research groups.

Second, the repetitive DNA sequences in genomes, such as transposons (mobile genetic elements), longer repeats (e.g., LINEs and SINEs, long interspersed nuclear elements and short interspersed nuclear elements, respectively) and short tandem repeats (STRs, pieces of DNA sequences entirely consisting of the same two, three, or four bases that are repeated many times), complicate the assembly process substantially. Such repetitive sequences can be highly similar or even completely identical and may occur in tandem (next to each other on the same chromosome) or spread throughout the entire genome (interspersed). A considerable proportion of a vertebrate genome can consist of repetitive sequences (2–5% of tandem repeats and 7–40% of repeats overall in some teleost genomes; Reichwald et al., 2009). Such similar DNA sequences that nonetheless originate from different locations within the genome will yield near identical sequencing reads. Therefore, those sequencing reads that originate from repetitive sequences longer than the read length of the sequencing technology cannot be uniquely placed within the assembly, and it is difficult or even impossible to reliably find their original location without additional information. The presence of these sequences induces gaps in the assembly, whereby sequences that can be reliably assembled (contigs) are interrupted by gaps often representing repetitive sequence. A genome with higher proportion of repeats is generally more difficult to assemble than a less repetitive one.

The challenges of assembling repetitive stretches of DNA can be partially addressed by preparing sequencing library types of different DNA fragment sizes. For instance, libraries can be generated starting from DNA fragments ranging from relatively short sizes of 300 bp to longer sizes of up to 20 thousand bp (kbp) in length. From these longer DNA fragments only the ends can be sequenced resulting in two separate DNA sequences (paired reads). The sum of the distance between the two reads and the read lengths is called "insert size." Importantly, the fact that these two sequences came from a single, larger molecule is crucial for the algorithms that are used to assemble a genome. By incorporating such information, unresolvable sections of the genome can be crossed, and stretches of continuously assembled DNA without any missing sequences can be oriented relative toward each other and linked in a larger data structure (a so-called *scaffold*), see Fig. 1.1b. In some cases, the repetitive sequence can be placed in the gap based on paired reads information, thereby closing the gap.

Another challenge is related to the biology of diploid organisms and the fact that these organisms inherit a copy of their genome from each parent. Each section of the genome is thus represented by two copies, one paternal and one maternal. These two copies are predominantly similar, yet can also contain divergent sequences. If sections of the paternal and maternal copy are divergent, it becomes more difficult to assemble such sections since most assembly programs presume a haploid genome. Assembly algorithms often can incorporate variation consisting of small *indels* (single-base or longer insertions and deletions) and *SNPs* (single nucleotide polymorphisms), but larger indels or *inversions* are harder to assemble reliably. Essentially, if the two parental copies differ significantly at a specific location, a so-called heterozygotic site, the assembly algorithm has two possible choices and it is not trivial to choose between them. Often such regions induce gaps in the assembly. Problems of this kind are even more prevalent in species that are polyploid, having more than two copies of the genome. For such species (such as many plant species and some teleosts such as salmonids), sequences may originate from four or more different copies that are more or less similar. The genomes of such species are notoriously difficult to assemble (Schatz et al., 2012). In contrast to genome size and repetitive content, it is possible to overcome this problem of heterozygous sections of the genome. For instance, genome-sequencing projects often use different methods to reduce or eliminate the number of such sections by generating *double-haploid* individuals [having one "genomic" parent, e.g., generated both for the zebrafish (*Danio rerio*) and Atlantic salmon (*Salmo salar*) genome sequencing projects; Howe et al., 2013; Lien et al., 2016] or create inbred strains with reduced levels of heterozygosity. The platyfish (*Xiphophorus maculatus*) genome sequencing project used a specimen from a family that was inbred for 104 generations using brother-sister crosses (Schartl et al., 2013). Such approaches are not always applicable for every species. The creation of double-haploid individuals is difficult and inbreeding requires substantial time and effort, and might not be successful at all. For instance, early attempts for obtaining a double-haploid specimen for the Atlantic cod genome project failed, though recent efforts have been more successful (Otterå et al., 2011).

Finally, all sequencing technologies provide data with some uncertainty, that is, the sequencing reads contain a certain percentage of erroneous bases. The different sequencing technologies each have their own specific error profiles. For example, indels are a specific characteristic of the 454 technology (the exact number of consecutive identical bases (homopolymers) cannot always be resolved), whereas substitutions (single-base changes) are more typical for the Illumina technology (DePristo et al., 2011; Balzer et al., 2010). Assembly algorithms need to cope with these errors, and assembly programs often include methods that aim to correct these errors before assembly as an integral part of the running of the program. Alternatively, separate software tools are available to perform error-correction, resulting in an error-corrected dataset that is subsequently used by an assembly program (e.g., Kelley et al., 2010).

Assembling the first version of the cod genome

The Atlantic cod genome project presented two alternative assemblies (Star et al., 2011), one obtained using the Celera Assembler (Miller et al., 2008), and another using the Newbler assembly program from Roche/454. In the early phases of the Atlantic cod-sequencing project, both programs struggled to cope with the large amount of data that had been generated. The bioinformaticians in the project worked with the software developers of the respective programs so that, in the end, assemblies could be obtained from both. Two assemblies were presented and made available because each had specific strengths and weaknesses. The Newbler assembly program produced longer scaffolds than the Celera Assembler, whereas the Celera Assembler produced longer contigs (Table 1.1).

For Atlantic cod, we prepared multiple large-insert libraries from DNA from the same specimen (Table 1.2) in order to address the problems caused by repetitive sequences.

Compared to other teleost genomes, the cod genome assemblies are relatively fragmented. For example, the Newbler assembly has 35% of its sequence in gaps and it has short contigs. Other teleost genome assemblies available at publication time of the

Table 1.1 Assembly statistics

	Number	Bases (Mbp)	N50 (bp)[a]	L50 (n)[b]	Max (bp)[c]
Newbler					
Contigs[d]	284,239	536	2,778	50,237	76,504
Scaffolds	6,467	611	687,709	218	4,999,318
Entire assembly[e]	157,887	753	459,495	344	4,999,318
Celera					
Contigs[d]	135,024	555	7,128	19,938	117,463
Scaffolds	3,832	608	488,312	373	2,810,583
Entire assembly[e]	17,039	629	469,840	395	2,810,583

[a] Length of contig or scaffold defined such that half the assembly consists of contigs or scaffold of that size or larger.
[b] These numbers of sequences are of length N50 or longer.
[c] Maximum length.
[d] Contigs longer than 500 bp.
[e] Scaffolds and unplaced contigs.
Source: Modified from Star et al. (2011).

Table 1.2 An overview of the different libraries sequenced, and the number of reads from each

Type of library	Libraries (n)	Unpaired reads[a] (millions)	Paired reads (millions)	Pair distance (bp)[b]	Coverage[c]
Unpaired	4	46.7			
1.0–2.3 kbp paired	4	8.8	7.0	1090–1791	12
3.0 kbp paired	2	2.1	3.6	2680	11.6
		1.7	2.7	2676	8.7
8.0 kbp paired	2	1.5	2.6	7146	22.2
		1.8	3.2	7686	29.3
20.0 kbp paired	2	0.7	0.9	21052	23.3
		0.2	0.2	21081	6.1

[a] The methodology for generating such paired fragments with the 454 technology produces libraries containing a mixture of unpaired and paired reads.
[b] Distance reported by the assembly program after assembly.
[c] Coverage is defined as the number of reads times the average insert size, divided by total genome size (so-called clone coverage). This number represents the coverage when the pair distance is taken as the read length.
Source: Modified from supplementary Table 1.1 of Star et al. (2011).

cod genome had 10–15% gap bases, for example, in fugu and tetraodon (*Tetraodon nigroviridis*) (Aparicio et al., 2002; Jaillon et al., 2004), with contig *N50* (see Glossary) 16 and 10 times longer, respectively, compared to cod. There are several causes for the higher fragmentation in the cod genome assemblies—a high level of heterozygosity, and a high abundance of short tandem repeats—issues complicating the assembly process (Star et al., 2011).

Annotating the first version of the cod genome

A genome assembly provides an estimate of the nucleotide order of an organism, but does not directly include the placement of different features of the genome. The location of these features, for instance where the different genes are located in the genome, needs to be determined by a process called genome *annotation*. This is a complex procedure, combining results from algorithms that *ab initio* predict the coding sequence of genes, with mapping known genes other genomes (e.g., human, mouse, or stickleback) to the assembly. The comparison with other genomes is largely driven by a fundamental assumption: important features in the genome are often evolutionarily conserved. These features can be recognized by sequence homology as genes from different organisms are often highly conserved, so that the amino acid composition is highly similar to those from known organisms. Moreover, genes need to be translated into functional proteins. This translation process requires that the nucleic acid composition of the genome conforms to specific characteristics following the universal codon structure of the genetic code. These patterns can be bioinformatically recognized. An annotated genome assembly is often presented in a so-called genome browser, such as the one provided by Ensembl (Flicek et al., 2013), a joint scientific project between the European Bioinformatics Institute and the Wellcome Trust Sanger Institute. A genome browser is a graphical interface for viewing and accessing genomic data for a species.

The Newbler assembly was chosen for annotation. Ensembl used their in-house developed automated pipeline to predict the locations of genes and annotate them. Due to the fragmented nature of the first Atlantic cod assembly, information from other annotated fish genomes, in particular stickleback (*Gasterosteus aculeatus*), was used to complete gene sequences in an approach that partially reordered and restructured the assembly. A total of 22,154 genes were annotated (Star et al., 2011). This number of genes is comparable to that of other teleost genomes (Table 1.3).

Table 1.3 Annotation statistics

Annotation	Number	Bases (Mbp)	Assembly (%)
Protein encoding genes	20,095	28.1[a]	5.27
Pseudogenes	518	0.38	0.07
rRNA	590	0.06	0.01
miRNA	414	0.03	0.01
snoRNA	382	0.05	0.01
snRNA	115	0.01	0.003
miscRNA	40	0.01	0.002
Transposable elements	614,494	137	19.74
Simple and low complexity	731,280	57	8.21

[a] excluding introns

Source: Modified from Star et al. (2011)

Synteny with other teleost genomes

The location of genes on genomes is not stable of over large evolutionary time periods. Despite this variation, the same types of features of a genome tend to cluster together on the same chromosomes between different species, a phenomenon called synteny. For example, if a set of genes is found close together in one species, the orthologs (homologs separated by a speciation event) of these genes are more likely to cluster together in another species than not. Synteny reflects the shared ancestry of living organisms. There is a high degree of synteny among vertebrates (Ohno, 1973; Masahiro Kasahara et al., 2007; Jaillon et al., 2004), such that most chromosomes, or large parts of chromosomes, can be identified by the genes located on them (Kikuta et al., 2007). Importantly, in species that are closely related, we expect to find a high degree of synteny, as their ancestor is evolutionary more recently compared to species that are distantly related. Patterns of synteny among chromosomes were fairly well known for several teleosts, and we could use this to assess the quality of the cod genome assembly by interspecies comparisons. For this, we utilized a linkage map for Atlantic cod consisting of 924 SNPs ordered in 23 so-called linkage groups (this number of linkage groups mirrors the number (23) of chromosomes in Atlantic cod) (Hubert et al., 2010). About half the Newbler assembly (332 Mbp) could be placed onto this linkage map, which means that we could infer which parts of this assembly were located on specific chromosomes. We subsequently compared the location of genes on the cod genome assembly to the location of orthologous genes on the genome assemblies of five other teleost fishes, namely, stickleback, tetraodon, fugu, medaka (*Oryzias latipes*), and zebrafish. We expected, based on earlier work on synteny in teleosts (Masahiro Kasahara et al., 2007), that a substantial number of genes mapped to the same linkage group in Atlantic cod, should co-occur on the same chromosome or linkage group in the other genomes. The results are shown in Fig. 1.2, showing that the highest level of synteny is with stickleback.

Ongoing efforts to improve the Atlantic cod genome assembly

The impact of combining different sequencing technologies

After the initial release of the sequencing machines by 454 in 2005 and Illumina in 2006, both platforms have steadily improved the read lengths and throughput, leading to increasingly affordable sequencing. Today, the reads from Illumina are long enough to be used for whole genome sequencing and subsequent assembly on their own (Gnerre et al., 2011; Li et al., 2010b). The now defunct 454 sequencing technology still had longer reads at its end of life (4–10 times depending on which versions of the different technologies that are being compared), thereby spanning longer repeats than the Illumina reads. Both Illumina reads and 454 reads have different systematic errors (Balzer et al., 2010; DePristo et al., 2011; Loman et al., 2012; Nakamura et al., 2011; Ross et al., 2013) and each technology has its weaknesses and strengths. By using a strategy that combines the longer 454 reads with the more affordable Illumina reads,

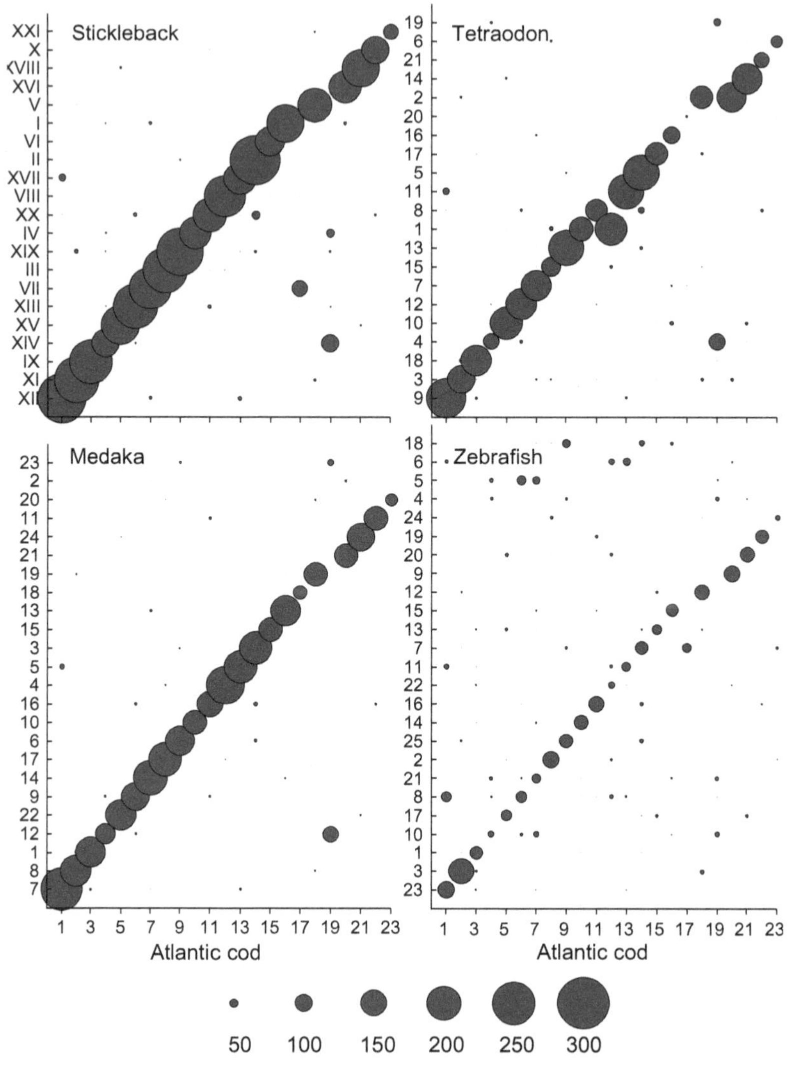

Figure 1.2 Synteny between Atlantic cod and selected teleosts. Sphere sizes indicate the number of syntenic genes, that is, the co-occurrence of orthologous genes, in the 23 Atlantic cod linkage groups (Hubert et al., 2010) (x-axis) with the chromosomes of four teleosts (y-axis). Apart from a strong pattern of conserved synteny among the teleost chromosomes, a lineage-specific chromosomal rearrangement in Atlantic cod compared to the other teleosts appears—indicated by a deviation from a linear order of the spheres, in particular with stickleback chromosome XIV, tetraodon chromosome 4 and medaka chromosome 12. Orthology is here defined as a minimum of 50% sequence identity over 50% of the alignment. Source: Modified from Star et al. (2011).

the strength of each platform can be used to an advantage, whereas their weaknesses can be negated (DiGuistini et al., 2009; Dalloul et al., 2010).

There are now several new long read sequencing technologies on the market—among those Oxford Nanopore (Eisenstein, 2012) and the single-molecule sequencing platform from Pacific Biosciences (Eid et al., 2009) are the most promising. While raw PacBio reads are only about 85–90% accurate, the technology does not have any systematic errors (Roberts et al., 2013; Ross et al., 2013), which means that the consensus accuracy with enough oversampling is up to 99.999% (Chin et al., 2013). This technology has been used in a multitude of genome sequencing projects, giving chromosome-length contigs without gaps also for large mammalian genomes (Berlin et al., 2015; Pendleton et al., 2015). The PacBio technology is steadily improving, with mean length about 10–15 kbp, a N50 read length of 14,000 on the current chemistry (P6-C4). As a consequence, PacBio reads span across much longer repeated sequence than any other technology—and at a steadily reduced cost.

Software developments for de novo assembly

Improvements in genome assembly are not only driven by new and improved sequencing technologies, but also by advances in the software used in assembly. While the Illumina sequencing platform was released in 2006 (Bentley et al., 2008), it was not until the end of 2009 that the first genome assemblies of vertebrates based on that sequencing technology were released (Li et al., 2010a,b). Indeed, the Atlantic cod genome assembly was the first sequenced and assembled with only 454 sequencing, and it was released in 2011 (Star et al., 2011), 6 years after the release of the 454 sequencing platform (Margulies et al., 2005). Initially, the read lengths of the early HTS technologies were too short to be useful for assembly of large genomes, but, over time, the read lengths have improved substantially. As an example, the assembler ALLPATHS was released in 2008 (Butler et al., 2008), producing genome assemblies based on 30 bp simulated Illumina reads, ALLPATHS 2 was released in 2009, using nonsimulated 36 bp reads from the Illumina platform (Maccallum et al., 2009), while ALLPATHS-LG was released in 2011 (Gnerre et al., 2011), using a specific set of sequencing libraries with the read length of 100 bp, to assemble mouse and human genomes to qualities approaching the previous Sanger-based assemblies (Lander et al., 2001; Venter et al., 2001).

Another example of continuous software improvements is the Celera Assembler. This program was used to assemble the first genome assembly of a multicellular organism sequenced with the Sanger technology (Myers et al., 2000). Later it was updated to handle 454 reads (Miller et al., 2008), and a subsequent unpublished version added support for Illumina reads. After the arrival of the PacBio sequencing platform, the program was adapted to utilize PacBio reads in combination with other sequencing technologies (Koren et al., 2012), and later created assemblies based on only PacBio sequencing data (Koren et al., 2013). Recently, it was updated to handle mammalian-size genomes at much improved speed (Berlin et al., 2015).

Goals for a new reference genome

Genome assemblies represent models for the actual genome—and thus are never perfect. A single assembly cannot represent all the diversity within populations of a species, and it is nearly impossible to eliminate all possible technological or algorithmical errors. Therefore, published genomes that have an active research community are continuously improved. For instance, in Dec. 2013, a new version of the human genome assembly was released (build 38), with several improvements compared to build 37, first released in 2009.

The first version of the cod genome has proven to be a valuable resource for the fish genomics community, and is frequently cited and downloaded. The fragmented nature of the Atlantic cod genome assembly however, poses limitations compared to some of the other available teleost genomes. For example, synteny analysis was only possible to a limited extent, since not all scaffolds have been placed inside the context of a larger chromosomal region. Moreover, the current number of gaps in the scaffolds results in the presence of relatively large parts of the genome that are solely available as short, unplaced pieces. We therefore continuously work toward an improved version of the genome (see Chapter 3).

Acknowledgments

We would like to thank Øystein Langangen, Helle Tessand Baalsrud, Paul R. Berg, and Michael Matschiner for critical reading and comments on the manuscript. The Atlantic cod genome sequencing and the ongoing work for improving the genome was supported by grants from the Research Council of Norway (the FUGE program and grants #181913/V00 and 187940/S10) to KSJ.

Glossary

Annotation Features on a genome assembly, such as the location of genes and their putative function. In addition to *de novo* prediction, different sources of data may be used as evidence for annotating a genomic region. For instance, the similarity of expressed genes or of genes from other species can be investigated and reported. Besides genes, other known elements such as promoter regions, and untranslated regions of mRNA (UTR's) are also annotated. An annotated genome allows for improved access to genes of interest.

Assembly The assembler software program identifies, reads with similar or identical, overlapping sequence, and joins these into longer stretches of nucleotides, often called *contigs*. The contigs can be ordered and oriented into even longer stretches of nucleotides called *scaffolds* by using paired reads information. Contigs often end where there is a lack of coverage (no reads cover that part of the genome), at genomic repeats that exceed the read length, and, for a diploid organism where haplotypes differ significantly, where two haplotypes occupy the same chromosomal position. Scaffolds connecting the contigs thus contain sequence gaps. The resulting assembly is a putative reconstruction of the actual genome.

cDNA Complementary DNA, created when sequencing RNA. Current technologies cannot directly sequence RNA, RNA needs to be reverse transcribed into DNA (cDNA) before sequencing.

Coverage In whole genome sequencing, the genome is oversampled, that is, each nucleotide site in the genome is sequenced several times, that is, represented by several reads. The average number of times each nucleotide is sequenced is called coverage.

Gap A region in a scaffold or assembly where the assembler inferred that there should be a sequence of nucleotides, but not precisely what sequence to place there. Gaps are caused by lack of coverage, repeats, or heterozygosity. Gaps are usually represented by stretches of Ns in the assembly.

High-throughput sequencing Also called *next-generation sequencing* and *second-generation sequencing*. The term reflects the massively parallel sequencing performed on the Roche/454, Illumina, Solid and Ion Torrent instruments. Pacific Biosciences (PacBio) and Oxford Nanopore sequencing technologies can be defined as a HTS, but are also regarded as *third-generation sequencing (single-molecule sequencing).*

Indel An insertion or deletion of one or more base pairs in a part of DNA sequence compared to a homologous part of DNA (from another individual or a homologous chromosome).

Inversion Describes the occurrence of one or more base pairs (often longer stretches) of DNA that are inversely oriented between two homologous stretches of DNA.

Library creation The biochemical process going from isolated DNA, via fragmentation, to DNA fragments with adaptors attached for HTS sequencing.

Linkage disequilibrium Two features of a genome (e.g., two alleles), that co-occur more often or less often than expected by random chance, are said to be in linkage disequilibrium.

N50 Length of contig or scaffold defined such that half the assembly consists of contigs or scaffold of that size or larger.

Paired reads Here we use paired reads for all pairs of reads with a known distance between them. When the distance is over 1 kbp, these are often referred to as "mate pairs." Paired reads (paired end sequencing) can be used to orient and order contigs into scaffolds, or detect rearrangements between genomes.

Read A read is the order of nucleotides produced by a sequencing instrument from a fragment of DNA. For example, a typical read from an Illumina sequencing instrument is 100 or 125 nucleotides long.

SNPs Single nucleotide polymorphisms, single nucleotide differences between two genomes (or homologous chromosomes).

References

Aparicio, S., Chapman, J., Stupka, E., Putnam, N., Chia, J.-M., Dehal, P., Christoffels, A., Rash, S., Hoon, S., Smit, A., Gelpke, M.D.S., Roach, J., Oh, T., Ho, I.Y., Wong, M., Detter, C., Verhoef, F., Predki, P., Tay, A., Lucas, S., Richardson, P., Smith, S.F., Clark, M.S., Edwards, Y.J.K., Doggett, N., Zharkikh, A., Tavtigian, S.V., Pruss, D., Barnstead, M., Evans, C., Baden, H., Powell, J., Glusman, G., Rowen, L., Hood, L., Tan, Y.H., Elgar, G., Hawkins, T., Venkatesh, B., Rokhsar, D., Brenner, S., 2002. Whole-genome shotgun assembly and analysis of the genome of *Fugu rubripes*. Science 297 (5585), 1301–1310.

Balzer, S., Malde, K., Lanzén, A., Sharma, A., Jonassen, I., 2010. Characteristics of 454 pyrosequencing data--enabling realistic simulation with flowsim. Bioinformatics 26 (18), i420–i425.

Bentley, D.R., Balasubramanian, S., Swerdlow, H.P., Smith, G.P., Milton, J., Brown, C.G., Hall, K.P., Evers, D.J., Barnes, C.L., Bignell, H.R., Boutell, J.M., Bryant, J., Carter, R.J., Keira Cheetham, R., Cox, A.J., Ellis, D.J., Flatbush, M.R., Gormley, N.A., Humphray, S.J., Irving, L.J., Karbelashvili, M.S., Kirk, S.M., Li, H., Liu, X., Maisinger, K.S., Murray, L.J., Obradovic, B., Ost, T., Parkinson, M.L., Pratt, M.R., Rasolonjatovo, I.M.J., Reed, M.T., Rigatti, R., Rodighiero, C., Ross, M.T., Sabot, A., Sankar, S.V., Scally, A., Schroth, G.P., Smith, M.E., Smith, V.P., Spiridou, A., Torrance, P.E., Tzonev, S.S., Vermaas, E.H., Walter, K., Wu, X., Zhang, L., Alam, M.D., Anastasi, C., Aniebo, I.C., Bailey, D.M.D., Bancarz, I.R., Banerjee, S., Barbour, S.G., Baybayan, P.A., Benoit, V.A., Benson, K.F., Bevis, C., Black, P.J., Boodhun, A., Brennan, J.S., Bridgham, J.A., Brown, R.C., Brown, A.A., Buermann, D.H., Bundu, A.A., Burrows, J.C., Carter, N.P., Castillo, N., Chiara, M., Catenazzi, E., Chang, S., Neil Cooley, R., Crake, N.R., Dada, O.O., Diakoumakos, K.D., Dominguez-Fernandez, B., Earnshaw, D.J., Egbujor, U.C., Elmore, D.W., Etchin, S.S., Ewan, M.R., Fedurco, M., Fraser, L.J., Fuentes Fajardo, K.V., Scott Furey, W., George, D., Gietzen, K.J., Goddard, C.P., Golda, G.S., Granieri, P.A., Green, D.E., Gustafson, D.L., Hansen, N.F., Harnish, K., Haudenschild, C.D., Heyer, N.I., Hims, M.M., Ho, J.T., Horgan, A.M., Hoschler, K., Hurwitz, S., Ivanov, D.V., Johnson, M.Q., James, T., Huw Jones, T.A., Kang, G.-D., Kerelska, T.H., Kersey, A.D., Khrebtukova, I., Kindwall, A.P., Kingsbury, Z., Kokko-Gonzales, P.I., Kumar, A., Laurent, M.A., Lawley, C.T., Lee, S.E., Lee, X., Liao, A.K., Loch, J.A., Lok, M., Luo, S., Mammen, R.M., Martin, J.W., McCauley, P.G., McNitt, P., Mehta, P., Moon, K.W., Mullens, J.W., Newington, T., Ning, Z., Ling Ng, B., Novo, S.M., O'Neill, M.J., Osborne, M.A., Osnowski, A., Ostadan, O., Paraschos, L.L., Pickering, L., Pike, A.C., Pike, A.C., Chris Pinkard, D., Pliskin, D.P., Podhasky, J., Quijano, V.J., Raczy, C., Rae, V.H., Rawlings, S.R., Chiva Rodriguez, A., Roe, P.M., Rogers, J., Rogert Bacigalupo, M.C., Romanov, N., Romieu, A., Roth, R.K., Rourke, N.J., Ruediger, S.T., Rusman, E., Sanches-Kuiper, R.M., Schenker, M.R., Seoane, J.M., Shaw, R.J., Shiver, M.K., Short, S.W., Sizto, N.L., Sluis, J.P., Smith, M.A., Ernest Sohna Sohna, J., Spence, E.J., Stevens, K., Sutton, N., Szajkowski, L., Tregidgo, C.L., Turcatti, G., Vandevondele, S., Verhovsky, Y., Virk, S.M., Wakelin, S., Walcott, G.C., Wang, J., Worsley, G.J., Yan, J., Yau, L., Zuerlein, M., Rogers, J., Mullikin, J.C., Hurles, M.E., McCooke, N.J., West, J.S., Oaks, F.L., Lundberg, P.L., Klenerman, D., Durbin, R., Smith, A.J., 2008. Accurate whole human genome sequencing using reversible terminator chemistry. Nature 456 (7218), 53–59.

Berlin, K., Koren, S., Chin, C.-S., Drake, J.P., Landolin, J.M., Phillippy, A.M., 2015. Assembling large genomes with single-molecule sequencing and locality-sensitive hashing. Nat. Biotechnol. 33 (6), 623–630.

Bradnam, K.R., Fass, J.N., Alexandrov, A., Baranay, P., Bechner, M., Birol, I., Boisvert, S., Chapman, J.A., Chapuis, G., Chikhi, R., Chitsaz, H., Chou, W.-C., Corbeil, J., Del Fabbro, C., Docking, T., Durbin, R., Earl, D., Emrich, S., Fedotov, P., Fonseca, N.A., Ganapathy, G., Gibbs, R.A., Gnerre, S., Godzaridis, É., Goldstein, S., Haimel, M., Hall, G., Haussler, D., Hiatt, J.B., Ho, I.Y., Howard, J., Hunt, M., Jackman, S.D., Jaffe, D.B., Jarvis, E.D., Jiang, H., Kazakov, S., Kersey, P.J., Kitzman, J.O., Knight, J.R., Koren, S., Lam, T.-W., Lavenier, D., Laviolette, F., Li, Y., Li, Z., Liu, B., Liu, Y., Luo, R., Maccallum, I., MacManes, M.D., Maillet, N., Melnikov, S., Naquin, D., Ning, Z., Otto, T.D., Paten, B., Paulo, O.S., Phillippy, A.M., Pina-Martins, F., Place, M., Przybylski, D., Qin, X., Qu, C., Ribeiro, F.J., Richards, S., Rokhsar, D.S., Ruby, J., Scalabrin, S., Schatz, M.C., Schwartz, D.C., Sergushichev, A., Sharpe, T., Shaw, T.I., Shendure, J., Shi, Y., Simpson, J.T., Song, H., Tsarev, F., Vezzi, F., Vicedomini, R., Vieira, B.M., Wang, J., Worley, K.C., Yin, S., Yiu, S.-M., Yuan, J., Zhang, G., Zhang, H., Zhou, S., Korf, I.F., 2013. Assemblathon 2: evaluating de novo methods of genome assembly in three vertebrate species. GigaScience 2 (1), 10.

Butler, J., Maccallum, I., Kleber, M., Shlyakhter, I.A., Belmonte, M.K., Lander, E.S., Nusbaum, C., Jaffe, D.B., 2008. ALLPATHS: de novo assembly of whole-genome shotgun microreads. Genome Res. 18, 810–820.

Chin, C.-S., Alexander, D.H., Marks, P., Klammer, A.A., Drake, J., Heiner, C., Clum, A., Copeland, A., et al., 2013. Nonhybrid, finished microbial genome assemblies from long-read SMRT sequencing data. Nat. Methods 10, 563–569.

Compeau, P.E.C., Pevzner, P.A., Tesler, G., 2011. How to apply de Bruijn graphs to genome assembly. Nat. Biotechnol. 29 (11), 987–991.

Dalloul, R.A., Long, J.A., Zimin, A.V., Aslam, L., Beal, K., Blomberg, L.A., Bouffard, P., Burt, D.W., Crasta, O., Crooijmans, R.P.M.A., Cooper, K., Coulombe, R.A., De, S., Delany, M.E., Dodgson, J.B., Dong, J.J., Evans, C., Frederickson, K.M., Flicek, P., Florea, L., Folkerts, O., Groenen, M.A.M., Harkins, T.T., Herrero, J., Hoffmann, S., Megens, H.-J., Jiang, A., de Jong, P., Kaiser, P., Kim, H., Kim, K.-W., Kim, S., Langenberger, D., Lee, M.-K., Lee, T., Mane, S., Marcais, G., Marz, M., McElroy, A.P., Modise, T., Nefedov, M., Notredame, C., Paton, I.R., Payne, W.S., Pertea, G., Prickett, D., Puiu, D., Qioa, D., Raineri, E., Ruffier, M., Salzberg, S.L., Schatz, M.C., Scheuring, C., Schmidt, C.J., Schroeder, S., Searle, S.M.J., Smith, E.J., Smith, J., Sonstegard, T.S., Stadler, P.F., Tafer, H., Tu, Z.J., Van Tassell, C.P., Vilella, A.J., Williams, K.P., Yorke, J.A., Zhang, L., Zhang, H.-B., Zhang, X., Zhang, Y., Reed, K.M., 2010. Multi-platform next-generation sequencing of the domestic turkey (*Meleagris gallopavo*): genome assembly and analysis. PLoS Biol. 8 (9), e1000475.

DePristo, M.A., Banks, E., Poplin, R., Garimella, K.V., Maguire, J.R., Hartl, C., Philippakis, A.A., del Angel, G., Rivas, M.A., Hanna, M., McKenna, A., Fennell, T.J., Kernytsky, A.M., Sivachenko, A.Y., Cibulskis, K., Gabriel, S.B., Altshuler, D., Daly, M.J., 2011. A framework for variation discovery and genotyping using next-generation DNA sequencing data. Nat. Genet. 43 (5), 491–498.

DiGuistini, S., Liao, N.Y., Platt, D., Robertson, G., Seidel, M., Chan, S.K., Docking, T.R., Birol, I., Holt, R.A., Hirst, M., Mardis, E., Marra, M.A., Hamelin, R.C., Bohlmann, J., Breuil, C., Jones, S.J.M., 2009. De novo genome sequence assembly of a filamentous fungus using Sanger, 454 and Illumina sequence data. Genome Biol. 10 (9), R94.

Earl, D., Bradnam, K., St John, J., Darling, A., Lin, D., Fass, J., Yu, H.O.K., Buffalo, V., Zerbino, D.R., Diekhans, M., Nguyen, N., Ariyaratne, P.N., Sung, W.-K., Ning, Z., Haimel, M., Simpson, J.T., Fonseca, N.A., Birol, I., Docking, T.R., Ho, I.Y., Rokhsar, D.S., Chikhi, R., Lavenier, D., Chapuis, G., Naquin, D., Maillet, N., Schatz, M.C., Kelley, D.R., Phillippy, A.M., Koren, S., Yang, S.-P., Wu, W., Chou, W.-C., Srivastava, A., Shaw, T.I., Ruby, J.G., Skewes-Cox, P., Betegon, M., Dimon, M.T., Solovyev, V., Seledtsov, I., Kosarev, P., Vorobyev, D., Ramirez-Gonzalez, R., Leggett, R., MacLean, D., Xia, F., Luo, R., Li, Z., Xie, Y., Liu, B., Gnerre, S., Maccallum, I., Przybylski, D., Ribeiro, F.J., Yin, S., Sharpe, T., Hall, G., Kersey, P.J., Durbin, R., Jackman, S.D., Chapman, J.A., Huang, X., DeRisi, J.L., Caccamo, M., Li, Y., Jaffe, D.B., Green, R.E., Haussler, D., Korf, I.F., Paten, B., 2011. Assemblathon 1: a competitive assessment of de novo short read assembly methods. Genome Res. 21 (12), 2224–2241.

Eid, J., Fehr, A., Gray, J., Luong, K., Lyle, J., Otto, G., Peluso, P., Rank, D., et al., 2009. Real-time DNA sequencing from single polymerase molecules. Science 323, 133–138.

Eisenstein, M., 2012. Oxford Nanopore announcement sets sequencing sector abuzz. Nat. Biotechnol. 30 (4), 295–296.

Flicek, P., Amode, M.R., Barrell, D., Beal, K., Billis, K., Brent, S., Carvalho-Silva, D., Clapham, P., Coates, G., Fitzgerald, S., Gil, L., Giron, C.G., Gordon, L., Hourlier, T., Hunt, S., Johnson, N., Juettemann, T., Kahari, A.K., Keenan, S., Kulesha, E., Martin, F.J., Maurel, T., McLaren, W.M., Murphy, D.N., Nag, R., Overduin, B., Pignatelli, M., Pritchard, B.,

Pritchard, E., Riat, H.S., Ruffier, M., Sheppard, D., Taylor, K., Thormann, A., Trevanion, S.J., Vullo, A., Wilder, S.P., Wilson, M., Zadissa, A., Aken, B.L., Birney, E., Cunningham, F., Harrow, J., Herrero, J., Hubbard, T.J.P., Kinsella, R., Muffato, M., Parker, A., Spudich, G., Yates, A., Zerbino, D.R., Searle, S.M.J., 2013. Ensembl 2014. Nucleic Acids Res. 42 (D1), D749–D755.

Gnerre, S., Maccallum, I., Przybylski, D., Ribeiro, F.J., Burton, J.N., Walker, B.J., Sharpe, T., Hall, G., Shea, T.P., Sykes, S., Berlin, A.M., Aird, D., Costello, M., Daza, R., Williams, L., Nicol, R., Gnirke, A., Nusbaum, C., Lander, E.S., Jaffe, D.B., 2011. High-quality draft assemblies of mammalian genomes from massively parallel sequence data. Proc. Natl. Acad. Sci. USA 108 (4), 1513–1518.

Howe, K., Clark, M.D., Torroja, C.F., Torrance, J., Berthelot, C., Muffato, M., Collins, J.E., Humphray, S., McLaren, K., Matthews, L., McLaren, S., Sealy, I., Caccamo, M., Churcher, C., Scott, C., Barrett, J.C., Koch, R., Rauch, G.-J., White, S., Chow, W., Kilian, B., Quintais, L.T., Guerra-Assunção, J.A., Zhou, Y., Gu, Y., Yen, J., Vogel, J.-H., Eyre, T., Redmond, S., Banerjee, R., Chi, J., Fu, B., Langley, E., Maguire, S.F., Laird, G.K., Lloyd, D., Kenyon, E., Donaldson, S., Sehra, H., Almeida-King, J., Loveland, J., Trevanion, S., Jones, M., Quail, M., Willey, D., Hunt, A., Burton, J., Sims, S., McLay, K., Plumb, B., Davis, J., Clee, C., Oliver, K., Clark, R., Riddle, C., Eliott, D., Threadgold, G., Harden, G., Ware, D., Mortimer, B., Kerry, G., Heath, P., Phillimore, B., Tracey, A., Corby, N., Dunn, M., Johnson, C., Wood, J., Clark, S., Pelan, S., Griffiths, G., Smith, M., Glithero, R., Howden, P., Barker, N., Stevens, C., Harley, J., Holt, K., Panagiotidis, G., Lovell, J., Beasley, H., Henderson, C., Gordon, D., Auger, K., Wright, D., Collins, J., Raisen, C., Dyer, L., Leung, K., Robertson, L., Ambridge, K., Leongamornlert, D., McGuire, S., Gilderthorp, R., Griffiths, C., Manthravadi, D., Nichol, S., Barker, G., Whitehead, S., Kay, M., Brown, J., Murnane, C., Gray, E., Humphries, M., Sycamore, N., Barker, D., Saunders, D., Wallis, J., Babbage, A., Hammond, S., Mashreghi-Mohammadi, M., Barr, L., Martin, S., Wray, P., Ellington, A., Matthews, N., Ellwood, M., Woodmansey, R., Clark, G., Cooper, J., Tromans, A., Grafham, D., Skuce, C., Pandian, R., Andrews, R., Harrison, E., Kimberley, A., Garnett, J., Fosker, N., Hall, R., Garner, P., Kelly, D., Bird, C., Palmer, S., Gehring, I., Berger, A., Dooley, C.M., Ersan-Ürün, Z., Eser, C., Geiger, H., Geisler, M., Karotki, L., Kirn, A., Konantz, J., Konantz, M., Oberländer, M., Rudolph-Geiger, S., Teucke, M., Osoegawa, K., Zhu, B., Rapp, A., Widaa, S., Langford, C., Yang, F., Carter, N.P., Harrow, J., Ning, Z., Herrero, J., Searle, S.M.J., Enright, A., Geisler, R., Plasterk, R.H.A., Lee, C., Westerfield, M., de Jong, P.J., Zon, L.I., Postlethwait, J.H., Nüsslein-Volhard, C., Hubbard, T.J.P., Crollius, H.R., Rogers, J., Stemple, D.L., 2013. The zebrafish reference genome sequence and its relationship to the human genome. Nature 496 (7446), 498–503.

Hubert, S., Higgins, B., Borza, T., Bowman, S., 2010. Development of a SNP resource and a genetic linkage map for Atlantic cod (*Gadus morhua*). BMC Genomics 11, 191.

Jaillon, O., Aury, J.-M., Brunet, F., Petit, J.-L., Stange-Thomann, N., Mauceli, E., Bouneau, L., Fischer, C., Ozouf-Costaz, C., Bernot, A., Nicaud, S., Jaffe, D., Fisher, S., Lutfalla, G., Dossat, C., Segurens, B., Dasilva, C., Salanoubat, M., Levy, M., Boudet, N., Castellano, S., Anthouard, V., Jubin, C., Castelli, V., Katinka, M., Vacherie, B., Biémont, C., Skalli, Z., Cattolico, L., Poulain, J., De Berardinis, V., Cruaud, C., Duprat, S., Brottier, P., Coutanceau, J.-P., Gouzy, J., Parra, G., Lardier, G., Chapple, C., McKernan, K.J., McEwan, P., Bosak, S., Kellis, M., Volff, J.-N., Guigó, R., Zody, M.C., Mesirov, J., Lindblad-Toh, K., Birren, B., Nusbaum, C., Kahn, D., Robinson-Rechavi, M., Laudet, V., Schachter, V., Quétier, F., Saurin, W., Scarpelli, C., Wincker, P., Lander, E.S., Weissenbach, J., Roest Crollius, H., 2004. Genome duplication in the teleost fish *Tetraodon nigroviridis* reveals the early vertebrate proto-karyotype. Nature 431 (7011), 946–957.

Kelley, D.R., Schatz, M.C., Salzberg, S.L., 2010. Quake: quality-aware detection and correction of sequencing errors. Genome Biol. 11 (11), R116.

Kikuta, H., Laplante, M., Navratilova, P., Komisarczuk, A.Z., Engstrom, P.G., Fredman, D., Akalin, A., Caccamo, M., Sealy, I., Howe, K., Ghislain, J., Pezeron, G., Mourrain, P., Ellingsen, S., Oates, A.C., Thisse, C., Thisse, B., Foucher, I., Adolf, B., Geling, A., Lenhard, B., Becker, T.S., 2007. Genomic regulatory blocks encompass multiple neighboring genes and maintain conserved synteny in vertebrates. Genome Res. 17 (5), 545–555.

Kjesbu, O.S., Bogstad, B., Devine, J.A., Gjosaeter, H., Howell, D., Ingvaldsen, R.B., Nash, R.D.M., Skjaeraasen, J.E., 2014. Synergies between climate and management for Atlantic cod fisheries at high latitudes. Proc. Natl. Acad. Sci. 111 (9), 3478–3483.

Koren, S., Schatz, M.C., Walenz, B.P., Martin, J., Howard, J.T., Ganapathy, G., Wang, Z., Rasko, D.A., et al., 2012. Hybrid error correction and de novo assembly of single-molecule sequencing reads. Nat. Biotechnol. 30, 693–700.

Koren, S., Harhay, G.P., Smith, T.P., Bono, J.L., Harhay, D.M., McVey, S.D., Radune, D., Bergman, N.H., et al., 2013. Reducing assembly complexity of microbial genomes with single-molecule sequencing. Genome Biol. 14, R101.

Lander, E.S., Linton, L.M., Birren, B., Nusbaum, C., Zody, M.C., Baldwin, J., Devon, K., Dewar, K., Doyle, M., FitzHugh, W., Funke, R., Gage, D., Harris, K., Heaford, A., Howland, J., Kann, L., Lehoczky, J., LeVine, R., McEwan, P., McKernan, K., Meldrim, J., Mesirov, J.P., Miranda, C., Morris, W., Naylor, J., Raymond, C., Rosetti, M., Santos, R., Sheridan, A., Sougnez, C., Stange-Thomann, N., Stojanovic, N., Subramanian, A., Wyman, D., Rogers, J., Sulston, J., Ainscough, R., Beck, S., Bentley, D., Burton, J., Clee, C., Carter, N., Coulson, A., Deadman, R., Deloukas, P., Dunham, A., Dunham, I., Durbin, R., French, L., Grafham, D., Gregory, S., Hubbard, T., Humphray, S., Hunt, A., Jones, M., Lloyd, C., McMurray, A., Matthews, L., Mercer, S., Milne, S., Mullikin, J.C., Mungall, A., Plumb, R., Ross, M., Shownkeen, R., Sims, S., Waterston, R.H., Wilson, R.K., Hillier, L.W., McPherson, J.D., Marra, M.A., Mardis, E.R., Fulton, L.A., Chinwalla, A.T., Pepin, K.H., Gish, W.R., Chissoe, S.L., Wendl, M.C., Delehaunty, K.D., Miner, T.L., Delehaunty, A., Kramer, J.B., Cook, L.L., Fulton, R.S., Johnson, D.L., Minx, P.J., Clifton, S.W., Hawkins, T., Branscomb, E., Predki, P., Richardson, P., Wenning, S., Slezak, T., Doggett, N., Cheng, J.F., Olsen, A., Lucas, S., Elkin, C., Uberbacher, E., Frazier, M., Gibbs, R.A., Muzny, D.M., Scherer, S.E., Bouck, J.B., Sodergren, E.J., Worley, K.C., Rives, C.M., Gorrell, J.H., Metzker, M.L., Naylor, S.L., Kucherlapati, R.S., Nelson, D.L., Weinstock, G.M., Sakaki, Y., Fujiyama, A., Hattori, M., Yada, T., Toyoda, A., Itoh, T., Kawagoe, C., Watanabe, H., Totoki, Y., Taylor, T., Weissenbach, J., Heilig, R., Saurin, W., Artiguenave, F., Brottier, P., Bruls, T., Pelletier, E., Robert, C., Wincker, P., Smith, D.R., Doucette-Stamm, L., Rubenfield, M., Weinstock, K., Lee, H.M., Dubois, J., Rosenthal, A., Platzer, M., Nyakatura, G., Taudien, S., Rump, A., Yang, H., Yu, J., Wang, J., Huang, G., Gu, J., Hood, L., Rowen, L., Madan, A., Qin, S., Davis, R.W., Federspiel, N.A., Abola, A.P., Proctor, M.J., Myers, R.M., Schmutz, J., Dickson, M., Grimwood, J., Cox, D.R., Olson, M.V., Kaul, R., Raymond, C., Shimizu, N., Kawasaki, K., Minoshima, S., Evans, G.A., Athanasiou, M., Schultz, R., Roe, B.A., Chen, F., Pan, H., Ramser, J., Lehrach, H., Reinhardt, R., McCombie, W.R., de la Bastide, M., Dedhia, N., Blöcker, H., Hornischer, K., Nordsiek, G., Agarwala, R., Aravind, L., Bailey, J.A., Bateman, A., Batzoglou, S., Birney, E., Bork, P., Brown, D.G., Burge, C.B., Cerutti, L., Chen, H.C., Church, D., Clamp, M., Copley, R.R., Doerks, T., Eddy, S.R., Eichler, E.E., Furey, T.S., Galagan, J., Gilbert, J.G., Harmon, C., Hayashizaki, Y., Haussler, D., Hermjakob, H., Hokamp, K., Jang, W., Johnson, L.S., Jones, T.A., Kasif, S., Kaspryzk, A., Kennedy, S., Kent, W.J., Kitts, P., Koonin, E.V., Korf, I., Kulp, D., Lancet, D., Lowe, T.M., McLysaght, A., Mikkelsen, T., Moran, J.V., Mulder, N., Pollara, V.J., Ponting, C.P.,

Schuler, G., Schultz, J., Slater, G., Smit, A.F., Stupka, E., Szustakowski, J., Thierry-Mieg, D., Thierry-Mieg, J., Wagner, L., Wallis, J., Wheeler, R., Williams, A., Wolf, Y.I., Wolfe, K.H., Yang, S.P., Yeh, R.F., Collins, F., Guyer, M.S., Peterson, J., Felsenfeld, A., Wetterstrand, K.A., Patrinos, A., Morgan, M.J., de Jong, P., Catanese, J.J., Osoegawa, K., Shizuya, H., Choi, S., Chen, Y.J., Szustakowki, J., International Human Genome Sequencing Consortium, 2001. Initial sequencing and analysis of the human genome. Nature 409 (6822), 860–921.

Li, R., Fan, W., Tian, G., Zhu, H., He, L., Cai, J., Huang, Q., Cai, Q., et al., 2010a. The sequence and de novo assembly of the giant panda genome. Nature 463, 311–317.

Li, R., Zhu, H., Ruan, J., Qian, W., Fang, X., Shi, Z., Li, Y., Li, S., Shan, G., Kristiansen, K., Li, S., Yang, H., Wang, J., Wang, J., 2010b. De novo assembly of human genomes with massively parallel short read sequencing. Genome Res. 20 (2), 265–272.

Lien, S., Koop, B.F., Sandve, S.R., Miller, J.R., Kent, M.P., Nome, T., Hvidsten, T.R., Leong, J., Minkley, D., Zimin, A., Grammes, F., Grove, H., Gjuvsland, A., Walenz, B., Hermansen, R.A., von Schalburg, K., Rondeau, E.B., Di Genova, A., Samy, J.K.A., Vik, J.O., Vigeland, M.D., Caler, L., Grimholt, U., Jentoft, S., Våge, D.I., de Jong, P., Moen, T., Baranski, M., Palti, Y., Smith, D.R., Yorke, J.A., Nederbragt, A.J., Tooming-Klunderud, A., Jakobsen, K.S., Jiang, X., Fan, D., Hu, Y., Liberles, D.A., Vidal, R., Iturra, P., Jones, S.J.M., Jonassen, I., Maass, A., Omholt, S.W., Davidson, W.S., 2016. The Atlantic salmon genome provides insights into rediploidization. Nature 533 (7602), 200–205.

Loman, N.J., Misra, R.V., Dallman, T.J., Constantinidou, C., Gharbia, S.E., Wain, J., Pallen, M.J., 2012. Performance comparison of benchtop high-throughput sequencing platforms. Nat. Biotechnol. 30 (5), 434–439.

Luo, R., Liu, B., Xie, Y., Li, Z., Huang, W., Yuan, J., He, G., Chen, Y., Pan, Q., Liu, Y., Tang, J., Wu, G., Zhang, H., Shi, Y., Liu, Y., Yu, C., Wang, B., Lu, Y., Han, C., Cheung, D.W., Yiu, S.-M., Peng, S., Xiaoqian, Z., Liu, G., Liao, X., Li, Y., Yang, H., Wang, J., Lam, T.-W., Wang, J., 2012. SOAPdenovo2: an empirically improved memory-efficient short-read de novo assembler. GigaScience 1 (1), 18.

Maccallum, I., Przybylski, D., Gnerre, S., Burton, J., Shlyakhter, I., Gnirke, A., Malek, J., McKernan, K., et al., 2009. ALLPATHS 2: small genomes assembled accurately and with high continuity from short paired reads. Genome Biol. 10, R103.

Margulies, M., Egholm, M., Altman, W.E., Attiya, S., Bader, J.S., Bemben, L.A., Berka, J., Braverman, M.S., Chen, Y.-J., Chen, Z., Dewell, S.B., Du, L., Fierro, J.M., Gomes, X.V., Godwin, B.C., He, W., Helgesen, S., Ho, C.H., Irzyk, G.P., Jando, S.C., Alenquer, M.L.I., Jarvie, T.P., Jirage, K.B., Kim, J.-B., Knight, J.R., Lanza, J.R., Leamon, J.H., Lefkowitz, S.M., Lei, M., Li, J., Lohman, K.L., Lu, H., Makhijani, V.B., McDade, K.E., McKenna, M.P., Myers, E.W., Nickerson, E., Nobile, J.R., Plant, R., Puc, B.P., Ronan, M.T., Roth, G.T., Sarkis, G.J., Simons, J.F., Simpson, J.W., Srinivasan, M., Tartaro, K.R., Tomasz, A., Vogt, K.A., Volkmer, G.A., Wang, S.H., Wang, Y., Weiner, M.P., Yu, P., Begley, R.F., Rothberg, J.M., 2005. Genome sequencing in microfabricated high-density picolitre reactors. Nature 437 (7057), 376–380.

Masahiro Kasahara, Naruse, K., Sasaki, S., Nakatani, Y., Qu, W., Ahsan, B., Yamada, T., Nagayasu, Y., Doi, K., Kasai, Y., Jindo, T., Kobayashi, D., Shimada, A., Toyoda, A., Kuroki, Y., Fujiyama, A., Sasaki, T., Shimizu, A., Asakawa, S., Shimizu, N., Hashimoto, S.-I., Yang, J., Lee, Y., Matsushima, K., Sugano, S., Sakaizumi, M., Narita, T., Ohishi, K., Haga, S., Ohta, F., Nomoto, H., Nogata, K., Morishita, T., Endo, T., Shin-I, T., Takeda, H., Morishita, S., Kohara, Y., 2007. The medaka draft genome and insights into vertebrate genome evolution. Nature 447 (7145), 714–719.

Miller, J.R., Delcher, A.L., Koren, S., Venter, E., Walenz, B.P., Brownley, A., Johnson, J., Li, K., Mobarry, C., Sutton, G.G., 2008. Aggressive assembly of pyrosequencing reads with mates. Bioinformatics 24 (24), 2818–2824.

Miller, J.R., Koren, S., Sutton, G.G., 2010. Assembly algorithms for next-generation sequencing data. Genomics 95 (6), 315–327.

Myers, E.W., 2005. The fragment assembly string graph. Bioinformatics 21 (Suppl. 2), ii79–ii85.

Myers, E.W., Sutton, G.G., Delcher, A.L., Dew, I.M., Fasulo, D.P., Flanigan, M.J., Kravitz, S.A., Mobarry, C.M., et al., 2000. A whole-genome assembly of Drosophila. Science 287, 2196–2204.

Nakamura, K., Oshima, T., Morimoto, T., Ikeda, S., Yoshikawa, H., Shiwa, Y., Ishikawa, S., Linak, M.C., Hirai, A., Takahashi, H., Altaf-Ul-Amin, M., Ogasawara, N., Kanaya, S., 2011. Sequence-specific error profile of Illumina sequencers. Nucleic Acids Res. 39 (13), e90–e190.

Nystedt, B., Street, N.R., Wetterbom, A., Zuccolo, A., Lin, Y.-C., Scofield, D.G., Vezzi, F., Delhomme, N., Giacomello, S., Alexeyenko, A., Vicedomini, R., Sahlin, K., Sherwood, E., Elfstrand, M., Gramzow, L., Holmberg, K., Hällman, J., Keech, O., Klasson, L., Koriabine, M., Kucukoglu, M., Käller, M., Luthman, J., Lysholm, F., Niittylä, T., Olson, Å., Rilakovic, N., Ritland, C., Rosselló, J.A., Sena, J., Svensson, T., Talavera-López, C., Theißen, G., Tuominen, H., Vanneste, K., Wu, Z.-Q., Zhang, B., Zerbe, P., Arvestad, L., Bhalerao, R., Bohlmann, J., Bousquet, J., Garcia Gil, R., Hvidsten, T.R., de Jong, P., MacKay, J., Morgante, M., Ritland, K., Sundberg, B., Lee Thompson, S., Van de Peer, Y., Andersson, B., Nilsson, O., Ingvarsson, P.K., Lundeberg, J., Jansson, S., 2013. The Norway spruce genome sequence and conifer genome evolution. Nature 497 (7451), 579–584.

Ohno, S., 1973. Ancient linkage groups and frozen accidents. Nature 244 (5414), 259–262.

Otterå, H., Thorsen, A., Peruzzi, S., Dahle, G., Hansen, T., Karlsen, Ø., 2011. Induction of meiotic gynogenesis in Atlantic cod, *Gadus morhua* (L.). J. Appl. Ichthyol. 27 (6), 1298–1302.

Ottersen, G., Bogstad, B., Yaragina, N.A., Stige, L.C., Vikebo, F.B., Dalpadado, P., 2014. A review of early life history dynamics of Barents Sea cod (*Gadus morhua*). ICES J. Mar. Sci. 71 (8), 2064–2087.

Pendleton, M., Sebra, R., Pang, A.W.C., Ummat, A., Franzen, O., Rausch, T., Stütz, A.M., Stedman, W., Anantharaman, T., Hastie, A., Dai, H., Fritz, M.H.-Y., Cao, H., Cohain, A., Deikus, G., Durrett, R.E., Blanchard, S.C., Altman, R., Chin, C.-S., Guo, Y., Paxinos, E.E., Korbel, J.O., Darnell, R.B., McCombie, W.R., Kwok, P.-Y., Mason, C.E., Schadt, E.E., Bashir, A., 2015. Assembly and diploid architecture of an individual human genome via single-molecule technologies. Nat. Methods 12 (8), 780–786.

Pevzner, P.A., Tang, H., Waterman, M.S., 2001. An Eulerian path approach to DNA fragment assembly. Proc. Natl. Acad. Sci. USA 98 (17), 9748–9753.

Reichwald, K., Lauber, C., Nanda, I., Kirschner, J., Hartmann, N., Schories, S., Gausmann, U., Taudien, S., Schilhabel, M.B., Szafranski, K., Glöckner, G., Schmid, M., Cellerino, A., Schartl, M., Englert, C., Platzer, M., 2009. High tandem repeat content in the genome of the short-lived annual fish *Nothobranchius furzeri*: a new vertebrate model for aging research. Genome Biol. 10 (2), R16.

Roberts, R.J., Carneiro, M.O., Schatz, M.C., 2013. The advantages of SMRT sequencing. Genome Biol. 14, 405.

Ross, M.G., Russ, C., Costello, M., Hollinger, A., Lennon, N.J., Hegarty, R., Nusbaum, C., Jaffe, D.B., 2013. Characterizing and measuring bias in sequence data. Genome Biol. 14 (5), R51.

Schartl, M., Walter, R.B., Shen, Y., Garcia, T., Catchen, J., Amores, A., Braasch, I., Chalopin, D., Volff, J.-N., Lesch, K.-P., Bisazza, A., Minx, P., Hillier, L., Wilson, R.K., Fuerstenberg,

S., Boore, J., Searle, S., Postlethwait, J.H., Warren, W.C., 2013. The genome of the platyfish, *Xiphophorus maculatus*, provides insights into evolutionary adaptation and several complex traits. Nat. Genet. 45 (5), 567–572.

Schatz, M.C., Witkowski, J., McCombie, W.R., 2012. Current challenges in de novo plant genome sequencing and assembly. Genome Biol. 13 (4), 243.

Star, B., Nederbragt, A.J., Jentoft, S., Grimholt, U., Malmstrøm, M., Gregers, T.F., Rounge, T.B., Paulsen, J., Solbakken, M.H., Sharma, A., Wetten, O.F., Lanzén, A., Winer, R., Knight, J., Vogel, J.-H., Aken, B., Andersen, Ø., Lagesen, K., Tooming-Klunderud, A., Edvardsen, R.B., Tina, K.G., Espelund, M., Nepal, C., Previti, C., Karlsen, B.O., Moum, T., Skage, M., Berg, P.R., Gjøen, T., Kuhl, H., Thorsen, J., Malde, K., Reinhardt, R., Du, L., Johansen, S.D., Searle, S., Lien, S., Nilsen, F., Jonassen, I., Omholt, S.W., Stenseth, N.C., Jakobsen, K.S., 2011. The genome sequence of Atlantic cod reveals a unique immune system. Nature 477 (7363), 1–4.

Venter, J.C., Adams, M., Myers, E., Li, P., Mural, R., 2001. The sequence of the human genome. Science 291 (5507), 1304–1351.

Zerbino, D.R., Birney, E., 2008. Velvet: algorithms for de novo short read assembly using de Bruijn graphs. Genome Res. 18 (5), 821–829.

The rainbow trout genome, an important landmark for aquaculture and genome evolution

2

Julien Bobe, Lucie Marandel**, Stéphane Panserat**, Pierre Boudinot[†], Camille Berthelot[‡,††], Edwige Quillet[§], Jean-Nicolas Volff[¶], Carine Genêt[§], Olivier Jaillon[††], Hugues Roest Crollius[‡‡], Yann Guiguen**

*INRA, Fish Physiology and Genomics, Rennes, France; **Nutrition, metabolism and aquaculture (NUMEA), INRA, Univ Pau & Pays Adour, Saint Pée sur Nivelle, France; [†]Molecular Virology and Immunology, INRA, Université Paris-Saclay, Jouy-en-Josas, France; [‡]European Molecular Biology Laboratory, European Bioinformatics Institute, Welcome Trust Genome Campus, Cambridge, United Kingdom; [§]Animal Genetics and Integrative Biology, INRA, AgroParisTech, Université Paris-Saclay, Jouy-en-Josas, France; [¶]Institute of Functional Genomics of Lyon (IGFL), Ecole Normale Supérieure de Lyon, CNRS, University of Lyon I, Lyon, France; [††]CEA-Genomic Institute, French National Sequencing Center, Evry Cedex, France; [‡‡]Ecole Normale Supérieure (ENS), ENS Biology Institute, IBENS, Paris, France

Rainbow trout

Rainbow trout and Salmonids commercial importance

Members of the salmonid family (Fig. 2.1) are present worldwide and many of them are species of major importance for aquaculture, wild stock fisheries or recreational sport fisheries. Their world aquaculture production was estimated over 3,000,000 metric tons (MT) with a global commercial value over US$ 17 billion in 2013, with Atlantic salmon being the most important farmed salmonid species with world aquaculture production accounting for 2,000,000 MT. Rainbow trout is a species native to western North America, but is considered as one of the most widely introduced fishes. According to FishBase (http://www.fishbase.org), rainbow trout is now recorded to be present in at least 70 different countries with a distribution in nearly all of the world's continents. Although its farming receives less attention than Atlantic salmon, the production data for rainbow trout are still quite impressive. Total world production reached 500,000 MT for the first time in 2001 and was recorded in 2013 as being over 800,000 MT with a total export value around US$ 3.5 billion. More than one third of this production currently comes from Asia (310,000 MT) with a significant production both in Iran (140,000 MT) and Turkey (130,000 MT), the remaining production being produced in America (220,000 MT) and Europe (280,000 MT). In Europe, Norway is the major producer of rainbow trout with 20% of the total European production

Genomics in Aquaculture. http://dx.doi.org/10.1016/B978-0-12-801418-9.00002-0

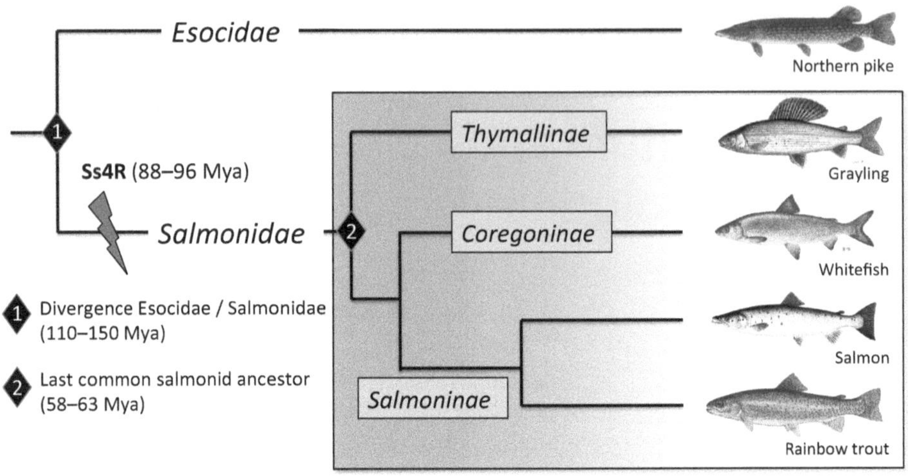

Figure 2.1 Salmonids evolution. The rainbow trout, *Oncorhynchus mykiss*, belongs to the salmonid (salmonidae) family that contains three sub-families, that is, the *Salmoninae* (mostly trouts, chars, and salmons), the *Thymallinae*, (graylings) and the *Coregoninae* (whitefish and ciscos). Salmonids diverged from their closest sister group, that is, *Esocidae* (pikes and mudminnows) 110–150 Mya (Near et al., 2012) and the last common ancestor of extant salmonids is dated around 58–63 Mya (Crête-Lafrenière et al., 2012). The specific whole genome duplication of salmonids (Ss4R) is dated between 88 Mya (Macqueen and Johnston, 2014) and 96 Mya (Berthelot et al., 2014).

(70,000 MT), followed by Italy (34,000 MT), France (29,000 MT), Russia (24,000), Spain (15,000 MT), United Kingdom (12,000 MT), and Finland (12,000 MT). More than 20 additional European countries share the remaining production. The majority of American production is localized in Chile (140,000 MT) and the production in the United States of America although rather modest (26,000 MT) was still worth $90 million in annual sales in 2013, and this does not include the industry aimed at restoration, conservation and recreation.

Rainbow trout: a freshwater laboratory rat

Beside its economic importance, rainbow trout has also been extensively used as a model of investigation in disciplines as diverse as cancer research, toxicology, comparative immunology, disease ecology, physiology, nutrition, and developmental and evolutionary biology (Thorgaard et al., 2002). *O. mykiss* is one the most studied fish species, with many experimental advantages including the availability of multiple natural populations (Behnke et al., 2002) and clonal lines (Komen and Thorgaard, 2007; Quillet et al., 2007), an easy breeding and the possibility to perform transgenesis (Thorgaard et al., 2002) and gene inactivation (Yano et al., 2014). Its relatively large body size compared to model fish like zebrafish or medaka makes rainbow trout a well suited alternative model to carry out biochemical and molecular studies on specific tissues or cells that are much more difficult to analyze in smaller fish models (Thorgaard et al., 2002). In agreement with its major relevance to biomedical research and aquaculture, many genomic resources have been developed during the last two decades, including genetic and physical maps, BAC libraries and BAC end sequences, as well as numerous expressed sequence tags (see later sections for a synopsis of the genomic resources available in rainbow trout). Most of the needed presequencing phase genomic resources were then available when the first *O. mykiss* whole genome-sequencing project was initiated in 2010.

Rainbow trout life cycle

Rainbow trout is a native species from the Northwest cost of North America where its natural distribution ranges from Alaska up to the north of Mexico (Behnke et al., 2002). It is mostly a freshwater species that inhabits lakes and rivers although there are populations, called steelhead trout that live in the ocean and have an anadromous migration back to freshwater to spawn. Freshwater populations usually inhabit well-oxygenated rivers or lakes with preferundum temperatures ranging from 10 to 16°C. Reproduction is generally taking place during spring with spawning occurring in cold well-oxygenated shallow rivers with fine gravel bottoms. One single female rainbow trout can produce 1–8 thousands eggs (3.5–6 mm) which are laid in one or multiple batches and immediately fertilized by sperm from a single male. Eggs fall between gravel spaces and usually hatch 4–5 weeks after fertilization. After vitellus resorption young fry emerge from the gravel substrate and start to feed essentially on zoobenthos and zooplankton prey. At the adult stage rainbow trout is an opportunistic feeder that relies on many different type of food ranging from small insects to crayfish.

Genomic aspects

The genome of rainbow trout per se is extremely interesting from an evolutionary genomics point of view. First of all, its genome is of respectable size, that is, 2.4×10^9 bp (Young et al., 1998), that is, 6 times bigger than the genome of the pufferfish (Roest Crollius et al., 2000), and even larger than that of the zebrafish, that is, 1.7×10^9 bp (Howe et al., 2013). Second, the genome of the rainbow trout is complex, with a chromosome arm number (NF) equal to 104 and a variable diploid set of chromosomes ranging from 58 to 64 due to Robertsonian rearrangements (Thorgaard, 1976). One major reason for this complexity is an event of tetraploidization (whole genome duplication, WGD), which occurred in the salmonid lineage (the Ss4R: Salmonid-specific fourth round of WGD). The genomes of salmonids are then assumed to still be undergoing a process of rediploidization and this involves a differential evolution or loss of gene duplicates, a process called "divergent resolution" (Taylor et al., 2001) that has been proposed to reduce hybrid fitness and favor speciation. Such a phenomenon might have played an important role in the amazing richness of species observed in fish (Volff, 2005). Indeed, divergent resolution has been already reported in teleosts subsequently to an ancestral genome duplication (third WGD or Teleost-specific third round of WGD = Ts3R) that took place between 225 and 330 million years ago (Mya) (Hurley et al., 2007; Near et al., 2012; Santini et al., 2009), but only very divergent species have been compared. Salmonid genomes provide a unique opportunity to analyze the differential evolution of tetraploidized genomes that diverged more recently (Allendorf and Thorgaard, 1984) and to assess the differential evolution of duplicated genes during early steps of rediploidization.

The rainbow trout genome: diversity, structure, organization

Salmonids combine both the essential characteristics of being very important commercial species and also species of considerable scientific importance with a key position in the ray-finned fish evolutionary tree (Fig. 2.1). These combined applied and scientific interests have deeply stimulated a well-structured international scientific community in developing a large number of genomics tools that are now available in many salmonid species.

Genomic resources preexisting before the first genome sequence in rainbow trout

The size of the rainbow trout genome is about 2.4×10^9 bp (Young et al., 1998) and its G + C content is 42% (Bernardi and Bernardi, 1990). One of the problems that plagued the initial sequencing and assembly of the zebrafish genome was the extent of polymorphism within and between the individuals of the pooled group that was used as the source of the DNA for the project. This problem has subsequently been overcome by using a single, doubled haploid fish that was produced by androgenesis. In rainbow trout, doubled haploid individuals were successfully obtained using gynogenesis or androgenesis (Chourrout, 1984; Parsons and Thorgaard, 1985; Quillet et al., 1991), and used to produce homozygous clonal lines (Scheerer et al., 1991; Quillet et al., 2007) that have been extensively characterized at genetic and phenotypic levels. Fully homozygous individuals were particularly desirable to develop genomic tools in rainbow trout, to deal with the high genome complexity that was anticipated because of the ancestral WGD in the species. The genomic DNA from a single doubled haploid fish (Swanson YY doubled haploid clonal line) was used to produce several rainbow trout bacterial artificial chromosomes (BAC) libraries. These BAC libraries have been fingerprinted to produce a first generation (Palti et al., 2011) and a second generation BAC physical map (Palti et al., 2012). Several linkage maps based on doubled haploid individuals and outbred populations have been constructed for rainbow trout using microsatellites and SNPs markers (Guyomard et al., 2012, 2006; Nichols et al., 2003; Rexroad et al., 2008; Sakamoto et al., 2000). A synthetic linkage map compiling previous ones was built (Guyomard et al., 2012) that comprised of 2,226 markers, with a medium density of 1 marker/cM.

In parallel, many different tissues and developmental stages cDNA libraries have been constructed and sequenced in the rainbow trout (Govoroun et al., 2006; Koop et al., 2008; Rexroad et al., 2003). As of Jun. 2015, 290,406 expressed sequence tags (ESTs) sequences are publicly available in NCBI GenBank and these ESTs have been the basis for several cDNA and oligos microarray platforms in rainbow trout (Canario et al., 2008). Sequences from both ends of 96,000 trout BACs were also available and these BAC-end sequences have been used to create a first trout repeat database (Genet et al., 2011), showing that repetitive DNA accounts for approximately 58% of the genome. These BAC-end sequences were screened for microsatellites and were used to produce the first and a second generation integrated maps (Palti et al., 2012, 2011).

Rainbow trout genome sequencing

The strategy developed for sequencing the rainbow trout genome was based on the NGS sequencing technologies available in 2010 at the beginning of this genome project. This genome was sequenced using a whole genome shotgun strategy (Berthelot et al., 2014) with genomic DNA from a unique doubled haploid (YY) rainbow trout male (Parsons and Thorgaard, 1985; Young et al., 1996). The availability of DNA from a totally homozygous individual has been a key resource that deeply facilitated the assembly of the rainbow trout genome. For the whole shotgun approach, single read, as well as 8 kb, 12 kb, and 20 kb mate-pairs libraries were sequenced using the 454 titanium technology up to a 20-fold sequence coverage with 454 Titanium and a 70-fold coverage using Solexa-Illumina sequences. Solexa-Illumina sequences were used to correct the contigs and scaffolds sequences resulting from the assembly of the 454 Titanium reads that are known to be error-rich especially in stretches of homopolymers (Gilles et al., 2011). This strategy was also complemented by a deep coverage of BAC-end sequences (Genet et al., 2011) that helped to scaffold contigs and to anchor the sequence onto the integrated maps. The total size of the resulting assembly of this first version of the rainbow trout genome (Berthelot et al., 2014) was 1.9 Gb with a scaffold N50 of 384 kb (half of the assembly is contained in 1,014 scaffolds longer than 384 kb). Using linkage and physical map information, scaffolds were anchored onto chromosomes at 898 distinct loci. Annotation of the genome sequence identified 46,585 protein-coding gene models with supporting protein evidence from other vertebrates and transcript evidence from 15 tissues of a doubled haploid rainbow trout adult obtained by RNA-seq. This high number of predicted protein-coding genes compared to other vertebrates is compatible with the recent Ss4R. 495 microRNA (miRNA) loci were also identified corresponding to 84 different families and 164 mature sequences. Transposable elements account for about 38% of the rainbow trout genome sequence.

Rainbow trout genome structure

The ancestral genome of teleost fish underwent a WGD event, termed here the teleost-specific third WGD (Ts3R) (Hurley et al., 2007; Near et al., 2012; Santini et al., 2009), approximately 300 million years ago (Mya). The signature of this Ts3R is still present in modern teleost genomes and comes in addition to two more ancestral WGD events common to all bony vertebrates (vertebrate genome duplications 1 and 2, VGD1 and VGD2) (Dehal and Boore, 2005). While WGD events are rare within animal lineages, they represent important evolutionary landmarks from which some major lineages have diversified. Duplicated genomes eventually retain only a small proportion of duplicated genes, while seemingly redundant copies are inactivated in a poorly understood process termed gene fractionation (Langham et al., 2004). The genome structure of the rainbow trout is of particular interest in this context because salmonids have undergone an additional relatively recent WGD event (the salmonid-specific fourth WGD or Ss4R) that has been initially dated between 25 and 100 Mya (Allendorf and Thorgaard, 1984). Gene fractionation may thus be ongoing in that species, providing a unique opportunity to better understand the main determinants that influence this rediploidization process.

By reconstructing the Ss4R paralogous regions, the genome of rainbow trout allowed the reconstruction of the ancestral karyotype of salmonids before the Ss4R duplication. The modern rainbow trout genome is organized in 38 major pairs of duplicated regions, and 14 out of 30 chromosomes result from the fusion of two different post-Ss4R chromosomes, in agreement with the known paralogies inferred from the trout linkage maps (Guyomard et al., 2012). Other chromosomes are more complex mosaics of different post-Ss4R chromosomes, reflecting additional inter-chromosomal rearrangements that have occurred since the Ss4R event (Fig. 2.2). Numerous ohnologs (i.e., paralogous genes formed by a WGD event) were identified in these Ss4R paralogous regions. The computed distribution of silent nucleotide substitutions (dS) among pairs of Ss4R ohnologs identified in rainbow trout, and between Atlantic salmon and rainbow trout pairs of orthologs (representative of the speciation divergence time, i.e., around 30 Mya, between these two species) allowed to estimate the date of the Ss4R at 96 Mya, in the upper range of the 25–100 Mya from previous estimation (Allendorf and Thorgaard, 1984) but in the same range (88 Mya) as another recent estimation (Macqueen and Johnston, 2014). These results contrast with the age of the Salmonidae family that has been estimated 50–60 Mya (Crête-Lafrenière et al., 2012), suggesting that the Ss4R occurred long before (> 30 Mya) the last common ancestor of extant salmonids (Fig. 2.1). This is consistent with the WGD Radiation Lag-Time Model (Schranz et al., 2012) that has been proposed in plants in which significant lag-times are proposed to be needed between WGDs and the subsequent adaptive radiations that are often associated with these WGD events.

Within a few million years after a WGD, genome evolution is thought to involve the loss of one gene copy of most ohnologous gene pairs by gene fractionation. This process has never been documented at the whole genome scale in any vertebrate because all WGD studied to date are too ancient to capture such information. The rainbow trout genome enabled that analysis as high confidence paralogous regions can be easily characterized due to the more recent age of the Ss4R. Across the entire genome, around 50% of the Ss4R duplicated gene pairs have undergone gene fractionation and returned to a single copy state, while the remaining retained both ohnologs. Gene fractionation is thus a relatively slow process in the trout genome, since genes were inactivated at an average rate of approximately 170 genes per million years. Additionally, the majority of singletons can still be paired with clear paralogous sequences

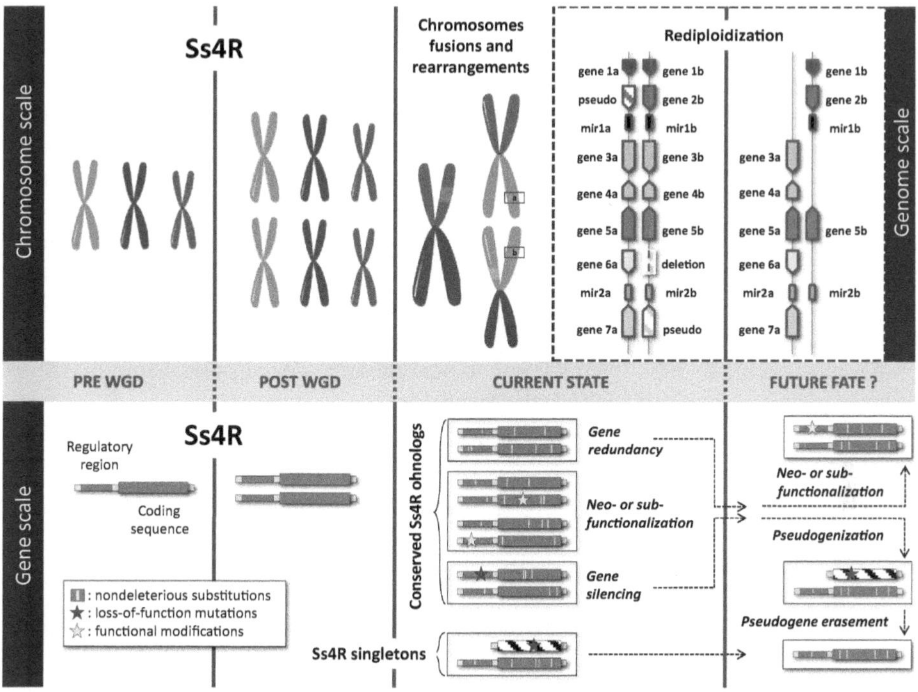

Figure 2.2 Schematic evolutionary history of the rainbow trout genome during the Ss4R WGD event. The Ss4R WGD resulted in a complete doubling of the genetic material, so every chromosome (and gene, associated with its regulatory regions) was present in two identical copies right after the Ss4R. The Ss4R WGD was followed by chromosome fusions and a few intrachromosomal rearrangements, bringing back the total number of chromosomes to 30, close to the preduplication karyotype. Rediploidization or gene fractionation, that is, the loss of duplicated genes after WGD by mutations and deletions to return to a mostly diploid state, is an ongoing process in the rainbow trout genome. At present, the duplicated regions within the genome (as exemplified by regions a and b in the boxed figure) are still highly collinear and retain both ohnologs for around 50% of the duplicated gene pairs. Ohnologs that are still present in both copies are slowly diverging at the sequence level, and may in the future acquire new or ancillary functions (neo- or sub-functionalization) or be lost via gene fractionation. Based on observations on older WGD events, only about 20–25% of duplicated genes are expected to be eventually retained in two copies (Howe et al., 2013).

stemming from the Ss4R, although the latter are largely nonfunctional (pseudogenes). In contrast to this 50% retention rate of ohnologous protein-coding genes, we found that the post-Ss4R conservation of genes encoding microRNAs is nearly complete, with almost all microRNA ohnologs being conserved as duplicated Ss4R copies. This higher conservation does not seem to be simply the outcome of the shorter length of sequences coding for miRNA and might result from more subtle selective processes. Altogether, the high retention rate of ohnologous copies and of pseudogenes suggests that the fractionation process is largely incomplete and still ongoing in the trout genome. At the genome organization level, the analysis of the Ss4R duplicated regions reveals a high colinearity between paralogous genomic sequences, consistent with a conserved order of ohnologs and no strong evidence of a clustering of singletons versus ohnologs throughout the genome. Together these observations show that gene fractionation does not involve many genomic rearrangements such as inversions or translocations that would modify the order of genes in the genome, or large deletions that would result in long clusters of singletons. The nucleotide sequence identity between paralogous genomic regions is still high (86%) and Ss4R ohnologous protein-coding sequences and microRNAs are also highly conserved at the sequence level with 92.9% amino-acid identity and 96.4% nucleotide identity, respectively. In addition, the identity between the Ss4R protein-coding singletons and their corresponding pseudogenes remains high (average amino acid identity 79.0%) suggesting that most of these gene inactivations took place recently.

A genome for genetic and functional investigations

Deciphering the genetic architecture of traits

Quantitative trait loci (QTLs) are genomic regions (loci) associated to the phenotypic variation of a trait. Typically, a QTL is linked to, or contains, gene(s) that control the target trait. Therefore, using the polymorphism (molecular or sequence information) associated to the QTL regions is a way to improve the efficiency of selective breeding by targeting the key regions that govern the variability of the target trait. Also, the identification of QTLs constitutes a step toward the molecular dissection of complex traits, the discovery of the genes that control functions of interest (causative genes) and their regulation.

Using the medium density genetic maps previously available (see earlier) and linkage association methodologies (LA) in family QTL designs, a number of QTL have been detected in rainbow trout for many traits, with a special focus on adaptation and disease resistance traits (Baerwald et al., 2011; Le-Bras et al., 2011; Hecht et al., 2012; Verrier et al., 2013; Liu et al., 2014; Quillet et al., 2014; Vallejo et al., 2014). However, the confidence intervals of QTL positions are usually wide, making their exploitation difficult.

Together with the rapid evolution of NGS technologies, high-density genotyping methods are now available. For instance, the restriction-site associated DNA tags (RAD-tags) sequencing (Etter et al., 2011) enables cost effective identification of thousands SNP. In rainbow trout, RAD-tags based linkage maps were produced resulting in an increased marker density (Miller et al., 2012). RAD-sequencing was also used in an extended SNP discovery study where nearly 145,000 SNPs were identified (Palti et al., 2014) allowing the development of high-density genotyping chips like the 57 K SNP chip recently developed for rainbow trout (Palti et al., 2015) that is now commercially available. In such studies, having a reference genome sequence as a template is of great interest to order the newly discovered SNPs into maps or build local haplotypes. These high-density genotyping tools allow a radical change in QTL mapping methods and enable the switch to methods exploiting linkage disequilibrium (LD) and association analyses at the population level (LDLA, Genome Wide Association Study or GWAS) (Liu et al, 2015). This opens new areas for deciphering complex traits, and for the identification of genomic regions under selection and the future implementation of genomic selection for aquaculture breeding.

Functional insights in Rainbow trout nutrition

Aquafeeds have been developed over the past thirty years for rainbow trout in aquaculture based on the well-known nutrient requirements identified for this species. In the context of the development of sustainable aquaculture, it is imperative to develop new aquafeeds in order to replace dietary fish meal and fish oil (originating from threatened marine resources) by either plant ingredients or other ingredients (Naylor et al., 2009). This evolution of dietary composition necessitates an extensive knowledge of the intermediary metabolism mainly based on molecular approaches (Panserat et al., 2009; Panserat and Kaushik, 2010).

Selected examples of nutritional regulation of transcript levels

Nutrition and flesh quality: molecular LC-PUFA biosynthesis pathway is functional

One of the most important challenges after suppression of fish oil in aquafeed is to maintain high level of long-chain polyunsaturated fatty acids (LC-PUFAs, i.e., the docosahexaenoic acid—DHA—and the eicosapentaenoic acid—EPA) in the flesh (Monroig et al., 2013). Numerous studies in rainbow trout demonstrated that this species has the molecular capacities to produce these LC-PUFAs from precursors found in vegetable oils. Indeed, higher levels of mRNAs coding for desaturase and elongase enzymes were observed following suppression of fish oil in the diets (Seiliez et al., 2011). However, this induction is not sufficient to obtain the same level of muscle LC-PUFAs compared to fish fed with fish oil. LC-PUFA biosynthesis in rainbow trout seems to substrate limited, even when some fish lines presented higher levels of fatty biosynthesis molecular capacity (Kamalam et al., 2012).

Functional but atypically regulated glucose metabolism

Rainbow trout is well known to be a poor user of dietary carbohydrates, limiting the incorporation of starch (main component of rich-carbohydrate plants) in aquafeeds (Polakof et al., 2012). Recent studies prove that rainbow trout has the inducible capacities to catabolize dietary glucose as suggested by levels of transcripts for glucose transporter in muscle (Diaz et al., 2009), for glucokinase enzyme in liver (Panserat et al., 2014) and for miRNAs involved in regulation of insulin signaling and metabolism (Mennigen et al., 2014, 2012) following carbohydrate intake. By contrast, the molecular regulation of hepatic glucose production seems to be atypical (see further sections) and could be one of the reasons for the poor metabolic use of carbohydrate in trout.

Linking nutrition to metabolism: hormonal and nutrient sensor roles

The decrease of feed intake associated with a low level of plasmatic insulin/glucagon ratio (mainly due to the pursuit of glucagon secretion) are often observed in rainbow trout following the intake of new aquafeeds, having potential strong impacts on feed efficiency (Panserat and Medale, 2013). Recent studies about the nutritional regulation of nutrient sensing pathway in hypothalamus (center of the feed intake regulation) and in endocrine pancreas (center of insulin and glucagon secretion linked to the nutritional status) demonstrated the existence of glucose and lipid sensing mechanisms in these tissues (Caruso and Sheridan, 2011). Fine nutritional regulations of these pathways at a molecular level are in progress in rainbow trout fed alternative diets.

The 21st century or the "omics" boom in nutritional studies

In the 2000s, the omics approaches have been introduced for the nutritional studies in rainbow trout. In particular, the hepatic transcriptome analysis became a new tool to study nutrition in rainbow trout. Indeed, because the liver is the center of the intermediary metabolism, postprandial hepatic transcriptomics were analyzed after fish oil suppression (Kolditz et al., 2008; Panserat et al., 2008), fish oil replacement by vegetable oils (Panserat et al., 2008), total fish oil and fish meal replacement (Overturf et al., 2012) and micronutrient deficiencies (Olsvik et al., 2013). These data confirmed studies linked to the intermediary metabolism but opened also new hypothesis related to health (immunology), xenobiotic metabolism, proteolysis and cell division. Some of the transcriptomics studies were completed by proteomics analysis (Kolditz et al., 2008; Martin et al., 2003), which seemed to confirm the transcript analysis but were limited by the current technical inability to identify proteins in rainbow trout.

Nutritional regulation of hepatic glucose-6-phosphatase genes expression in rainbow trout: how the trout genome sequencing provides new answers to old questions

The rainbow trout is considered to be a glucose-intolerant species due mainly to persistent hyperglycemia after intake of carbohydrate-enriched meal(s) or glucose tolerance tests (Polakof et al., 2012). One hypothesis to explain such a phenotype is an insufficient inhibition of endogenous glucose production via gluconeogenesis pathway (Panserat et al., 2000). Hepatic glucose-6-phosphatase (G6pc), the last gluconeogenic enzyme catalyzing the hydrolysis of glucose-6-phosphate (G6P) in glucose, was proposed to be a major actor involved in blood glucose enrichment under hyperglycemic conditions [Type 2 diabetes (Rooney et al., 1993)] or after carbohydrate intake [in seabass (Viegas et al., 2015)] by establishing a futile glucose/G6P cycle together with glucokinase, the first glycolytic enzyme catalyzing the phosphorylation of glucose in G6P. A gene coding for G6pc in rainbow trout was for the first time partially sequenced in 2000 (Panserat et al., 2000) and the nutritional/hormonal regulation of its expression was then deeply investigated first by northern blot analysis (Panserat et al., 2002, 2000) progressively replaced by contemporary real-time quantitative PCR analysis. *g6pc* mRNA level was thus shown to be up-regulated by high dietary lipids but was surprisingly not regulated by high carbohydrate diet intake containing digestible starch (Panserat et al., 2008). In vivo and in vitro analysis also demonstrated that *g6pc* mRNA level was positively correlated to protein (Kirchner et al., 2003a) and amino acids (AA) pool (Lansard et al., 2010) levels respectively. Moreover specific AA were involved in the regulation of *g6pc* gene such as alanine (Kirchner et al., 2003b), one gluconeogenic dispensable AA (G-DAA), or leucine and methionine (Lansard et al., 2011). Finally insulin regulation of this gene was investigated and demonstrated to have a down-regulation effect on *g6pc* mRNA level both in vivo (Polakof et al., 2009) and in vitro (Plagnes-Juan et al., 2008) and to be able to counteract the increase of mRNA level induced by glucose addition in the cell cultured medium or in trout fed a high carbohydrate diet. However, some of these expression results remained questioning when putting into perspective of enzyme activity results. For instance insulin seemed to act at the molecular level to down-regulate *g6pc* mRNA level but had no effect on G6pc hepatic and gut activity (Polakof et al., 2011). In the same way, *g6pc* mRNA level was down-regulated by only one G-DAA (alanine), whereas the enzyme activity was decreased in trout fed a three G-DAA substituted diets (alanine or aspartic acid or glutamic acid, (Kirchner et al., 2003b)). In addition, no clear reason could be given to explain why G6pc activity did not change when *g6pc* was repressed by a single meal with glucose (Panserat et al., 2000) until the identification of a second gene coding for G6pc in EST databases in 2008 (Plagnes-Juan et al., 2008). Indeed the analysis of this new gene brought new information to interpret data presented above and let sense that the complexity of rainbow trout genome has to be considered to better understand nutritional regulation. In particular, data showed that both *g6pc* genes were differentially regulated by nutritional status (Mennigen et al., 2013) but also displayed a contrasting regulation regarding the relative proportion of carbohydrate in the diet (Kamalam et al., 2012; Seiliez et al., 2011). With the recent sequencing of the rainbow trout genome (Berthelot et al., 2014) molecular comprehension of *g6pc*/G6pc nutritional regulation moved on to a next step. Actually in silico investigations revealed that in fact five paralogous genes coding for G6pc were retained in the rainbow trout genome after Ss4R and that two of them, never identified before, displayed an unexpected up-regulation of their mRNA levels in trout fed a high carbohydrate diet (Marandel et al., 2015). Overall enzyme activity measured in the liver of these fish suggested that proteins translated from the latter two genes might be involved in this activity. It was hypothesized that these two genes may contribute to the hyperglycemic phenotype by releasing glucose in blood via the establishment of a futile cycle together with glucokinase. Old results concerning *g6pc* genes regulation by insulin or G-DAA can be now reinterpreted in the light of this new discovery and by considering that *g6pc* paralogs are potentially differentially regulated but that they all may contribute to overall enzyme activity.

The example of *g6pc* genes strongly illustrates how advances in genomics can serve our understanding of nutritional regulation of metabolism and provide new answers to old questions. But as usual, close a window often opens a door and many questions related to the role of differential regulation of *g6pc* paralogous genes and about their related contribution to one or another phenotype arise.

Trout genomics for health

The rainbow trout immunome: the added value of a full repertoire of fish immunity genes

Toward the definition of a trout immunome

When the human genome was sequenced, a quasicomplete repertoire of proteins involved in immunity—the "Immunome"—became available (Ortutay and Vihinen, 2009). Lots of data were already available at the molecular, structural, and cellular levels, in both normal and diseased states regarding genes involved in mouse and human immunity; these data were integrated in databases and tools dedicated to immune genes and pathways. When the genome of another vertebrate is sequenced, its immunome can be annotated using three main approaches: (1) the functional knowledge available for immune genes of other species like human or mouse can be attached to their orthologs, considering that phylogenetic orthologs in different species often have similar roles, (2) datasets from high-throughput approaches of gene expression in infectious and other pathological states can identify genes modulated in diseases and during responses (3) at last, direct genomic and functional studies of genes or gene families can be undertaken from the newly available genome.

The annotation of the rainbow trout immunome is not an easy task and cannot be achieved by a direct transposition of the knowledge available in mouse and human databases; indeed, as mentioned above, two cycles of WGD occurred during the evolution of salmonids, providing a very large number of paralogous genes susceptible to sub functionalization. Additionally, several gene families such as chemokines and trims underwent considerable specific expansion in salmonids (Boudinot et al., 2011; Chen et al., 2013; Marancik et al., 2014). Hence, a well-assembled genome in which closely related duplicates are predicted constitutes a pivotal resource to define the immunome, especially from high-throughput transcriptomes.

An advanced trout immunome will provide a good basis for further developments such as the analysis of genetic polymorphism and epigenetic approaches, which will be instrumental for marker based selection programs.

The case of antigen specific receptors, Igs and TCRs

As new sequencing technologies gave access to the quasi-complete diversity of immunoglobulin (Ig) or T cell receptor (TCR) sequences expressed in a tissue—or even in a whole individual—studies of immune repertoires have greatly expanded in the last five years (Castro et al., 2013; Weinstein et al., 2009). These studies provide a detailed account of the adaptive immune responses against pathogens, and assess the status of the immune system in different pathologies. Such approaches have been developed in rainbow trout and zebrafish based on Ig and TCR sequences available in these species. The availability of a full trout genome sequence, and the potential possibility to obtain a detailed annotation of TCR and Ig loci in this species, will provide researchers with a much-empowered system to access the complete diversity of immune repertoires and responses. Additionally, the modelization approaches that apply to repertoire data using theoretical physics tools (Mora et al., 2010) can be developed only when the sequences of all V, D, and J gene segments to recombine are available. The rainbow trout genome represents the first step—and a necessary resource—for the development of such approaches that will be instrumental for evaluation of future vaccines.

Interest of a genome assembly for linkage studies: gene clusters and linkage studies

The presence and functional importance of gene clusters involved in immunity remain open questions. Several selection pressures may account for such clusters, including local dynamics of gene duplication and coregulation of expression. The trout genome provides an interesting opportunity to address such issues, one of which refers to the history of the major histocompatibility complex (MHC). TCRs recognize antigens as small peptides (9–22 amino acids) bound to membrane proteins called MHC class I or class II molecules, that are expressed at the surface of "antigen presenting cells." The TCR recognizes both the MHC protein and the peptide antigen presented by the MHC molecule, a phenomenon known as "MHC restriction." The genetic "MHC" is a region of the genome defined by the presence of MHC class I and II molecules, and genes involved in the preparation and loading of the peptide antigen for MHC presentation. In fact, the MHC encodes many genes (more than 100 genes in human), approximately half of which are implicated in immunity (The_MHC_sequencing_consortium, 1999); thus the genes of this region bring an important contribution to the immunome and defense mechanisms. The origin and evolutionary pathway of the MHC remains incompletely understood. MHC is found in all jaw vertebrates from sharks to mammals. As mentioned earlier, it is now generally accepted that according to Ohno's theory (Ohno et al., 1968), two rounds of whole genome duplication occurred after the emergence of urochordates and before the radiation of jawed vertebrates, leading to four sets of paralogous regions. Such tetrads were identified for several key genetic regions, including the HOX and MHC complexes. Two additional tetrads were identified across vertebrate genomes, that apparently derived from neighboring regions on the same ancestral chromosome containing the "proto-MHC" of common ancestors of vertebrates and other phyla (Flajnik et al., 2012); such proto-MHC regions were, for example, identified in the placozoan *Trichoplax adhaerens*, which is considered as a basal bilaterian branch (Suurvali et al., 2014). Altogether, these three MHC related tetrads encode a large number of genes involved in immunity in vertebrates, including several major gene complexes such as the natural killer receptor complex and the leukocyte receptor complex; they also count many important immune gene families as B7 receptors, TRIMs, etc.

The MHC was conserved as a defined genomic entity across vertebrates from sharks to mammals, with the exception of fishes in which the regions encoding MHC class I and MHC class II are found on different chromosomes (Flajnik and Kasahara, 2010). The functional significance of this conservation, and of the different pattern observed in fish, remains elusive. Coregulation of many genes involved in immunity is a tempting hypothesis, but it has received little direct support to date and is challenged by the pattern observed in fish. Fishes represent interesting models to address such questions and get insights into the immunological impact of a "broken" MHC. Fish groups that have lost MHC class II genes and classical T helper responses—such as cods and pipefishes (Star and Jentoft, 2012)—are good examples of potential repercussions; Salmonids, with additional cycles of genome duplication and ongoing rediploidization, represent another context of great interest to challenge and understand the fate of the MHC region and its paralogs or related regions.

The rainbow trout genome constitutes a great resource for fish and comparative immunologists. As a basis to define a complete immunome, it is a very important resource for future developments requiring monitoring of immune responses, including selection of more robust animals. Additionally, with the genome of the closely related Atlantic salmon, it also represents a unique opportunity to address generic and evolutionary questions for which teleosts harbor unique features. Finally, a good quality genome assembly will also be very important for future studies of the trout gut (or other) microbiota.

Rainbow trout reproduction with a genome sequence

A search in the Web of Science database (Science Citation Expanded 1975–present) yielded over 8,000 articles dealing with various aspects of rainbow trout reproduction. In contrast, a similar search for zebrafish—a widely used biological model—reproduction only yielded 3,500 references. This illustrates the importance of the scientific community working in the field of reproduction in rainbow trout and the potential impact of the rainbow trout genome and related genomic resources to study reproduction in this species. As for other disciplinary fields, the recent release of the rainbow trout genome now offers new possibilities to study gene synteny and to elucidate previously unresolved evolutionary history of genes playing important roles in reproduction as illustrated hereafter.

Sex hormone-binding globulins (Shbg) are carrier blood proteins involved in the transport of sex steroids in plasma and the regulation of their availability to target organs. Two genes—*shbga* and *shbgb*—exist in fish, *shbga* being the orthologs of the *SHBG* human gene. Shbgb has been found in rainbow trout and salmon but never described in any nonsalmonid teleost species to date. While this could suggest that *shbga* and *shbgb* originate from Ss4R WGD, the topology of the Shbg tree (Bobe et al., 2010) as well as recent analyses using the rainbow trout genome sequence suggest that this duplication is much more ancient than initially thought.

Gonadal soma-derived growth factor (Gsdf) is a recently identified member of the TGF-beta superfamily (Sawatari et al., 2007) that is believed to play an important role for proliferation of primordial germ cells and spermatogonia. It is also spatially and temporally correlated with early testicular differentiation in medaka (Shibata et al., 2010). Within medaka gonads, *gsdf* gene expression is restricted to Sertoli and granulosa cells. The fate of *gsdf* after Ss4R WGD remains however uncharacterized but recent preliminary analyses suggest that Ss4R yielded two *gsdf* genes in rainbow trout with possible different expression patterns (Fig. 2.3). The rainbow trout genome will allow thorough characterization of *gsdf* ohnologs including any gene silencing or sub-/neo-functionalization after duplication.

With the genome sequence of rainbow trout, the use of genome editing will also become possible without relying on existing resources (e.g., BAC libraries) to target

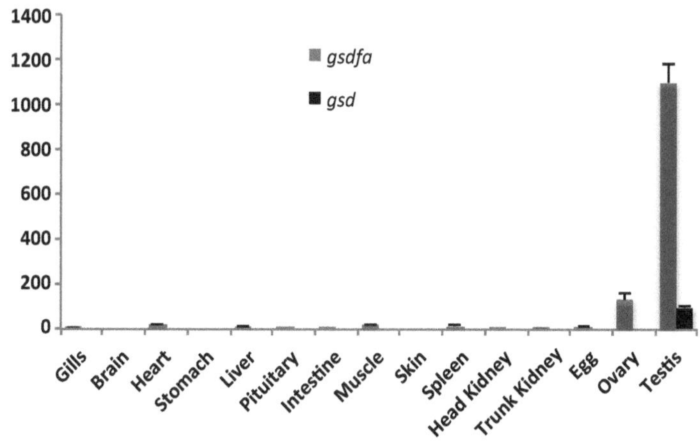

Figure 2.3 Example of evolution of gene expression after Ss4R gene duplication. Expression levels of *gsdf* Ss4R ohnologs in different rainbow trout tissues, showing that one of the two *gsdf* ohnolog is predominantly expressed even if the expression profiles of these two genes are still identical. mRNA levels (mean, SD) were measured in tissues originating from three individuals. Expression was arbitrarily set to 100 for *gsdfb* expression in testis.

important genes and describe the phenotypes associated with corresponding knockouts. While rainbow trout has a relatively long lifecycle, in comparison to other fish model species (e.g., zebrafish, medaka), gene knockout is sometimes necessary to decipher the function of salmonid specific genes such as the sex determining gene *sdY* (Yano et al., 2012). Because of its long-term use as a model for reproduction, rainbow trout could be an interesting model to perform genome editing in a well-characterized teleost model exhibiting group-synchronous oogenetic development. A recent study showed that the promising and now widely used Crispr/CAS9 technology can successfully be used in salmonids (Edvardsen et al., 2014).

In summary, the recent publication of the rainbow trout genome offers new possibilities in the field of reproductive biology with major outcomes related to the control of sex-ratio, puberty, fecundity, and sterility.

Future directions and concluding remarks: resources are still expending

Next generation sequencing (NGS) is still a rapidly evolving field, and second-generation of NGS technologies are now available with the emergence of longer-read length technologies and higher volumes of produced sequence information. Using NGS in rainbow trout, transcriptome resources have been extended by numerous studies that have sequenced cDNAs in different tissues, stages of development, or conditions of treatments (Ali et al., 2014; Liu et al., 2014; Marancik et al., 2014; Salem et al., 2015). In addition to cDNA transcripts, deep microRNA sequencing has also been carried out in rainbow trout (Farlora et al., 2015; Juanchich et al., 2013; Ma et al., 2015; Salem et al., 2010) leading to detailed catalogs of microRNA repertoires. With new NGS technologies resequencing of the whole genome of new rainbow trout individuals from differing genetic origins, including doubled haploids individuals, will also be possible at limited cost. Therefore, new markers and new polymorphisms will be discovered, that will facilitate further QTL detection and refine functional analyses.

A new updated version of the rainbow trout genome has been announced to be released in 2016 based on recent NGS improvements of the Illumina sequencing technique allowing the production of 2×250 bp reads in large amounts (USDA initiative, Unpublished). The total size of the resulting assembly of this updated version of the rainbow trout genome is announced to be 2,17 Gb with a scaffold N50 of 1,700 kb. All these quickly growing resources will now have to be integrated to produce reference comprehensive maps of all functional elements in the rainbow trout genomes for instance through international initiative like the Functional Annotation of Animal Genomes (Andersson et al., 2015), which is the equivalent of the ENCODE Human genome. Altogether these joined efforts will lead to a better understanding of the rainbow trout genome allowing the use of these integrated resources for whole genome genetic selection, investigation of epigenetic regulations on a genome-scale (Baerwald et al., 2015) and all kinds of whole genome analysis.

References

Ali, A., Rexroad, C.E., Thorgaard, G.H., Yao, J., Salem, M., 2014. Characterization of the rainbow trout spleen transcriptome and identification of immune-related genes. Front. Genet. 5, 348.

Allendorf, F.W., Thorgaard, G.H., 1984. Tetraploidy and the evolution of salmonid fishes. In: Turner, B.J. (Ed.), Evolutionary Genetics of Fishes, Monographs in Evolutionary Biology. Springer, US, pp. 1–53.

Andersson, L., Archibald, A.L., Bottema, C.D., Brauning, R., Burgess, S.C., Burt, D.W., Casas, E., Cheng, H.H., Clarke, L., Couldrey, C., Dalrymple, B.P., Elsik, C.G., Foissac, S., Giuffra, E., Groenen, M.A., Hayes, B.J., Huang, L.S., Khatib, H., Kijas, J.W., Kim, H., Lunney, J.K., McCarthy, F.M., McEwan, J.C., Moore, S., Nanduri, B., Notredame, C., Palti, Y., Plastow, G.S., Reecy, J.M., Rohrer, G.A., Sarropoulou, E., Schmidt, C.J., Silverstein, J., Tellam, R.L., Tixier-Boichard, M., Tosser-Klopp, G., Tuggle, C.K., Vilkki, J., White, S.N., Zhao, S., Zhou, H., FAANG Consortium, 2015. Coordinated international action to accelerate genome-to-phenome with FAANG, the functional annotation of animal genomes project. Genome Biol. 16, 57.

Baerwald, M.R., Meek, M.H., Stephens, M.R., Nagarajan, R.P., Goodbla, A.M., Tomalty, K.M.H., Thorgaard, G.H., May, B., Nichols, K.M., 2015. Migration-related phenotypic divergence is associated with epigenetic modifications in rainbow trout. Mol. Ecol. 25 (8), 1785–1800.

Baerwald, M.R., Petersen, J.L., Hedrick, R.P., Schisler, G.J., May, B., 2011. A major effect quantitative trait locus for whirling disease resistance identified in rainbow trout (*Oncorhynchus mykiss*). Heredity 106, 920–926.

Behnke, R.J., Tomelleri, J., Proebstel, D.S., 2002. Trout and Salmon of North America, first ed. Chanticleer Press, Free Press, New York.

Bernardi, G., Bernardi, G., 1990. Compositional transitions in the nuclear genomes of cold-blooded vertebrates. J. Mol. Evol. 31, 282–293.

Berthelot, C., Brunet, F., Chalopin, D., Juanchich, A., Bernard, M., Noël, B., Bento, P., Da Silva, C., Labadie, K., Alberti, A., Aury, J.-M., Louis, A., Dehais, P., Bardou, P., Montfort, J., Klopp, C., Cabau, C., Gaspin, C., Thorgaard, G.H., Boussaha, M., Quillet, E., Guyomard, R., Galiana, D., Bobe, J., Volff, J.-N., Genêt, C., Wincker, P., Jaillon, O., Roest Crollius, H., Guiguen, Y., 2014. The rainbow trout genome provides novel insights into evolution after whole-genome duplication in vertebrates. Nat. Commun. 5, 3657.

Bobe, J., Guiguen, Y., Fostier, A., 2010. Diversity and biological significance of sex hormone-binding globulin in fish, an evolutionary perspective. Mol. Cell. Endocrinol. 316, 66–78.

Boudinot, P., van der Aa, L.M., Jouneau, L., Pasquier, L. Du, Pontarotti, P., Briolat, V., Benmansour, A., Levraud, J.P., 2011. Origin and evolution of TRIM proteins: new insights from the complete TRIM repertoire of zebrafish and pufferfish. PLoS One 6, e22022.

Canario, A.V.M., Bargelloni, L., Volckaert, F., Houston, R.D., Massault, C., Guiguen, Y., 2008. Genomics toolbox for farmed fish. Rev. Fish. Sci. 16, 3–15.

Caruso, M.A., Sheridan, M.A., 2011. New insights into the signaling system and function of insulin in fish. Gen. Comp. Endocrinol. 173, 227–247.

Castro, R., Jouneau, L., Pham, H.P., Bouchez, O., Giudicelli, V., Lefranc, M.P., Quillet, E., Benmansour, A., Cazals, F., Six, A., Fillatreau, S., Sunyer, O., Boudinot, P., 2013. Teleost fish mount complex clonal IgM and IgT responses in spleen upon systemic viral infection. PLoS Pathog. 9, e1003098.

Chen, J., Xu, Q., Wang, T., Collet, B., Corripio-Miyar, Y., Bird, S., Xie, P., Nie, P., Secombes, C.J., Zou, J., 2013. Phylogenetic analysis of vertebrate CXC chemokines reveals novel lineage specific groups in teleost fish. Dev. Comp. Immunol. 41, 137–152.

Chourrout, D., 1984. Pressure-induced retention of second polar body and suppression of first cleavage in rainbow trout: production of all-triploids, all-tetraploids, and heterozygous and homozygous diploid gynogenetics. Aquaculture 36, 111–126.

Crête-Lafrenière, A., Weir, L.K., Bernatchez, L., 2012. Framing the Salmonidae family phylogenetic portrait: a more complete picture from increased taxon sampling. PloS One 7, e46662.

Dehal, P., Boore, J.L., 2005. Two rounds of whole genome duplication in the ancestral vertebrate. PLoS Biol. 3, e314.

Diaz, M., Vraskou, Y., Gutierrez, J., Planas, J.V., 2009. Expression of rainbow trout glucose transporters GLUT1 and GLUT4 during in vitro muscle cell differentiation and regulation by insulin and IGF-I. Am. J. Physiol. Regul. Integr. Comp. Physiol. 296, R794–R800.

Edvardsen, R.B., Leininger, S., Kleppe, L., Skaftnesmo, K.O., Wargelius, A., 2014. Targeted mutagenesis in Atlantic salmon (*Salmo salar* L.) using the CRISPR/Cas9 system induces complete knockout individuals in the F0 generation. PloS One 9, e108622.

Etter, P.D., Bassham, S., Hohenlohe, P.A., Johnson, E.A., Cresko, W.A., 2011. SNP discovery and genotyping for evolutionary genetics using RAD sequencing. Methods Mol. Biol. Clifton NJ 772, 157–178.

Farlora, R., Valenzuela-Miranda, D., Alarcón-Matus, P., Gallardo-Escárate, C., 2015. Identification of microRNAs associated with sexual maturity in rainbow trout brain and testis through small RNA deep sequencing. Mol. Reprod. Dev. 82 (9), 651–662.

Flajnik, M.F., Kasahara, M., 2010. Origin and evolution of the adaptive immune system: genetic events and selective pressures. Nat. Rev. Genet. 11, 47–59.

Flajnik, M.F., Tlapakova, T., Criscitiello, M.F., Krylov, V., Ohta, Y., 2012. Evolution of the B7 family: co-evolution of B7H6 and NKp30, identification of a new B7 family member, B7H7, and of B7's historical relationship with the MHC. Immunogenetics 64, 571–590.

Genet, C., Dehais, P., Palti, Y., Gao, G., Gavory, F., Wincker, P., Quillet, E., Boussaha, M., 2011. Analysis of BAC-end sequences in rainbow trout: content characterization and assessment of synteny between trout and other fish genomes. BMC Genomics 12, 314.

Gilles, A., Meglécz, E., Pech, N., Ferreira, S., Malausa, T., Martin, J.-F., 2011. Accuracy and quality assessment of 454 GS-FLX Titanium pyrosequencing. BMC Genomics 12, 245.

Govoroun, M., Le Gac, F., Guiguen, Y., 2006. Generation of a large scale repertoire of expressed sequence tags (ESTs) from normalised rainbow trout cDNA libraries. BMC Genomics 7, 196.

Guyomard, R., Mauger, S., Tabet-Canale, K., Martineau, S., Genet, C., Krieg, F., Quillet, E., 2006. A type I and type II microsatellite linkage map of rainbow trout (*Oncorhynchus mykiss*) with presumptive coverage of all chromosome arms. BMC Genomics 7, 302.

Guyomard, R., Boussaha, M., Krieg, F., Hervet, C., Quillet, E., 2012. A synthetic rainbow trout linkage map provides new insights into the salmonid whole genome duplication and the conservation of synteny among teleosts. BMC Genet. 13, 15.

Hecht, B.C., Thrower, F.P., Hale, M.C., Miller, M.R., Nichols, K.M., 2012. Genetic architecture of migration-related traits in rainbow and steelhead trout, *Oncorhynchus mykiss*. Genes Genom. Genet. 2, 1113–1127.

Howe, K., Clark, M.D., Torroja, C.F., Torrance, J., Berthelot, C., Muffato, M., Collins, J.E., Humphray, S., McLaren, K., Matthews, L., McLaren, S., Sealy, I., Caccamo, M., Churcher, C., Scott, C., Barrett, J.C., Koch, R., Rauch, G.-J., White, S., Chow, W., Kilian, B., Quintais, L.T., Guerra-Assunção, J.A., Zhou, Y., Gu, Y., Yen, J., Vogel, J.-H., Eyre, T., Redmond, S., Banerjee, R., Chi, J., Fu, B., Langley, E., Maguire, S.F., Laird, G.K., Lloyd, D., Kenyon, E., Donaldson, S., Sehra, H., Almeida-King, J., Loveland, J., Trevanion, S., Jones, M., Quail, M., Willey, D., Hunt, A., Burton, J., Sims, S., McLay, K., Plumb, B., Davis, J., Clee, C., Oliver, K., Clark, R., Riddle, C., Elliot, D., Eliott, D., Threadgold, G., Harden, G., Ware, D., Begum, S., Mortimore, B., Mortimer, B., Kerry, G., Heath, P., Phillimore, B., Tracey, A., Corby, N., Dunn, M., Johnson, C., Wood, J., Clark, S., Pelan, S., Griffiths, G., Smith, M., Glithero, R., Howden, P., Barker, N., Lloyd, C., Stevens, C., Harley, J., Holt, K., Panagiotidis, G., Lovell, J., Beasley, H., Henderson, C., Gordon, D., Auger, K., Wright, D., Collins, J., Raisen, C., Dyer, L., Leung, K., Robertson, L., Ambridge, K., Leongamornlert, D., McGuire, S., Gilderthorp, R., Griffiths, C., Manthravadi, D., Nichol, S., Barker, G., Whitehead, S., Kay, M., Brown, J., Murnane, C., Gray, E., Humphries, M., Sycamore, N., Barker, D., Saunders, D., Wallis, J., Babbage, A., Hammond, S., Mashreghi-Mohammadi, M., Barr, L., Martin, S., Wray, P., Ellington, A., Matthews, N., Ellwood, M., Woodmansey, R., Clark, G., Cooper, J.D., Cooper, J., Tromans, A., Grafham, D., Skuce, C., Pandian, R., Andrews, R., Harrison, E., Kimberley, A., Garnett, J., Fosker, N., Hall, R., Garner, P., Kelly, D., Bird, C., Palmer, S., Gehring, I., Berger, A., Dooley, C.M., Ersan-Ürün, Z., Eser, C., Geiger, H., Geisler, M., Karotki, L., Kirn, A., Konantz, J., Konantz, M., Oberländer, M., Rudolph-Geiger, S., Teucke, M., Lanz, C., Raddatz, G., Osoegawa, K., Zhu, B., Rapp, A., Widaa, S., Langford, C., Yang, F., Schuster, S.C., Carter, N.P., Harrow, J., Ning, Z., Herrero, J., Searle, S.M.J., Enright, A., Geisler, R., Plasterk, R.H.A., Lee, C., Westerfield, M., de Jong, P.J., Zon, L.I., Postlethwait, J.H., Nüsslein-Volhard, C., Hubbard, T.J.P., Roest Crollius, H., Rogers, J., Stemple, D.L., 2013. The zebrafish reference genome sequence and its relationship to the human genome. Nature 496, 498–503.

Hurley, I.A., Mueller, R.L., Dunn, K.A., Schmidt, E.J., Friedman, M., Ho, R.K., Prince, V.E., Yang, Z., Thomas, M.G., Coates, M.I., 2007. A new time-scale for ray-finned fish evolution. Proc. Biol. Sci. 274, 489–498.

Juanchich, A., Le Cam, A., Montfort, J., Guiguen, Y., Bobe, J., 2013. Identification of differentially expressed miRNAs and their potential targets during fish ovarian development. Biol. Reprod. 88, 128.

Kamalam, B.S., Medale, F., Kaushik, S., Polakof, S., Skiba-Cassy, S., Panserat, S., 2012. Regulation of metabolism by dietary carbohydrates in two lines of rainbow trout divergently selected for muscle fat content. J. Exp. Biol. 215, 2567–2578.

Kirchner, S., Kaushik, S., Panserat, S., 2003a. Low protein intake is associated with reduced hepatic gluconeogenic enzyme expression in rainbow trout (*Oncorhynchus mykiss*). J. Nutr. 133, 2561–2564.

Kirchner, S., Kaushik, S., Panserat, S., 2003b. Effect of partial substitution of dietary protein by a single gluconeogenic dispensable amino acid on hepatic glucose metabolism in rainbow trout (*Oncorhynchus mykiss*). Comp. Biochem. Physiol. Mol. Integr. Physiol. 134, 337–347.

Kolditz, C.I., Paboeuf, G., Borthaire, M., Esquerre, D., SanCristobal, M., Lefevre, F., Medale, F., 2008. Changes induced by dietary energy intake and divergent selection for muscle fat content in rainbow trout (*Oncorhynchus mykiss*), assessed by transcriptome and proteome analysis of the liver. BMC Genomics 9, 506.

Komen, H., Thorgaard, G.H., 2007. Androgenesis, gynogenesis and the production of clones in fishes: a review. Aquaculture 269, 150–173.

Koop, B.F., von Schalburg, K.R., Leong, J., Walker, N., Lieph, R., Cooper, G.A., Robb, A., Beetz-Sargent, M., Holt, R.A., Moore, R., Brahmbhatt, S., Rosner, J., Rexroad, C.E., McGowan, C.R., Davidson, W.S., 2008. A salmonid EST genomic study: genes, duplications, phylogeny and microarrays. BMC Genomics 9, 545.

Langham, R.J., Walsh, J., Dunn, M., Ko, C., Goff, S.A., Freeling, M., 2004. Genomic duplication, fractionation and the origin of regulatory novelty. Genetics 166, 935–945.

Lansard, M., Panserat, S., Plagnes-Juan, E., Seiliez, I., Skiba-Cassy, S., 2010. Integration of insulin and amino acid signals that regulate hepatic metabolism-related gene expression in rainbow trout: role of TOR. Amino Acids 39, 801–810.

Lansard, M., Panserat, S., Plagnes-Juan, E., Dias, K., Seiliez, I., Skiba-Cassy, S., 2011. L-leucine, L-methionine, and L-lysine are involved in the regulation of intermediary metabolism-related gene expression in rainbow trout hepatocytes. J. Nutr. 141, 75–80.

Le-Bras, Y., Dechamp, N., Krieg, F., Filangi, O., Guyomard, R., Boussaha, M., Bovenhuis, H., Pottinger, T.G., Prunet, P., Le-Roy, P., Quillet, E., 2011. Detection of QTL with effects on osmoregulation capacities in the rainbow trout (*Oncorhynchus mykiss*). BMC Genet. 12, 46.

Liu, S., Gao, G., Palti, Y., Cleveland, B.M., Weber, G.M., Rexroad, C.E., 2014. RNA-seq analysis of early hepatic response to handling and confinement stress in rainbow trout. PloS One 9, e88492.

Liu, S.X., Vallejo, R.L., Gao, G.T., Palti, Y., Weber, G.M., Hernandez, A., Rexroad, C.E., 2015. Identification of Single-Nucleotide Polymorphism Markers Associated with Cortisol Response to Crowding in Rainbow Trout. Mar. Biotechnol. 17 (3), 328–337.

Ma, H., Weber, G.M., Hostuttler, M.A., Wei, H., Wang, L., Yao, J., 2015. MicroRNA expression profiles from eggs of different qualities associated with post-ovulatory ageing in rainbow trout (*Oncorhynchus mykiss*). BMC Genomics 16, 201.

Macqueen, D.J., Johnston, I.A., 2014. A well-constrained estimate for the timing of the salmonid whole genome duplication reveals major decoupling from species diversification. Proc. Biol. Sci. 281, 20132881.

Marancik, D., Gao, G., Paneru, B., Ma, H., Hernandez, A.G., Salem, M., Yao, J., Palti, Y., Wiens, G.D., 2014. Whole-body transcriptome of selectively bred, resistant-, control-, and susceptible-line rainbow trout following experimental challenge with *Flavobacterium psychrophilum*. Front Genet. 5, 453.

Marandel, L., Seiliez, I., Veron, V., Skiba-Cassy, S., Panserat, S., 2015. New insights into the nutritional regulation of gluconeogenesis in carnivorous rainbow trout (*Oncorhynchus mykiss*): a gene duplication trail. Physiol. Genomics 47 (7), 253–263.

Martin, S.A., Vilhelmsson, O., Medale, F., Watt, P., Kaushik, S., Houlihan, D.F., 2003. Proteomic sensitivity to dietary manipulations in rainbow trout. Biochim. Biophys. Acta 1651, 17–29.

Mennigen, J.A., Panserat, S., Larquier, M., Plagnes-Juan, E., Medale, F., Seiliez, I., Skiba-Cassy, S., 2012. Postprandial regulation of hepatic microRNAs predicted to target the insulin pathway in rainbow trout. PLoS One 7, e38604.

Mennigen, J.A., Skiba-Cassy, S., Panserat, S., 2013. Ontogenetic expression of metabolic genes and microRNAs in rainbow trout alevins during the transition from the endogenous to the exogenous feeding period. J. Exp. Biol. 216, 1597–1608.

Mennigen, J.A., Martyniuk, C.J., Seiliez, I., Panserat, S., Skiba-Cassy, S., 2014. Metabolic consequences of microRNA-122 inhibition in rainbow trout, *Oncorhynchus mykiss*. BMC Genomics 15, 70.

Miller, M.R., Brunelli, J.P., Wheeler, P.A., Liu, S., Rexroad, C.E., Palti, Y., Doe, C.Q., Thorgaard, G.H., 2012. A conserved haplotype controls parallel adaptation in geographically distant salmonid populations. Mol. Ecol. 21, 237–249.

Monroig, O., Tocher, D.R., Navarro, J.C., 2013. Biosynthesis of polyunsaturated fatty acids in marine invertebrates: recent advances in molecular mechanisms. Mar. Drugs 11, 3998–4018.

Mora, T., Walczak, A.M., Bialek, W., Callan, C.G., 2010. Maximum entropy models for antibody diversity. Proc. Natl. Acad. Sci. USA 107, 5405–5410.

Naylor, R.L., Hardy, R.W., Bureau, D.P., Chiu, A., Elliott, M., Farrell, A.P., Forster, I., Gatlin, D.M., Goldburg, R.J., Hua, K., Nichols, P.D., 2009. Feeding aquaculture in an era of finite resources. Proc. Natl. Acad. Sci. USA 106, 15103–15110.

Near, T.J., Eytan, R.I., Dornburg, A., Kuhn, K.L., Moore, J.A., Davis, M.P., Wainwright, P.C., Friedman, M., Smith, W.L., 2012. Resolution of ray-finned fish phylogeny and timing of diversification. Proc. Natl. Acad. Sci. USA 109, 13698–13703.

Nichols, K.M., Young, W.P., Danzmann, R.G., Robison, B.D., Rexroad, C., Noakes, M., Phillips, R.B., Bentzen, P., Spies, I., Knudsen, K., Allendorf, F.W., Cunningham, B.M., Brunelli, J., Zhang, H., Ristow, S., Drew, R., Brown, K.H., Wheeler, P.A., Thorgaard, G.H., 2003. A consolidated linkage map for rainbow trout (*Oncorhynchus mykiss*). Anim. Genet. 34, 102–115.

Ohno, S., Wolf, U., Atkin, N.B., 1968. Evolution from fish to mammals by gene duplication. Hereditas 59, 169–187.

Olsvik, P.A., Hemre, G.I., Waagbo, R., 2013. Exploring early micronutrient deficiencies in rainbow Trout (*Oncorhynchus mykiss*) by next-generation sequencing technology--from black box to functional genomics. PLoS One 8, e69461.

Ortutay, C., Vihinen, M., 2009. Immunome knowledge base (IKB): an integrated service for immunome research. BMC Immunol. 10, 3.

Overturf, K., Vallejo, R.L., Palti, Y., Barrows, F.T., Parsons, J.E., 2012. Microarray analysis of differential utilization of plant-based diets by rainbow trout. Aquac. Int. 20, 213–232.

Palti, Y., Genet, C., Luo, M.-C., Charlet, A., Gao, G., Hu, Y., Castaño-Sánchez, C., Tabet-Canale, K., Krieg, F., Yao, J., Vallejo, R.L., Rexroad, C.E., 2011. A first generation integrated map of the rainbow trout genome. BMC Genomics 12, 180.

Palti, Y., Genet, C., Gao, G., Hu, Y., You, F.M., Boussaha, M., Rexroad, C.E., Luo, M.-C., 2012. A second generation integrated map of the rainbow trout (*Oncorhynchus mykiss*) genome: analysis of conserved synteny with model fish genomes. Mar. Biotechnol. 14, 343–357.

Palti, Y., Gao, G., Miller, M.R., Vallejo, R.L., Wheeler, P.A., Quillet, E., Yao, J., Thorgaard, G.H., Salem, M., Rexroad, C.E., 2014. A resource of single-nucleotide polymorphisms for rainbow trout generated by restriction-site associated DNA sequencing of doubled haploids. Mol. Ecol. Resour. 14, 588–596.

Palti, Y., Gao, G., Liu, S., Kent, M.P., Lien, S., Miller, M.R., Rexroad, C.E., Moen, T., 2015. The development and characterization of a 57K single nucleotide polymorphism array for rainbow trout. Mol. Ecol. Resour. 15, 662–672.

Panserat, S., Kaushik, S.J., 2010. Regulation of gene expression by nutritional factors in fish. Aquac. Res. 41, 751–762.

Panserat, S.K.S., Medale, F., 2013. Rainbow trout as a model for nutrition and nutrient metabolism studies. Trout Physiol. Conserv, 8, -Nova Sci. Publ., Hauppauge, USA, p. 22.

Panserat, S., Medale, F., Blin, C., Breque, J., Vachot, C., Plagnes-Juan, E., Gomes, E., Krishnamoorthy, R., Kaushik, S., 2000. Hepatic glucokinase is induced by dietary carbohydrates in rainbow trout, gilthead seabream, and common carp. Am. J. Physiol. Regul. Integr. Comp. Physiol. 278, R1164–R1170.

Panserat, S., Perrin, A., Kaushik, S., 2002. High dietary lipids induce liver glucose-6-phosphatase expression in rainbow trout (*Oncorhynchus mykiss*). J. Nutr. 132, 137–141.

Panserat, S., Ducasse-Cabanot, S., Plagnes-Juan, E., Srivastava, P.P., Kolditz, C., Piumi, F., Esquerre, D., Kaushik, S., 2008. Dietary fat level modifies the expression of hepatic genes in juvenile rainbow trout (*Oncorhynchus mykiss*) as revealed by microarray analysis. Aquaculture 275, 235–241.

Panserat, S., Hortopan, G.A., Plagnes-Juan, E., Kolditz, C., Lansard, M., Skiba-Cassy, S., Esquerre, D., Geurden, I., Medale, F., Kaushik, S., Corraze, G., 2009. Differential gene expression after total replacement of dietary fish meal and fish oil by plant products in rainbow trout (*Oncorhynchus mykiss*) liver. Aquaculture 294, 123–131.

Panserat, S., Rideau, N., Polakof, S., 2014. Nutritional regulation of glucokinase: a cross-species story. Nutr. Res. Rev. 27, 21–47.

Parsons, J., Thorgaard, G., 1985. Production of androgenetic diploid rainbow-trout. J. Hered. 76, 177–181.

Plagnes-Juan, E., Lansard, M., Seiliez, I., Medale, F., Corraze, G., Kaushik, S., Panserat, S., Skiba-Cassy, S., 2008. Insulin regulates the expression of several metabolism-related genes in the liver and primary hepatocytes of rainbow trout (*Oncorhynchus mykiss*). J. Exp. Biol. 211, 2510–2518.

Polakof, S., Skiba-Cassy, S., Panserat, S., 2009. Glucose homeostasis is impaired by a paradoxical interaction between metformin and insulin in carnivorous rainbow trout. Am. J. Physiol. Regul. Integr. Comp. Physiol. 297, R1769–R1776.

Polakof, S., Moon, T.W., Aguirre, P., Skiba-Cassy, S., Panserat, S., 2011. Glucose homeostasis in rainbow trout fed a high-carbohydrate diet: metformin and insulin interact in a tissue-dependent manner. Am. J. Physiol. Regul. Integr. Comp. Physiol. 300, R166–R174.

Polakof, S., Panserat, S., Soengas, J.L., Moon, T.W., 2012. Glucose metabolism in fish: a review. J. Comp. Physiol. B 182, 1015–1045.

Quillet, E., Garcia, P., Guyomard, R., 1991. Analysis of the production of all homozygous lines of rainbow trout by gynogenesis. J. Exp. Zool. 257, 367–374.

Quillet, E., Dorson, M., Le-Guillou, S., Benmansour, A., Boudinot, P., 2007. Wide range of suceptibility to rhabdoviruses in homozygous clones of rainbow trout, Fish Shellfish. Immunol. 22, 510–519.

Quillet, E., Krieg, F., Dechamp, N., Hervet, C., Bérard, A., Le-Roy, P., Guyomard, R., Prunet, P., Pottinger, T.G., 2014. Quantitative trait loci for magnitude of the plasma cortisol response to confinement in rainbow trout. Anim. Genet. 45, 223–234.

Rexroad, C.E., Lee, Y., Keele, J.W., Karamycheva, S., Brown, G., Koop, B., Gahr, S.A., Palti, Y., Quackenbush, J., 2003. Sequence analysis of a rainbow trout cDNA library and creation of a gene index. Cytogenet. Genome Res. 102, 347–354.

Rexroad, C.E., Palti, Y., Gahr, S.A., Vallejo, R.L., 2008. A second generation genetic map for rainbow trout (*Oncorhynchus mykiss*). BMC Genet. 9, 74.

Roest Crollius, H., Jaillon, O., Bernot, A., Dasilva, C., Bouneau, L., Fischer, C., Fizames, C., Wincker, P., Brottier, P., Quétier, F., Saurin, W., Weissenbach, J., 2000. Estimate of human gene number provided by genome-wide analysis using Tetraodon nigroviridis DNA sequence. Nat. Genet. 25, 235–238.

Rooney, D.P., Neely, R.D., Beatty, O., Bell, N.P., Sheridan, B., Atkinson, A.B., Trimble, E.R., Bell, P.M., 1993. Contribution of glucose/glucose 6-phosphate cycle activity to insulin resistance in type 2 (non-insulin-dependent) diabetes mellitus. Diabetologia 36, 106–112.

Sakamoto, T., Danzmann, R.G., Gharbi, K., Howard, P., Ozaki, A., Khoo, S.K., Woram, R.A., Okamoto, N., Ferguson, M.M., Holm, L.E., Guyomard, R., Hoyheim, B., 2000. A microsatellite linkage map of rainbow trout (*Oncorhynchus mykiss*) characterized by large sex-specific differences in recombination rates. Genetics 155, 1331–1345.

Salem, M., Xiao, C., Womack, J., Rexroad, C.E., Yao, J., 2010. A microRNA repertoire for functional genome research in rainbow trout (*Oncorhynchus mykiss*). Mar. Biotechnol. 12, 410–429.

Salem, M., Paneru, B., Al-Tobasei, R., Abdouni, F., Thorgaard, G.H., Rexroad, C.E., Yao, J., 2015. Transcriptome assembly, gene annotation and tissue gene expression atlas of the rainbow trout. PloS One 10, e0121778.

Santini, F., Harmon, L.J., Carnevale, G., Alfaro, M.E., 2009. Did genome duplication drive the origin of teleosts? A comparative study of diversification in ray-finned fishes. BMC Evol. Biol. 9, 194.

Sawatari, E., Shikina, S., Takeuchi, T., Yoshizaki, G., 2007. A novel transforming growth factor-beta superfamily member expressed in gonadal somatic cells enhances primordial germ cell and spermatogonial proliferation in rainbow trout (*Oncorhynchus mykiss*). Dev. Biol. 301, 266–275.

Scheerer, P.D., Thorgaard, G.H., Allendorf, F.W., 1991. Genetic analysis of androgenetic rainbow trout. J. Exp. Zool. 260, 382–390.

Schranz, M.E., Mohammadin, S., Edger, P.P., 2012. Ancient whole genome duplications, novelty and diversification: the WGD radiation lag-time model. Curr. Opin. Plant Biol. 15, 147–153.

Seiliez, I., Panserat, S., Lansard, M., Polakof, S., Plagnes-Juan, E., Surget, A., Dias, K., Larquier, M., Kaushik, S., Skiba-Cassy, S., 2011. Dietary carbohydrate-to-protein ratio affects TOR signaling and metabolism-related gene expression in the liver and muscle of rainbow trout after a single meal. Am. J. Physiol. Regul. Integr. Comp. Physiol. 300, R733–R743.

Shibata, Y., Paul-Prasanth, B., Suzuki, A., Usami, T., Nakamoto, M., Matsuda, M., Nagahama, Y., 2010. Expression of gonadal soma derived factor (GSDF) is spatially and temporally correlated with early testicular differentiation in medaka. Gene Expr. Patterns GEP 10, 283–289.

Star, B., Jentoft, S., 2012. Why does the immune system of Atlantic cod lack MHC II? BioEssays News Rev. Mol. Cell. Dev. Biol. 34, 648–651.

Suurvali, J., Jouneau, L., Thepot, D., Grusea, S., Pontarotti, P., Pasquier, L. Du, Ruutel Boudinot, S., Boudinot, P., 2014. The proto-MHC of placozoans, a region specialized in cellular stress and ubiquitination/proteasome pathways. J. Immunol. 193, 2891–2901.

Taylor, J.S., Van de Peer, Y., Meyer, A., 2001. Genome duplication, divergent resolution and speciation. Trends Genet. 17, 299–301.

The_MHC_sequencing_consortium, 1999. Complete sequence and gene map of a human major histocompatibility complex. Nature 401, 921–923.

Thorgaard, G.H., 1976. Robertsonian polymorphism and constitutive heterochromatin distribution in chromosomes of the rainbow trout (*Salmo gairdneri*). Cytogenet. Cell Genet. 17, 174–184.

Thorgaard, G.H., Bailey, G.S., Williams, D., Buhler, D.R., Kaattari, S.L., Ristow, S.S., Hansen, J.D., Winton, J.R., Bartholomew, J.L., Nagler, J.J., Walsh, P.J., Vijayan, M.M., Devlin, R.H., Hardy, R.W., Overturf, K.E., Young, W.P., Robison, B.D., Rexroad, C., Palti, Y., 2002. Status and opportunities for genomics research with rainbow trout. Comp. Biochem. Physiol. B Biochem. Mol. Biol. 133, 609–646.

Vallejo, R.L., Palti, Y., Liu, S.X., Evenhuis, J.P., Gao, G.T., Rexroad, C.E., Wiens, G.D., 2014. Detection of QTL in Rainbow Trout Affecting Survival When Challenged with *Flavobacterium psychrophilum*. Mar. Biotechnol. 16 (3), 349–360.

Verrier, E.R., Dorson, M., Mauger, S., Torhy, C., Ciobotaru, C., Hervet, C., Dechamp, N., Genêt, C., Boudinot, P., Quillet, E., 2013. Resistance to a rhabdovirus (VHSV) in rainbow trout: identification of a major QTL related to innate mechanisms. PLoS ONE 8 (2), e55302.

Viegas, I., Rito, J., Jarak, I., Leston, S., Caballero-Solares, A., Metón, I., Pardal, M.A., Baanante, I.V., Jones, J.G., 2015. Contribution of dietary starch to hepatic and systemic carbohydrate fluxes in European seabass (*Dicentrarchus labrax* L.). Br. J. Nutr. 113, 1345–1354.

Volff, J.-N., 2005. Genome evolution and biodiversity in teleost fish. Heredity 94, 280–294.

Weinstein, J.A., Jiang, N., Fisher, D.S., Quake, S.R.S., White, III, R.A., 2009. High-throughput sequencing of the zebrafish antibody repertoire. Science 324, 807–811.

Yano, A., Guyomard, R., Nicol, B., Jouanno, E., Quillet, E., Klopp, C., Cabau, C., Bouchez, O., Fostier, A., Guiguen, Y., 2012. An immune-related gene evolved into the master sex-determining gene in rainbow trout, Oncorhynchus mykiss. Curr. Biol. 22, 1423–1428.

Yano, A., Nicol, B., Jouanno, E., Guiguen, Y., 2014. Heritable targeted inactivation of the rainbow trout (*Oncorhynchus mykiss*) master sex-determining gene using zinc-finger nucleases. Mar. Biotechnol. 16, 243–250.

Young, W.P., Wheeler, P.A., Fields, R.D., Thorgaard, G.H., 1996. DNA fingerprinting confirms isogenicity of androgenetically derived rainbow trout lines. J. Hered. 87, 77–80.

Young, W.P., Wheeler, P.A., Coryell, V.H., Keim, P., Thorgaard, G.H., 1998. A detailed linkage map of rainbow trout produced using doubled haploids. Genetics 148, 839–850.

An improved version of the Atlantic cod genome and advancements in functional genomics: implications for the future of cod farming

3

Ole Kristian Tørresen, Matthew L. Rise**, Xingkun Jin*, Bastiaan Star*, Simon MacKenzie†, Kjetill S. Jakobsen*, Alexander J. Nederbragt*,‡, Sissel Jentoft*,§*

*Department of Biosciences, Centre for Ecological and Evolutionary Synthesis (CEES), University of Oslo, Oslo, Norway; **Department of Ocean Sciences, Memorial University of Newfoundland, St. John's, NL, Canada; †Institute of Aquaculture, University of Stirling, Stirling, United Kingdom; ‡Department of Informatics, Biomedical Informatics Research Group, University of Oslo, Oslo, Norway; §Department of Natural Sciences, University of Agder, Kristiansand, Norway

The Atlantic cod fisheries—and the rebirth of cod aquaculture

From a historical perspective Atlantic cod (*Gadus morhua*) has been, and still is, one of the most important commercial fish species in the north Atlantic. The large traditional cod fisheries date back centuries and have been vital for settlements along the coast of Northern Europe as well as North America's eastern coast. Due to heavy overfishing the northwest Atlantic cod collapsed in the early 1990s—and has still not yet recovered. Compared to the northwest Atlantic cod the northeast Atlantic cod is still viable and thriving—having the largest cod population worldwide—and by far the most abundant is the Northeast Arctic cod (NEAC). In these areas, the northeast Arctic area and around Iceland, the fisheries industry of cod is still profitable—and is an essential marine resource.

From the newest report from FAO (http://www.fao.org/), there is an increasing demand for food. The world is consuming more fish than before, without catching more than recorded in the mid 1990s. This has become the reality due to fish farming. Aquaculture supplies 42% of the world fish production. Since farmed fish is looked upon as a healthy product—also in regard to food safety—the prognosis is

Genomics in Aquaculture. http://dx.doi.org/10.1016/B978-0-12-801418-9.00003-2

that the aquaculture industry will keep growing in the future. To be able to meet this future food demand, there is a need for a successful as well as diversifying aquaculture industry that includes new species such as Atlantic cod. Cod aquaculture had its first attempts in the 1980s—but due to low profitability the enterprises were closed down. During the early 2000s, a second wave of efforts was initiated. An emerging and in some areas rapidly growing cod farming industry was then found in Norway, Canada and the United Kingdom as well as in Iceland and the northeast USA. In Norway, the production rate was steadily increasing until the collapse in 2011—and the total production of cod was 3770 tons in 2013, which was a decrease of 62.4% from 2012 and comprising only 0.3% of the market share of aquaculture production in total (FAO; http://www.fao.org/). The main obstacles for a cost-efficient cod aquaculture is mainly linked to precocious sexual maturation and disease that reduce flesh quality or increase mortality, causing serious economic harm (Almeida et al., 2009; Skjæraasen et al., 2004). Additionally, the release of gametes and escaped fish from sea cages may result in interbreeding of farmed fish with wild stocks, potentially reducing the "genetic health" of these wild stocks (Jørstad et al., 2008; Meager et al., 2009; Skjæraasen et al., 2009). New insight—how to solve the aforementioned and other issues such as diet and stress effects on cod health—provided by the extremely rapid advancements in genomic research may facilitate the rebirth of cod aquaculture, making it a profitable and viable industry.

Immunogenomics—the unusual immune system of Atlantic cod

The vertebrate adaptive immune system (AIS) is one of the most complex response systems in nature, and is responsible for the recognition of pathogens. The AIS emerged over 500 million years ago in the ancestral lineage leading to all jawed vertebrates (Flajnik and Kasahara, 2010). Basic components of the AIS, the major histocompatibility complex (MHC) I and II, were thought to be conserved since their emergence, and thus believed to be central to the survival of vertebrates (Flajnik and Kasahara, 2001; Litman et al., 2010). Initial attempts to characterize different components of the Atlantic cod immune system provided some tantalizing clues about a divergent immune functionality in this species compared to other fishes. For instance, measurable specific antibody responses were difficult to obtain (Espelid et al., 1991) and the absence was subject of debate (Pilström et al., 2005). Moreover, in comparison to other studied teleosts, Atlantic cod has abundant phagocytic neutrophils in its peripheral blood and high levels of IgM (Rønneseth et al., 2007; Øverland et al., 2010; Pilström et al., 2005). While other experiments did detect specific antibody production (Schrøder et al., 2009), vaccination also proved effective in the absence of a measurable specific immune response (Mikkelsen et al., 2011), suggesting that mechanisms other than the classical adaptive immune response provide protection. Finally, a quantification of expressed MHC I genes indicated that these genes were expanded in cod compared to other fishes (Persson et al., 1999; Miller et al., 2002).

The first genome assembly of Atlantic cod provided an excellent opportunity to investigate its immune gene architecture and thus address these observations of a divergent immune functionality. While most genes involved in the vertebrate immune response were present, we did not find genes for MHC II, their assembly and trafficking chaperone Ii (invariant chain/CD74), and the MHC II-interacting protein CD4, which is essential for helper T-cell activation. Importantly, Atlantic cod lacks antigen presentation through CD4+ T-cell responses without these genes. We ensured that the Atlantic cod genome assembly did not fail to reconstruct the region where these genes were located through extensive, local gene synteny analyses. In other words, based on the location of genes that are usually colocalized with MHC II, Ii and CD4 in other fishes, we could identify the regions where these should have been located in Atlantic cod, but were not. In a search for expression of MHC II using cDNA, we determined that the loss of MHC II expression is likely shared among at least some gadoid species, haddock (*Melanogrammus aeglefinus*), whiting (*Merlangius merlangus*) and burbot (*Lota lota*) (Star et al., 2011), though it is currently not clear why, or when such loss occurred (Star and Jentoft, 2012). We also quantified the number of MHC I copies in the genome more accurately and found that Atlantic cod has about 100 classical MHC I loci, which is a highly expanded number compared to other teleosts (Fig. 3.1A). A phylogenetic analysis of these genes supports the existence of two clades of MHC I in cod (Fig. 3.1B). Interestingly, MHC I genes belonging to one of these clades have MHC II-like sorting signals, indicating that loss of MHC II functionality has coincided with a more versatile usage of the cytosolic pathway of MHC I (Malmstrøm et al., 2013).

Apart from these observations that show alterations affecting AIS, we also found an unusual composition of genes involved the innate immune response. For instance, some families of highly conserved Toll-like receptors (TLRs) have markedly expanded while others are absent. Even so, the Atlantic cod has the highest number of TLRs found in a teleost so far (Solbakken et al., 2016; Star et al., 2011). Interestingly, high copy numbers of TLRs have been reported in sea urchin (222 TLRs) and amphioxus (77 TLRs) and these numbers have been discussed in light of their absent adaptive immune system (Rast et al., 2006; Hibino et al., 2006; Huang et al., 2008). The high number of (some) TLR gene families indicate that Atlantic cod particularly relies on these genes for its initial detection of pathogens (Solbakken et al., 2016).

Overall, the loss of MHC II function and lack of a CD4+ T-cell response represent a fundamentally different manner of how the adaptive immune system is initiated and regulated in Atlantic cod. The marked expansion of MHC I genes and unusual TLR composition signify a shift of its immune system in handling microbial pathogens. Importantly, these findings change fundamental assumptions regarding the evolution of the vertebrate immune system.

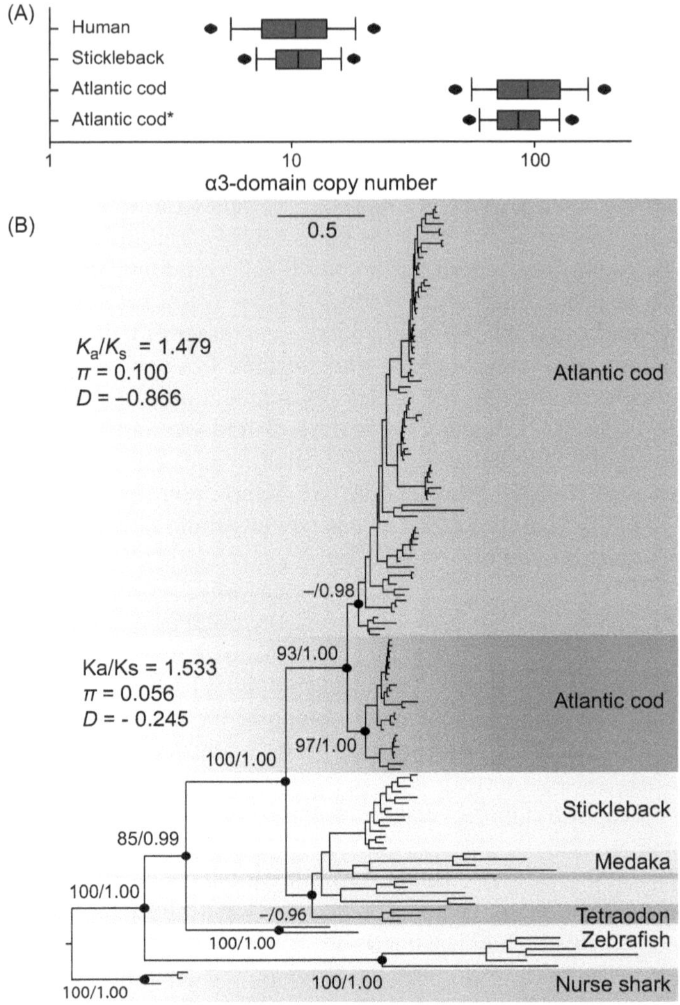

Figure 3.1 Copy numbers and diversity of MHC I genes in Atlantic cod. (A) Estimates of the copy-number of the MHC I α3 domain in human, three-spined stickleback and Atlantic cod. The estimates are based on qPCR ratios of the MHC I α3 domain and single-copy reference genes. β2-microglobulin was used as a reference gene for human and stickleback, while β2-microglobulin and topoisomerase III-α (*) were used as reference genes for Atlantic cod. The estimates for human and stickleback agree with the expected number of α3 domains found in both reference genomes (9 and 13, respectively). Black dots indicate 95% confidence intervals calculated by bootstrapping (n = 50,000). (B) Phylogeny based on the amino acid sequences of the MHC I α1–α3 domains in selected teleosts. The sequences for Atlantic cod are derived from cDNA, while the other teleost sequences were taken from Ensembl and NCBI. The main branches are supported by Maximum likelihood (ML) and Bayesian posterior probabilities, ML/Bayesian, shown by the dots. The scale bar shows the distance in number of substitutions per site. Also shown for the two main clades in Atlantic cod, is the ratio of nonsynonymous to synonymous variable sites (K_a/K_s), the average nucleotide diversity per site (π) and Tajima's D (D). Source: The figure is modified from Star et al. (2011).

Atlantic cod functional genomics research

Functional genomics research on Atlantic cod has focused on large-scale transcript expression (i.e., transcriptome) profiling to study the genes and molecular pathways involved in biological processes such as reproduction and development, and responses to pathogens (or other immune stimuli), environmental stressors (e.g., toxicants, elevated temperature), and/or various diets. Part of the impetus for studying cod functional genomics lies in its potential as an aquaculture species. Indeed, many of the functional genomics resources for cod [e.g., cDNA libraries, expressed sequence tag (EST) databases, DNA microarrays, transcriptome] were built to improve our understanding of the genetic basis of traits that are important for cod aquaculture (e.g., antiviral or antibacterial responses; responses to environmental stressors such as elevated temperature; growth, and development). A solid foundation of functional genomics resources has been built for cod. In addition to aquaculture related applications, these tools are being used for research in other areas such as toxicogenomics.

With regard to the molecular techniques utilized, cod functional genomics research may be divided into three areas: (1) suppression subtractive hybridization (SSH) based targeted gene discovery; (2) DNA microarray development, quality testing, and application; and (3) transcriptome assembly and RNA deep sequencing (RNAseq). Real-time quantitative PCR (qPCR) is a ubiquitous technique in functional genomics laboratories, and is often used to confirm or validate the results arising from the use of the aforementioned techniques. Guidelines have been developed for microarray [MIAME: minimum information about a microarray experiment (Brazma et al., 2001)] and qPCR [MIQE: minimum information for publication of qPCR experiments (Bustin et al., 2009)] experiments to ensure transparency and rigor. Details such as RNA integrity and biological versus technical replicate numbers must be considered when designing, running, and reporting the results from a functional genomics experiment. Also, the data from functional genomics experiments should be made publically available [e.g., archiving microarray data in NCBI's Gene Expression Omnibus (GEO), or RNAseq data in NCBI's Sequence Read Archive (SRA) or equivalent databases]. These and other aspects of functional genomics experiments will be highlighted subsequently in the context of Atlantic cod functional genomics literature (Tables 3.1 and 3.2).

With regard to areas of study, much of the published cod functional genomics research may be divided into the following areas: (1) immune and/or stress related; (2) reproduction and development; (3) nutrigenomics; and (4) toxicogenomics. We will consider examples from the literature in each of these areas.

Immunity and/or stress

SSH libraries and DNA microarrays have been used to study the genes and pathways involved in cod defense (i.e., immune and/or stress) responses. The method for SSH library construction was first described in Diatchenko et al. (1996). It generally involves the creation of a reciprocal set of cDNA libraries, one to identify upregulated genes and the other to identify downregulated genes. Immune and stress relevant cod SSH

Table 3.1 Examples of microarray-based functional genomics studies on Atlantic cod

Microarray platform (reference)	Tissue: overview of findings
Immune-related studies	
20K custom-built 50-mer oligo (platform GEO accession number GPL10532; series GEO accession number GSE22312) (Booman et al., 2011).	Spleen: 82 probes were identified using Significance Analysis of Microarrays (SAM) (Tusher et al., 2001) as significantly differentially expressed (FDR 0.01) between fish that were stimulated with bacterial antigens (formalin-killed atypical *Aeromonas salmonicida*) 24 h postinjection (HPI) and two sets of control fish [preinjection, and 24 HPI phosphate buffered saline (PBS) injected control]. A selection of microarray-identified genes was validated by qPCR.
~1K custom-built cDNA focused on genes involved in stress and immune responses (ArrayExpress accession number E-MTAB-1058) (Holen et al., 2012).	Head kidney leukocytes: responses to single doses of lipopolysaccharide (LPS), the viral mimic polyinosinic acid:polycytidylic acid [poly(I:C)], and phytohaemoagglutinin (PHA) were evaluated. Rank Product analysis (Breitling et al., 2004) identified 12 genes responsive to LPS, 4 genes responsive to poly(I:C), and 4 genes responsive to PHA. All 20 microarray-identified genes were further studied using qPCR.
20K custom-built 50-mer oligo (GEO series accession number GSE27299) (Hori et al., 2012)	Spleen: The impact of moderately elevated temperature (from 10 to 16°C) on the spleen transcriptome response to intraperitoneal (IP) injection with the viral mimic poly(I:C) was evaluated. Many antiviral transcripts were shown using SAM (FDR 0.01) to be upregulated by poly(I:C) earlier at the elevated temperature. Several microarray-identified genes were validated by qPCR.
44K 60-mer oligo array from Agilent (Nofima's Atlantic Cod Immunity and Quality array version 1, ACIQ-1) (Krasnov et al., 2013)	Brain: identification of brain transcripts that respond to nodavirus infection at three time points (5, 11, and 25 days) after intramuscular injection of the pathogen. 1179 upregulated and 454 downregulated transcripts were identified using t-tests with $p < 0.01$ and fold-change threshold of 2. Fifteen microarray-identified (i.e., nodavirus-responsive in brain) transcripts were included in a qPCR experiment involving a cod cell line (ACL cells) exposed to poly(I:C).
Reproduction and development studies	
16K cDNA representing 7000 genes (ArrayExpress accession number E-TABM-1156) (Drivenes et al., 2012)	Pools representing 6 developmental stages: oocyte; 2-cell and blastula pool; gastrula and 50% epiboly pool; segmentation (20, 30, 40, and 50 somites) pool; golden eye and hatching pool; first-feeding. Rank Product analysis (q value cutoff 0.05) was used to identify differentially expressed genes between stages. For example, 281 genes were upregulated in the golden eye and hatching pool compared with the segmentation pool.

Table 3.1 Examples of microarray-based functional genomics studies on Atlantic cod (*cont.*)

Microarray platform (reference)	Tissue: overview of findings
44K 60-mer oligo array from Agilent, representing ~24,000 unique genes (ArrayExpress accession number E-MTAB-1296) (Kleppe et al., 2013)	Eggs and embryos: to investigate the potential influence of maternal stress on offspring, the influence of cortisol treatment (broodstock implanted with osmotic pumps with or without cortisol, and untreated controls) on unfertilized egg, blastula, and gastrula global transcript expression was evaluated. Rank Products and SAM were used to identify egg and embryo transcripts that respond to elevated cortisol, with Rank Products giving much larger genes lists. Several genes identified by both Rank Products and SAM as cortisol-responsive were further studied using qPCR.
44K 60-mer oligo array from Agilent (Nofima's ACIQ-2) (GEO accession number GSE58392) (Škugor et al., 2014a)	Unfertilized oocytes, fertilized oocytes (2 h postfertilization, to assess potential sperm transcript contribution to the fertilized egg), midblastula transition, and segmentation stage embryos: transcripts with greater than twofold differential expression between egg or embryonic stages and the reference sample (adult tissues) were identified. Developmentally regulated genes identified in this study include genes involved in metabolic processes (e.g., lipid and cholesterol metabolism) and many genes with known roles in development (e.g., notch, frizzled 8a). Eight microarray-identified genes were selected for qPCR validation.
20K custom-built 50-mer oligo (series GEO accession number GSE54233) (Rise et al., 2014).	Fertilized eggs (7 h postfertilization, ~2-cell stage): 43 probes consistently differentially expressed in both of the lowest quality egg samples compared with the highest quality egg sample; qPCR validated genes included *kpna7* and *hacd1* (both of which were more highly expressed in the highest quality female).
44K 60-mer oligo array from Agilent (Nofima's ACIQ-1) (Škugor et al., 2014b)	Hatched larvae: the impact of morpholino-based knockdown of Dead end (DnD, a gene that is required for germ cell viability) on the hatchling transcriptome was investigated. Over 2000 genes were identified as differentially expressed (*t* test $p < 0.05$, fold change > 1.6) between DnD knockdown and control larvae at hatch, and several of these genes are involved in lipid and steroid metabolism.
44K oligo array from Agilent (ArrayExpress accession number E-MTAB-2170) (Kleppe et al., 2014)	Ovarian follicles from various stages, ovulated eggs, and embryos (blastula, gastrula): transcriptome dynamics during oogenesis and early embryonic development were investigated. In total, more than 3000 genes were identified as differentially expressed during vitellogenesis or early embryogenesis. Seven microarray-identified genes were validated using qPCR.

(*Continued*)

Table 3.1 **Examples of microarray-based functional genomics studies on Atlantic cod (*cont.*)**

Microarray platform (reference)	Tissue: overview of findings
Nutrigenomics studies	
16K cDNA (Morais et al., 2012)	Intestine: 289 features significantly differentially expressed between cod fed C0 (full fish oil) and cod fed C66 (66% of fish oil replaced with camelina oil) diets. Biological processes represented by differentially expressed genes include: lipid metabolism; cell proliferation, differentiation and apoptosis; immune response. A selection of microarray-identified genes were validated by qPCR.
20K custom-built 50-mer oligo (series GEO accession number GSE54358) (Booman et al., 2014).	Spleen: impacts of diet and poly(I:C) on the spleen transcriptome were evaluated. Fish from 3 diets (fish oil control, FO; 40CO and 80CO (40 and 80% of added fish oil replaced with camelina oil) were injected with poly(I:C) or PBS and sampled 24 HPI. Fish were also sampled preinjection to evaluate the effect of the test diets on the constitutive spleen transcriptome. Over 500 significantly poly(I:C)-responsive genes were identified using SAM (FDR 0.01, fold change cutoff of 2) for each diet group; no significant differences in gene expression were identified between the poly(I:C)-stimulated fish from the different diets. Only one gene was identified as significantly influenced by diet in the preinjection samples. These camelina oil-containing diets had minimal influence on the basal or antiviral spleen transcriptome. Selected microarray-identified genes were qPCR validated.
20K custom-built 50-mer oligo (series GEO accession number GSE68792) (Rise et al., 2015).	Whole larvae: transcriptome responses to dietary zooplankton (5–10% of live prey items) were investigated. Over 300 genes were differentially expressed between the zooplankton-fed group and two other diet groups. Gene ontology (GO) terms that were overrepresented in the zooplankton-responsive gene set included terms related to oxidation–reduction, cell division, and response to selenium. A selection of microarray-identified genes were qPCR validated.
Toxicogenomics studies	
Small-scale (746 cDNA) custom-built array focused on genes involved in stress and immune responses (ArrayExpress accession number E-BASE-11) (Lie et al., 2009a)	Liver: gene expression was analyzed in cod sampled from two contaminated sites, one unpolluted reference site, and an aquaculture site. Rank Products analysis identified genes that were differentially expressed between sites. For example, 18 genes (5 in males, 13 in females) were found to be differentially expressed between fish from one of the contaminated sites and the fish from the reference site. Eight microarray-identified genes were subjected to qPCR validation.

Table 3.1 Examples of microarray-based functional genomics studies on Atlantic cod (*cont.*)

Microarray platform (reference)	Tissue: overview of findings
744 cDNA custom-built array focused on stress and immune response genes (ArrayExpress accession number E-TABM-676) (Lie et al., 2009b)	Liver (female only): transcript expression responses to 20-week oral exposure to 17B-estradiol, produced water (discharged from offshore oil fields), or low or high dose of alkylphenols (major components of produced water) were assessed. Rank Products ($q < 0.05$) identified 46 transcripts as significantly responsive to the low dose alkylphenols compared with controls; smaller numbers of genes were identified for the other treatments. Nine genes were qPCR validated.
135K Nimblegen with three different 60mer probes designed for each transcript (ArrayExpress accession number E-MTAB-1372) (Olsvik et al., 2012)	Larvae: impacts of three levels of chemically dispersed oil or three levels of mechanically dispersed oil on the whole larval transcriptome were evaluated. ANOVA analysis of microarray data showed that mechanically dispersed oil exposure influenced a larger number of genes than exposure to chemically dispersed oil. A selection of microarray-identified genes were qPCR validated.
44K oligonucleotide array from Agilent (Bratberg et al., 2013)	Liver: impacts of oral exposure to a mixture of persistent organic pollutants (POPs) and/or weathered crude oil on the liver transcriptome were evaluated. SAM and Rank Product analysis (Breitling et al., 2004) were used to identify differentially expressed genes, with Rank Product analysis identifying much larger gene lists. Five microarray-identified CYP450 family genes were selected for qPCR.
135K Nimblegen oligo with 125,825 cod probes (at least 3 probes per cDNA) and 11,779 control probes (NCBI GEO accession number GSE43733) (Yadetie et al., 2014).	Liver: transcript expression responses to polychlorinated biphenyls (PCBs) were evaluated. PCBs (0.5, 2, or 8 mg/kg body weight) or vehicle control administered by intraperitoneal injections. SAM (FDR 10%) identified 160 genes differentially expressed (139 upregulated, 21 downregulated) between the highest PCB dose and the control samples, and no genes differentially expressed between the other two PCB dose groups and the control fish. Ten microarray-identified genes (all upregulated by PCBs) were further studied using qPCR.
Other study	
16K cDNA from Institute for Marine Research (IMR) (Edvardsen et al., 2011)	Pituitary gland, spleen, pylorus, and testis: SAM was used to identify transcripts that were differentially expressed between tissues as an initial test of the platform. Four microarray-identified genes were further studied using Northern blots.

GEO, Gene Expression Omnibus (NCBI) (GEO accession numbers are provided); FDR, false discovery rate.

Table 3.2 **Examples of RNA sequencing (RNAseq) based functional genomics studies on Atlantic cod**

Sequencing platform (reference)	Tissue: overview of findings
Developmental studies	
Roche 454, Illumina GA, and ABI SOLiD (Johansen et al., 2011)	Various developmental stages from zygote to larvae, as well as adult tissues: Three different sequencing technologies were used for transcriptome studies involving nuclear and mitochondrial mRNAs, nonprotein-coding RNAs (ncRNAs), small nuclear RNAs (snRNAs), small nucleolar RNAs (snoRNAs), and microRNAs (miRNAs). Over 200 conserved miRNAs were identified in embryos, and expression profiles for 30 miRNAs in 4 developmental stages (blastula, gastrula, 15 somite, and golden eye) are presented.
Roche 454 GS-FLX pyrosequencing (Lanes et al., 2013) (SRA accession number SRA058818)	Fertilized eggs (2- to 8-cell stage): 238 transcripts were differentially expressed between wild broodstock and farmed broodstock. Biological processes represented by DEGs include: fructose metabolism; oxidative phosphorylation; fatty acid metabolism; glycerophospholipid metabolism.
ABI SOLiD (Bizuayehu et al., 2015) (Sequencing data submitted to miRBase)	Various developmental stages from zygote to larvae, as well as adult tissues: The impact of temperature (4 vs 9.5°C) during early development on miRNA expression in early and later life stage cod was investigated. Selected miRNAs were validated using qPCR.
Nutrigenomics studies	
Illumina (Penglase et al., 2015) (SRA acc. number: SRP056073)	Whole larvae: transcriptome responses to dietary copepods (marine zooplankton) versus enriched rotifers were investigated. Over 40 redox-relevant genes were identified as copepod responsive and potentially involved in enhanced growth.
Other studies	
Roche 454 pyrosequencing (Johansen et al., 2009)	Various tissues and developmental stages: approximately 400,000 reads from 454 sequence analyses contributed to characterization of the Atlantic cod transcriptome; microRNA characterization was achieved through RNA sequencing.
Roche 454 GS-FLX pyrosequencing (Małachowicz et al., 2015) (SRA accession number SRP052904)	Gill: this study involved transcriptome characterization of gill tissues from Atlantic cod sampled from the Baltic Sea.

DEG, differentially expressed gene; SRA, Sequence Read Archive of NCBI [SRA accession numbers (IDs) are provided].

libraries were constructed as part of the targeted gene discovery efforts of the Atlantic Cod Genomics and Broodstock Development Project (CGP) based in Canada. These libraries were Sanger sequenced, and a list of libraries with number of ESTs generated for each library is provided in Bowman et al. (2011). For example, reciprocal SSH libraries were constructed to identify: (1) cod spleen and brain transcripts that respond to the viral mimic polyriboinosinic polyribocytidylic acid (poly(I:C)) or nodavirus carrier state (Rise et al., 2008, 2010); (2) cod spleen and head kidney transcripts responsive to bacterial (formalin-killed atypical *Aeromonas salmonicida*) agonist (antigens and/or pathogen-associated molecular patterns) (Feng et al., 2009); and (3) cod liver, head kidney, and skeletal muscle transcripts responsive to heat shock (Hori et al., 2010). For each of these studies, qPCR was used to confirm the expression responses of several SSH-identified transcripts. SSH has proven to be a reliable method for the targeted discovery of defense relevant cod transcripts, although of which the depth of sequencing was much lower than that seen in next-generation sequencing-based RNAseq (e.g., thousands of ESTs vs millions of reads).

At least three different microarray platforms have been used for functional genomics research on Atlantic cod immune responses. As part of the CGP, a 20,000 feature (20K) oligonucleotide (oligo) microarray was designed, built, and quality tested. The 50mer oligo probes for this microarray platform (GEO accession number GPL10532) were designed based on a database containing over 150,000 ESTs from a total of 42 cDNA libraries (23 normalized, 19 SSH) representing various tissues (e.g., brain, gill, ovary, liver, muscle, head kidney, spleen, pyloric caecum), treatments (e.g., stimulated with viral mimic, bacterial antigens, thermal stress), and life stages (e.g., embryonic, larval, adult) (Bowman et al., 2011). Booman et al. (2011) provided details on EST assembly, selection of contiguous sequences (contigs) for 20K probe design, microarray production, and quality testing. The initial experiment with this 20K microarray aimed to identify genes that were significantly [false discovery rate (FDR) 0.01] differentially expressed between cod that were stimulated by intraperitoneal (IP) injection with formalin-killed atypical *A. salmonicida* 24 h postinjection (HPI) versus two sets of control fish [0 h preinjection, and 24 HPI phosphate-buffered saline (PBS) IP injected control]. Details regarding the 20K microarray data analysis methods, including the use of the Bioconductor (siggenes package) implementation of the Significance Analysis of Microarrays (SAM) algorithm (Tusher et al., 2001; Schwender et al., 2006), may be found in Booman et al. (2011). This microarray experiment identified many known immune-relevant genes as *A. salmonicida* responsive (e.g., encoding cathelicidin 1, hepcidin, interleukin-8 variant 5, bactericidal permeability increasing protein/lipopolysaccharide binding protein variant b, and several CC chemokines), and several microarray-identified genes were validated by qPCR. This experiment demonstrated the utility of the 20K platform for immune-related studies, and also provided new transcriptomic information on the cod antibacterial response.

Since little is known regarding the effects of environmental stressors on cod immune responses, the 20K microarray platform was used to study how a moderate increase in ambient temperature (from 10 to 16°C) influences the cod spleen transcriptome response to viral or bacterial agonist (Hori et al., 2012, 2013). Hori et al.

(2012, 2013) used the synthetic dsRNA viral mimic poly(I:C) to elicit the cod antiviral response, and formalin-killed typical *A. salmonicida* (Asal) to elicit the antibacterial response. The gradual rise in temperature was designed to simulate the temperature change experienced by farmed cod in sea-cages in Newfoundland during the summer months, where temperatures can reach 20°C and high levels of mortality have been observed (Hori et al., 2012; Gollock et al., 2006; Pérez-Casanova et al., 2008). The 6°C increase in water temperature did not influence plasma cortisol levels, and had little effect on the basal spleen transcriptome [i.e., 6 upregulated and 0 downregulated (FDR 0.01) microarray probes in nonimmune stimulated fish at 16°C compared with 10°C]. However, the microarray results of Hori et al. (2012) revealed that moderately elevated temperature modulated the expression of many poly(I:C)-responsive transcripts; some of these transcripts (e.g., *tlr9*, *pkr*, *dhx58*, *stat1*, *il8*) were upregulated by the viral mimic earlier at the elevated temperature. As is customary for microarray studies, Hori et al. (2012) presented the results of hierarchical clustering of samples and differentially expressed genes using heatmaps. The heatmaps showed that the poly(I:C) 6HPI at 10°C (i.e., poly(I:C)-stimulated cod 6 h postinjection at the lower temperature) fish had a nonstimulated gene expression signature (i.e., grouping with fish that were not poly(I:C) stimulated), while poly(I:C) 6HPI at 16°C had a stimulated gene expression signature (i.e., grouping with poly(I:C)-stimulated cod 24 HPI at both temperatures) (Hori et al., 2012). Several pathogen detection (e.g., *tlr9*, *pkr*), signal transduction/transcription control (e.g., *irf1*, *irf7*, *irf10*, *stat1*), and immune effector (e.g., *isg15*, *sacs*, *rsad2*) relevant genes, that were microarray-identified as differentially expressed between poly(I:C)-responsive fish at 10 and 16°C, were validated by qPCR (Hori et al., 2012) (Table 3.1). While SAM (FDR 0.01) identified hundreds of genes as differentially expressed between poly(I:C) stimulated cod at 10 versus 16°C, less than 20 genes were identified as differentially expressed between Asal stimulated cod at the two temperatures using SAM with the same FDR cutoff; therefore, moderately elevated temperature may have a greater overall impact on the cod antiviral transcriptomic response compared with the antibacterial transcriptomic response (Hori et al., 2013).

In addition to the aforementioned studies involving the CGP 20K microarray, at least two other microarray platforms have been used for immune-related research of Atlantic cod. One of these platforms is the ~1K custom-built cDNA microarray that is focused on genes involved in stress and immune responses (ArrayExpress accession number E-MTAB-1058) (Holen et al., 2012). This platform was used to study cod head kidney leukocyte responses to single doses of lipopolysaccharide (LPS), poly(I:C), or phytohaemoagglutinin (PHA). Rank products analysis (Breitling et al., 2004) as implemented in J-Express was used to identify 12 genes responsive to LPS (e.g., *CC chemokine type 2* and *leukocyte cell-derived chemotaxin 2*), 4 genes responsive to poly(I:C) [e.g., *interferon stimulated gene 15* (*ISG15*) and *bloodthirsty*], and 4 genes responsive to PHA (e.g., *CC chemokine type 2* and *scavenger receptor class B, member 2*); all 20 microarray-identified genes were further studied using qPCR (Holen et al., 2012). A 44K Agilent 60mer oligo microarray platform (Nofima's Atlantic Cod Immunity and Quality array version 1, ACIQ-1) was used to study the cod brain transcriptome response to nodavirus infection at three time points (5, 11, and 25 days) after intramuscular (IM) injection of the pathogen (Krasnov et al., 2013). Using *t*-tests with p-value less than 0.01 and fold-change threshold of 2, Krasnov et al. (2013) identified 1179 upregulated and 454 downregulated nodavirus-responsive transcripts. Among the nodavirus-responsive genes identified in this study, significantly enriched pathways included "Toll-like receptor signalling" and "Natural killer cell mediated cytotoxicity" (Krasnov et al., 2013). Fifteen microarray-identified transcripts were included in a qPCR experiment involving a cod cell line (ACL cells) exposed to pIC (Krasnov et al., 2013) (Table 3.1).

Functional genomics research has identified a catalogue of defense-relevant genes, including suites of genes that respond to pathogens (e.g., nodavirus) or pathogen-associated molecular patterns (PAMPs) such as poly(I:C) and Asal. This information can be valuable in a variety of subsequent studies. For example, qPCR assays or other mRNA-based methods (e.g., in situ hybridization) may be used to further study the expression of SSH-, microarray-, or RNAseq-identified immune-relevant genes [e.g., to study host responses to pathogens, or the influence of vaccines, therapeutants, or diet on cod immune responses (see subsequent section on nutrigenomics)]. In addition, defense-relevant genes identified in functional genomics studies may be candidates for single nucleotide polymorphism (SNP) discovery and association studies for developing markers for selection of disease-resistant broodstock (Booman and Rise, 2012).

Reproduction and development

SSH, microarrays, and RNAseq have all been used in research related to Atlantic cod reproduction and development. Kortner et al. (2008) used SSH and microarrays to study the effects of androgens on cod oocyte development. In this study, reciprocal SSH cDNA libraries were created using mRNA from androgen-treated and control ovary samples, and 150 ESTs were generated and assembled; a small targeted cDNA array was built and used to study androgen-responsive transcript expression in previtellogenic cod oocyte cultures, and selected genes were further studied using qPCR (5 genes) and in situ hybridization (2 genes) (Kortner et al., 2008).

In addition, several larger-scale microarray platforms have been used for reproduction and development studies in cod. For example, a 16K cDNA array representing 7000 genes, built utilizing ESTs from cDNA libraries representing several different tissues (Edvardsen et al., 2011), was used to study transcriptome changes occurring during cod embryogenesis (Drivenes et al., 2012). In this study, Rank Product analysis (q value cutoff 0.05) identified differentially expressed genes between the following groups: maternal (oocyte); premidblastula transition (2-cell and blastula); postmidblastula transition (gastrula and 50% epiboly); segmentation (20, 30, 40, and 50 somites); hatching (golden eye and hatching); and first-feeding; for example, 281 genes were upregulated in hatching pool compared with the segmentation pool (Drivenes et al., 2012). The data from this microarray experiment are publically available (ArrayExpress accession number E-TABM-1156) (Table 3.1).

The CGP 20K oligo microarray platform was used to compare the transcriptomes of fertilized eggs (7 h postfertilization, ~2-cell stage) from different females (Rise et al., 2014) (Table 3.1). Eggs from 15 females representing 11 CGP families were fertilized by sperm from one male representing a 12th CGP family. The highest and lowest quality females were identified using total mortality at 7 days postfertilization (7 dpf) as an index of egg quality. Forty-three probes were consistently differentially expressed between the highest quality egg samples and both of the lowest quality egg samples. qPCR validated genes included *kpna7* and *hacd1* (more highly expressed in eggs from the highest quality female) and *dcbld1* and *ddc* (more highly expressed in eggs from the two lowest quality females and potentially influenced by family) (Rise et al., 2014). While some genes identified in this study may be useful for identifying extremes in egg quality, there was a lack of significant correlation between gene expression and egg quality when all females were considered. Rise et al. (2014) suggested that future studies involving multiplex qPCR and suites of candidate biomarkers (including genes associated with extremes in egg quality, e.g., *kpna7*, *ddc*) could be conducted to determine if expression of multiple biomarkers is more predictive of egg quality (i.e., developmental potential) than expression of a single gene. This study also identified many immune-relevant genes [e.g., *interferon gamma receptor 1* and *double-stranded RNA activated protein kinase (PKR) type* 2] that were expressed at high levels in cod eggs regardless of quality (Rise et al., 2014). This microarray data set is publically available (NCBI GEO series accession number GSE54233).

Several development-related studies have been conducted on Atlantic cod using 44K 60mer oligo microarrays from Agilent. Kleppe et al. (2013) used a 44K microarray experiment to investigate the potential influence of maternal stress on cod offspring. Females were implanted with osmotic pumps with cortisol (to mimic maternal stress) or osmotic pumps without cortisol (sham control); another control group was left untreated (i.e., no implanted pump). The effect of cortisol treatment on unfertilized egg, blastula, and gastrula global transcript expression was evaluated. Rank Products and SAM were used to identify egg and embryo transcripts that respond to elevated cortisol, with rank products analysis giving much larger gene lists (Kleppe et al., 2013) (Table 3.1). Eight genes were identified by both rank products and SAM (*Q*-value cutoff 0.05) as differentially expressed in maternal stages (egg and blastula) in response to cortisol treatment of broodstock females; these and some additional genes (e.g., glucocorticoid receptor-like transcripts) were studied using qPCR. Among the differentially expressed (i.e., cortisol-responsive) genes in maternal stages identified by rank products and/or SAM, significantly overrepresented GO terms included "response to stress", "response to wounding", and "hematopoietic progenitor cell differentiation" (Kleppe et al., 2013). While there was no visible phenotypic effect of elevated cortisol on cod early embryos, this microarray study revealed that genes and pathways in the egg and early embryo respond to maternal cortisol (mimicking stress); the long-term effects of these gene expression responses are not known (Kleppe et al., 2013). This microarray data set may be accessed through ArrayExpress (accession number E-MTAB-1296). Details on other examples of cod reproduction and development studies using microarrays are shown in Table 3.1 (Škugor et al., 2014a,b; Kleppe et al., 2014).

RNAseq based developmental studies of Atlantic cod are listed in Table 3.2. For example, Johansen et al. (2011) used three different sequencing technologies (Roche 454 pyrosequencing, Illumina GA, and ABI SOLiD) for RNAseq studies involving adult tissues as well as various developmental stages from zygote to larvae (Table 3.2). Johansen et al. (2011) identified over 200 conserved miRNAs in embryos, and presented miRNA expression profiles in blastula, gastrula, 15 somite, and golden eye stage embryos. Lanes et al. (2013) used Roche 454 pyrosequencing-based RNAseq to identify fertilized egg transcripts that were differentially expressed between wild and farmed broodstock, while Bizuayehu et al. (2015) used ABI SOLiD-based RNAseq to study the influence of rearing temperature on early and later life stage miRNA expression. Additional details on these cod RNAseq studies may be found in Table 3.2.

Nutrigenomics

Animal health and welfare is greatly influenced by different diet/food intake, via a wide range of regulatory mechanisms driven by correlated gene regulation. High-throughput "omics" approaches have facilitated the study of molecular regulatory mechanisms under different dietary formula, hence identifying the metabolic pathways that are affected by a certain nutrient (Conceicao et al., 2010). Nutrigenomics therefore is an approach to study at the molecular-level the interaction between nutritive materials and the host genomic response. In reference to fish and in this chapter cod this approach aims to optimize nutrition in feed formulae (specific nutrient regime) and sustainability in diets for aquaculture. Taking this into account, and the expensive and limited marine resources such as fish meal, inclusion of alternatives including plant ingredients is desirable (Tingbo et al., 2012).

Lilleeng et al., 2007, employing the suppression subtractive hybridization (SSH) method identified differentially expressed genes in the Atlantic cod intestine (distal part of midintestine) after two different dietary treatments, that is, fish meal (FM) and dehulled and extracted soybean meal (SBM). In this study 192 clones were isolated and sequenced and 157 of those clones had predicted BLAST annotations. Of these 12 clones were selected for further study by qPCR and upregulation of mRNA transcripts for amino peptidase N, transcobalamin I precursor, cytochrome P450 3A40, ras-related nuclear protein and fatty acid binding protein was observed in the SBM treated group compared to the FM diet (Lilleeng et al., 2007). Another study using a custom-made 744 EST microarray (CodStress array) concentrating upon the response of the liver to FM replacement was reported where following hybridizations it was concluded that replacement had a minimal effect upon liver function. A set of mRNA transcripts including insulin-like growth factor II (IGF-II), insulin-like growth factor receptors I and II (IGF-IR, IGF-IIR), cathepsinD (CatD), cathepsinF (CatF) and calpain 2 were further studied by qPCR and were shown to exhibit moderate changes as a result of meal replacement. This led the authors to suggest that replacement of fish meal in cod diet with plant protein would not affect its overall health and welfare (Lie et al., 2011). In line with these earlier studies several further studies addressed the inclusion of oil from *Camelina sativa* as a replacement for fish oil in cod diets. A 16K cDNA microarray study suggested that Camelina oil (CO) affected the transcriptome by impacting upon cellular proliferation and death processes as well as the cellular structure in intestine although no metabolic changes were identified (Morais et al., 2012) (Table 3.1). In a later study using juvenile cod and CO diets spleen transcriptomes were compared after dietary adaptation using an immune challenge model [virus mimic poly(I:C)]. In this study GO analysis demonstrated a significant enrichment of immune related genes was observable after poly(I:C) challenge. Further validation confirmed that five transcripts related to immune function (LGP2, STAT1, IRF1, ISG15, and viperin) were identically induced either in the replacement diet group or the fish oil feed group. Such comprehensive data implied that fish oil replacement with camelina oil would not have significant effect on functional immunity in the cod (Booman et al., 2014) (Table 3.1). A further study addressing immune function in the cod intestine based upon qPCR data looked at the impact of two separate purified yeast-derivatives: mannan oligosaccharide product (YM) or a β-glucan (BG) product upon the subsequent response to a *Vibrio anguillarum* exposure. Cytokine mRNA regulation was analyzed after bacterial challenge including inflammatory markers; il1b, il8, and ifng and the antiinflammatory marker il10 in different intestinal segments. Differential mRNA abundances related to dietary additives were reported leading the authors to suggest that the microbial products used had significant potential as immunomodulatory treatments for cod (Lokesh et al., 2012). Further examples of Atlantic cod nutrigenomics studies using microarrays (Rise et al., 2105) and RNAseq (Penglase et al. 2015) are provided in Tables 3.1 and 3.2, respectively.

Toxicogenomics

A wide range of contaminants including industrial oil, produced water (a byproduct of the oil industry), heavy metals, and organic compounds exist in the aquatic environment. Several studies have looked in detail to the impact of such compounds in the cod using a combination of array types and qPCR expression analyses. Lie et al., 2009a used the CodStress array (746 ESTs) that represented a set of stress-responsive and immune-relevant ESTs to examine the effects of contaminated waters including on both natural and farmed cod populations (Table 3.1). Results from liver transcriptome analyses suggested that significant generalized effects mediated through the stress responsive pathways were evident (Lie et al., 2009a). Using the same CodStress array the major component of produced water (PW), alkylphenols (AP), was shown to impact upon glycolysis, apoptosis and the general stress response in cod (Lie et al., 2009b) (Table 3.1). Significant upregulation of the mRNA abundance of beta-2-microglobulin, immunoglobulin-M light chain and interleukins-1 beta and 8 was also reported after PW exposure whereas interferon stimulated gene 15 expression was downregulated. This qPCR based study further validated the significant impact of produced water upon the immune system of juvenile cod (Pérez-Casanova et al., 2010). A parallel study also identified an AhR-responsive gene (UDP-GT) and genes relevant to immune function (complement C3, MHC 1), apoptosis (PERP), and oxidative stress (hepcidin, serotransferrin, glutathione peroxidase) to be modified by PW exposure (Holth et al., 2010). Using a genome-wide microarray Olsvik et al., 2012 demonstrated that dispersed oil was a strong inducer of genes regulating DNA replication, chromatin formation, cell proliferation, and protein degradation (Table 3.1). Further oil exposure experiments have described how genes involved in bone resorption were increased, whilst transcripts related to bone formation are decreased (Olsvik et al., 2011b). Interestingly many of the earlier studies and an additional related study using a 135k oligonucleotide array to look at the impact of methylmercury exposure upon the cod liver have highlighted common pathway regulation to environmental contamination.

In a mercury-enriched environment, some immune system related mRNAs were found to be significantly upregulated in cod gills including calreticulin, HSP70, and heme-oxygenase whereas glutathione peroxidase 4B and zona pellucida 3 were significantly downregulated in liver. Distinct genes expression patterns signified the antitoxin genes were to some extent tissue-specific (Olsvik et al., 2011a). Across these studies many potential biomarker candidate genes were identified which are specific to environmental pollution (Yadetie et al., 2013). In particular, the detoxification enzyme, cytochrome P450 1A (Cyp1a) is widely used as an key environmental indicator, and was found to be dramatically upregulated by most toxins, such as oil droplets (Olsvik et al., 2011b), produced water (Lie et al., 2009b), dispersants (Olsvik et al., 2012), mercury (Olsvik et al., 2011a), beta-naphthoflavone (BNF), 1,2,3,7,8-polychlorinated dibenzo-p-dioxin (PCDD), and cadmium (Cd) (Softeland et al., 2010), but downregulated by some phenol derivatives (Olsvik et al., 2009). Many other mRNA transcripts were observed to be regulated by various toxins; in cod hepatocytes upon 1,2,3,7,8-polychlorinated dibenzo-p-dioxin (PCDD) and cadmium (Cd) challenge respectively, metallothionein (MT), aryl hydrocarbon receptor 2 (AhR2), UDP-glucuronosyltransferase (UGT), glutathione S-transferase (GST), vitellogenin B (VTGB), hypoxia-inducible factor 1 (HIFI), heme oxygenase 1 (HO-1), transferrin, glutathione peroxidase (GPx), and heat shock protein 70 (HSP70) were all upregulated (Softeland et al., 2010).

Generation of a new version of the Atlantic cod genome

Despite the fragmented nature of the first version of the Atlantic cod genome, exclusively based on high-throughput 454 pyrosequencing reads, it has proven to be a valuable resource for the research community. Nevertheless, recent advancements within bioinformatics tools available as well as new sequencing technologies—such as the introduction of long-read sequencing technology (the PacBio platform) —have made it possible to substantially improve the original version. To create a better and more contiguous cod genome assembly, we combined sequencing data from various sequencing platforms (454, Illumina and PacBio) and used different assemblers to create several new assemblies. None of these assemblies (even with the combined sequencing platform data), however, were satisfactory with regard to gene content as well as contig and scaffold lengths, even if some of them displayed improved statistics for certain parameters. By development and utilization of a novel assembly reconciliation method we were able to merge the different assemblies, which resulted in one final assembly where 93% of the scaffolds were ordered and oriented into 23 linkage groups (Table 3.3) (Tørresen et al., 2016).

The contig N50 of the new assembly (gadMor2) is 50-fold larger than in the first (gadMor1) and the scaffold N50 is 8 times larger (prior to placement on the linkage map). This increased contiguity is significant: while the gadMor1 assembly consisted of more than 550,000 contigs, this is now reduced to only 25,000 in gadMor2. The average gene length in teleosts is around 10–20,000 bp (Xu et al., 2014), which means that very few of the genes in gadMor1 would be contained within one contig. In gadMor2, however, most genes are contained on one contig (Tørresen et al., 2016).

Table 3.3 **Assembly statistics**

Assembly	Total assembly size (Mbp)	Bases in contigs (Mbp)	N50[a] contig (kbp)	N50[a] scaffold (Mbp)	Percentage gap bases	BUSCO[b]
gadMor1	832	608	2.3	0.14	26.9	3308 (89.4%)
gadMor2	643	633	116	1.15[c]	1.69	3447 (93.2%)

[a] Length of contig or scaffold defined such that half the assembly consists of contigs or scaffold of that size or larger.
[b] Amount found of total 3698 conserved actinopterygii genes (percentage).
[c] This is before ordering and orienting to the linkage map.
Source: Modified from Star et al. (2011) and Tørresen et al. (2016).

Annotation using MAKER2

The significantly improved assembly (gadMor2) with respect to contiguity was automatically annotated using the program MAKER2. This resulted in a final set of 23,243 predicted (protein coding) genes, whereas the first version of the assembly (gadMor1)—annotated at Ensembl—provided a set of 20,095 (protein coding) genes (Star et al., 2011). Comparing different annotations is nontrivial. While more genes are annotated in gadMor2 than in gadMor1 (Table 3.4), the exact number of genes is not known. In general, the gene models in gadMor2 are longer than in gadMor1 (Tørresen et al., 2016), most likely reflecting the better contiguity in gadMor2. The contigs of gadMor1 were gene-model optimized during manual annotation according to three-spined stickleback annotated proteins, compensating the fragmented nature of that assembly. It is conceivable that such an approach may have lead to an underestimation of the number of protein-coding sequences (Tørresen et al., 2016).

Tandem repeat content—assembly and biological implications

The lack of sequence contiguity in the first version of the cod genome was hypothesized to be due to high content of short-tandem repeats (STRs). This idea was supported by the findings from the original assembly where it was noted that many contigs

Table 3.4 **Assembly characteristics**

Assembly	Number of protein coding genes	Number of TRs	Mean length TRs (sd) (bp)	Density of TRs (% of assembly)
gadMor1	20,095	970,798	56.50 (45.17)	8.8
gadMor2	23,246	876,691	84.32 (121.86)	10.9

TR, Tandem repeats (total number in assembly).
Source: Modified from Tørresen et al. (2016).

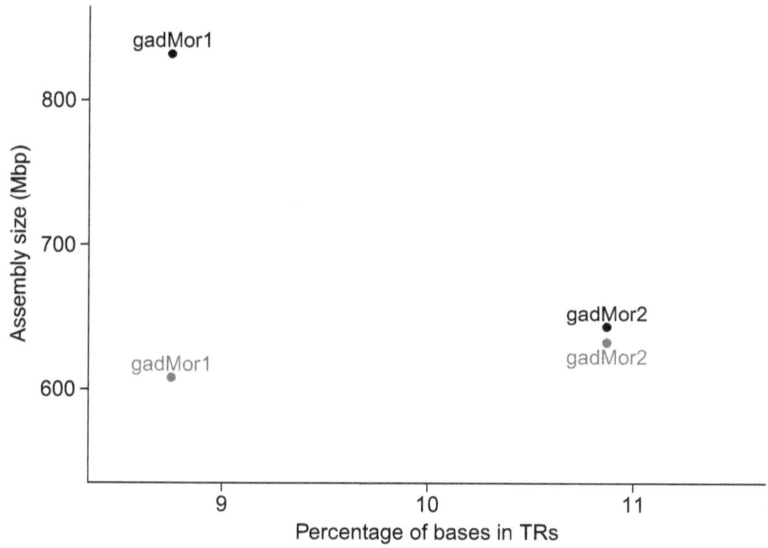

Figure 3.2 The density of tandem repeats (TRs) and the size of the assemblies for gadMor1 and gadMor2 (the original and new assemblies, respectively). The contig sequences for both assemblies add up to a bit less than 650 Mbp (gray dots), while the sequences included in scaffolds differ substantially between gadMor1 and 2 (black dots). The vertical distance between the pair of points equals the amount of sequence in gaps.

ended in such repeats (Star et al., 2011). When comparing gadMor1 and gadMor2 with regards to tandem repeats (TRs), gadMor2 has on average much longer TRs (Table 3.4), suggesting that these are captured in their whole (and exact) lengths in gadMor2. While the scaffolds of gadMor1 add up to more sequence than in gadMor2 (Fig. 3.2), the sequences captured in contigs are less (Tørresen et al., 2016). Some of this difference might be due to TRs, since a lesser density is found in gadMor1. Even so, cod was identified as an outlier with regard to the density of STRs compared to other marine animals based on gadMor1 (Jiang et al., 2014). Furthermore, in connection with amplification of repetitive DNA in ancient DNA samples, we found that cod has up to an order of magnitude more AC and AG tandem repeats than other vertebrates (Star et al., 2016). Comparing the gadMor2 assembly with the whole genome sequenced fish species available at Ensembl, we found that the cod genome has up to twice the density of TRs compared to other fish (Fig. 3.3). Half the TRs are dinucleotide repeats, which alone make up 5.7 % of the gadMor2 assembly. On average, the length of dinucleotide repeats is 84.4 ± 87.2 bp. Most notably, there is also a larger amount of TRs in promoters and coding sequences, compared to the other sequenced teleosts (Fig. 3.3). Moreover, within the coding regions the trinuclotide repeats are far most the abundant ones (Fig. 3.4), but there is also a large amount of dinucleotide repeats present (Tørresen et al., 2016). This is likely reflective of the large amount of dinucleotide repeats in the overall genome. Since STRs in general mutate several

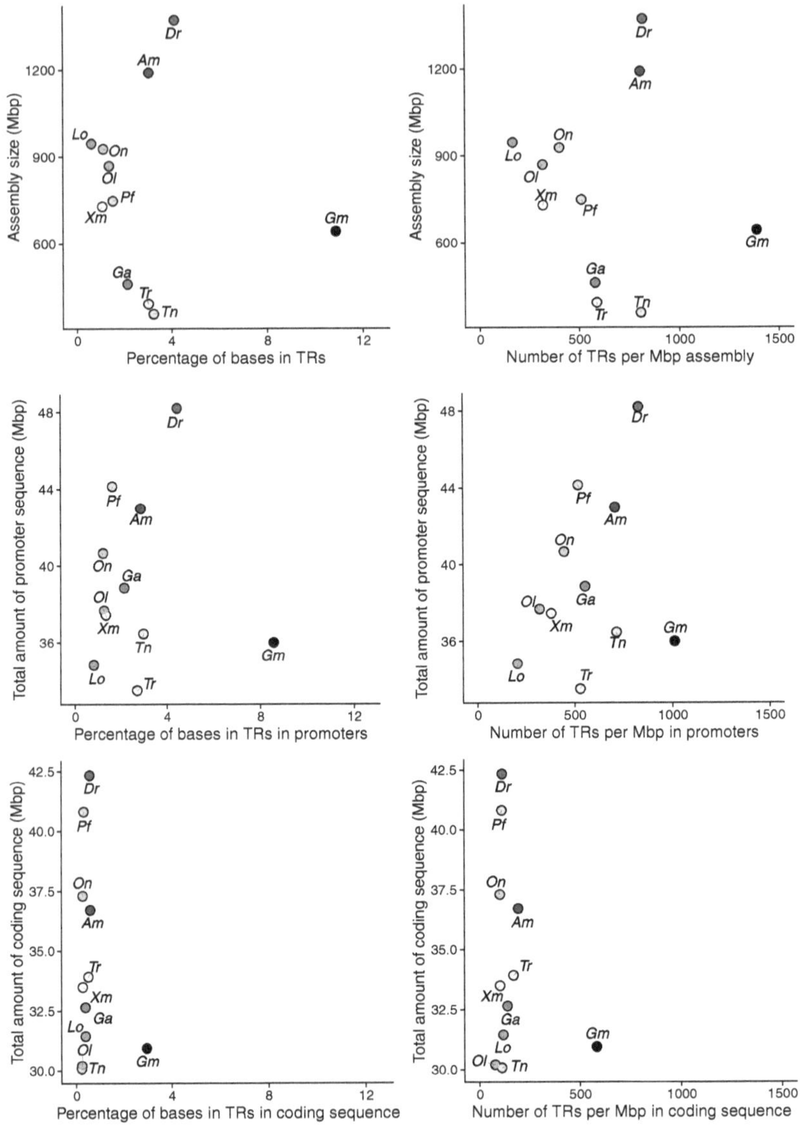

Figure 3.3 The density and frequency of tandem repeats in the whole genome, coding regions and promoters. The assemblies shown here are fish genomes from Ensembl release 81, excluding gadMor1, but including gadMor2. The left panels show the *density* of TRs in the whole assemblies, coding regions and promoter regions, respectively, while the right panels show the *frequency* of TRs in the same regions. Promoter regions are conservatively defined as 2 kb upstream of the coding regions (translation start). The species marked are Am (*Astyanax mexicanus*, cave fish), Dr (*Danio rerio*, zebrafish), Gm (*G. morhua*, Atlantic cod, gadMor2), Ga (*Gasterosteus aculeatus*, three-spined stickleback), Lo (*Lepisosteus oculatus*, spotted gar), On (*Oreochromis niloticus*, tilapia), Ol (*Oryzias latipes*, medaka), Pf (*Poecilia formosa*, Amazon molly), Tr (*Takifugu rubripes*, fugu), Tn (*Tetraodon nigroviridis*, tetraodon), and Xm (*Xiphophorus maculatus*, platyfish). Atlantic cod (in black) is an outlier with respect to the whole genome, promoter sequences and coding regions.

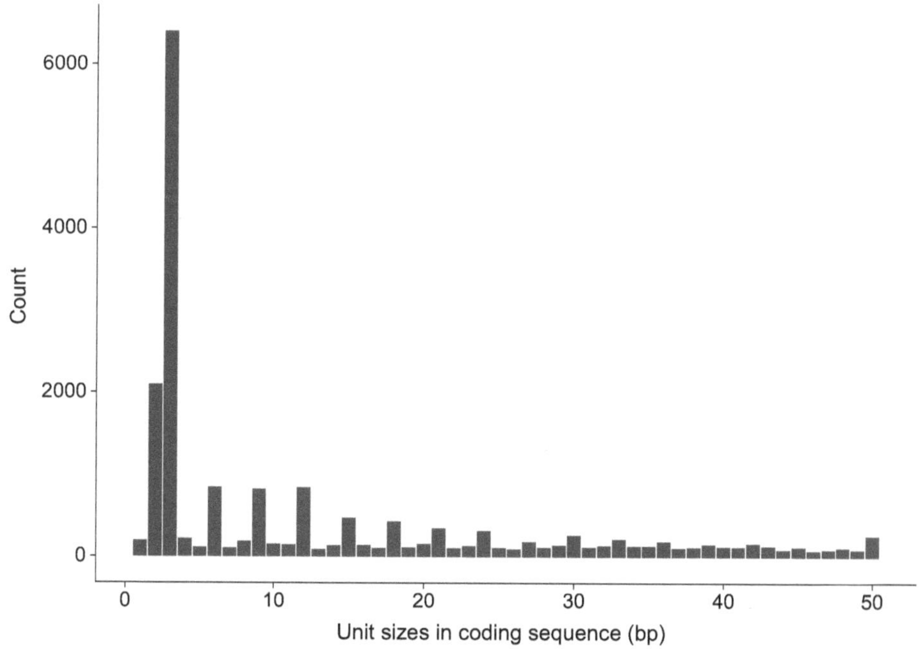

Figure 3.4 The amount of tandem repeats (TRs) in coding sequence in gadMor2 and their unit sizes. The vertical axis shows the number of TRs of different unit sizes in the amino acid coding sequences (count). For the horizontal axis unit size denotes mononucleotide, dinucleotide, trinucleotide repeats etc. Unit sizes of 50 or more are included in the single last column.

orders of magnitude more often than other sequences (Gemayel et al., 2010), these are likely to be a significant source of variation affecting gene regulation and functional properties of proteins.

An additional perspective on the TR distribution is that a high proportion (40 %) of genes in cod contains one or more TRs (Fig. 3.5). Between 5% and 25 % of the genes of the other fish contain one or more TRs, as observed elsewhere (Gemayel et al., 2010). Thus, Atlantic cod is an outlier compared to the other teleosts. Taking into account the large effective population size in cod, it is likely that many of these TRs are polymorphic. Estimations based on the new reference genome suggest that at least one third of the TRs are polymorphic in this single northeast Arctic cod specimen (Tørresen et al., 2016). Polymorphic TRs in promoters and coding sequence have been associated with a variety of phenomena, from behavior in voles (Hammock and Young, 2005) to clock genes in salmon (O'Malley et al., 2010). It is tempting to speculate that the high frequency of TRs in the cod genome implies that TRs are involved in several aspects of cod biology including adaptations to fluctuating environments. With the new reference genome at hand, future population genomics research will be directed to address such questions.

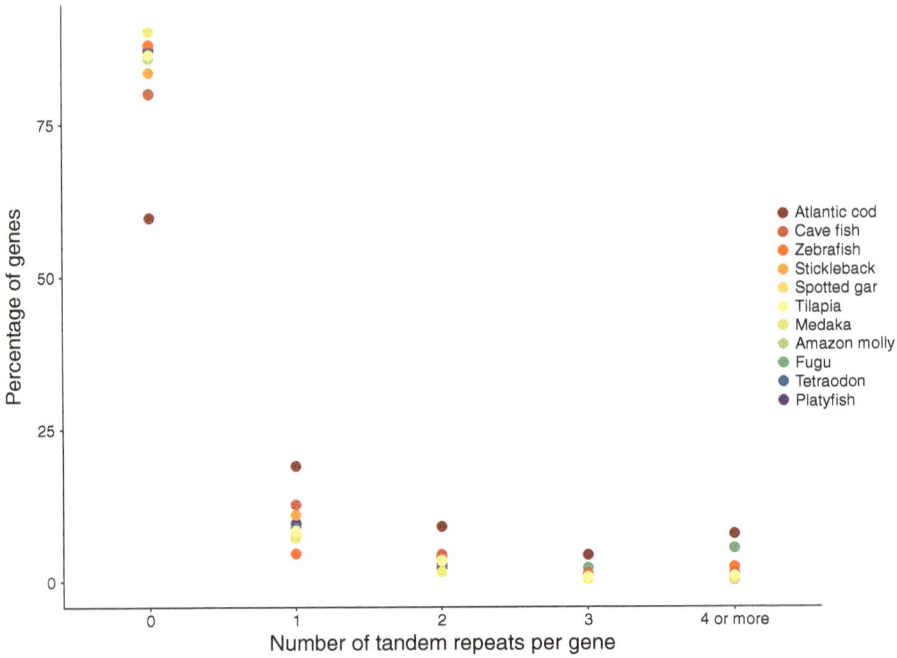

Figure 3.5 The number of tandem repeats (TRs) per gene in selected teleosts. The horizontal axis shows how many TRs per gene found (as percentage of number of genes), while the vertical axis shows the percentage for the different species within each category (i.e., 0, 1, 2, 4 or more individual TRs within the genes). The fish species included are: Atlantic cod, cave fish, zebrafish, stickleback, spotted gar, tilapia, medaka, Amazon molly, fugu, tetraodon, and platyfish. The color codes for each species are depicted in the figure. Atlantic cod (in dark red) is an outlier for each within-gene TR-number category.

An improved Atlantic cod genome—a valuable resource for biological inferences, fisheries management and the future cod aquaculture

The new and improved cod reference genome assembly and annotation will become an important resource for future research by facilitating in-depth functional genomics fundamental for biological inference. Additionally, comprehensive analysis of whole genome resequencing data from contemporary specimens and their historical counterparts would enlighten our understanding of the genomic basis of natural selection and how Atlantic cod adapt to changing environments. This is information that will aid management programs and bring conservation genetics to the next level. Further, the new reference genome assembly and annotation will also be essential for addressing genomic effects of artificial selection related to breeding programs and thus, enhance the efficiency of selecting family material for improved growth, disease resistance, delayed sexual maturation, and other economically important traits for the aquaculture industry. The goal should be to facilitate breeding programs that build upon utilizing the genome information from both wild and captive populations. For a viable future for an emerging cod aquaculture these approaches will be inherent components towards a successful breeding program for Atlantic cod.

References

Almeida, F.F.L., Taranger, G.L., Norberg, B., Karlsen, Ø., Bogerd, J., Schulz, R.W., 2009. Photoperiod-modulated testis maturation in Atlantic cod (*Gadus morhua*, L.). Biol. Reprod. 80 (4), 631–640.

Bizuayehu, T.T., Johansen, S.D., Puvanendran, V., Toften, H., Babiak, I., 2015. Temperature during early development has long-term effects on microRNA expression in Atlantic cod. BMC Genomics 16, 305.

Booman, M., Rise, M.L., 2012. Genomic tools for understanding the molecular basis of production-relevant traits in finfish. In: Fletcher, G.L., Rise, M.L. (Eds.), Aquaculture Biotechnology. Wiley-Blackwell, Ames, IA, pp. 3–20.

Booman, M., Borza, T., Hori, T.S., Feng, C.Y., Higgins, B., Culf, A., Leger, D., Chute, I., Hall, J.R., Belkaid, A., Rise, M., Gamperl, A.K., Hubert, S., Kimball, J., Ouelette, R., Johnson, S.C., Bowman, S., Rise, M.L., 2011. Development and experimental validation of a 20K Atlantic cod (*Gadus morhua*) oligonucleotide microarray based on a collection of over 150,000 ESTs. Mar. Biotechnol. 13, 733–750.

Booman, M., Xu, Q., Rise, M.L., 2014. Evaluation of the impact of camelina oil-containing diets on the expression of genes involved in the innate anti-viral immune response in Atlantic cod (*Gadus morhua*). Fish Shellfish Immunol.Immunol. 41 (1), 52–63.

Bowman, S., Hubert, S., Higgins, B., Stone, C., Kimball, J., Borza, T., Bussey, J.T., Simpson, G., Hall, J.R., Hori, T.S., Feng, C.Y., Gamperl, A.K., Booman, M., Rise, M., Symonds, J., Johnson, S.C., Rise, M.L., 2011. An integrated approach to gene discovery and marker development in Atlantic cod (*Gadus morhua*). Mar. Biotechnol. 13, 242–255.

Bratberg, M., Olsvik, P.A., Edvardsen, R.B., Brekken, H.K., Vadla, R., Meier, S., 2013. Effects of oil pollution and persistent organic pollutants (POPs) on glycerophospholipids in liver and brain of male Atlantic cod (*Gadus morhua*). Chemosphere 90, 2157–2171.

Brazma, A., Hingamp, P., Quackenbush, J., Sherlock, G., Spellman, P., Stoeckert, C., et al., 2001. Minimum information about a microarray experiment (MIAME)—toward standards for microarray data. Nat. Genet. 29, 365–371.

Breitling, R., Armengaud, P., Amtmann, A., Herzyk, P., 2004. Rank products: a simple, yet powerful, new method to detect differentially regulated genes in replicated microarray experiments. FEBS Lett. 573, 83–92.

Bustin, S.A., Benes, V., Garson, J.A., Hellemans, J., Huggett, J., Kubista, M., et al., 2009. The MIQE guidelines: minimum information for publication of quantitative real-time PCR experiments. Clin. Chem. 55, 611–622.

Conceicao, L.E.C., Aragao, C., Richard, N., Engrola, S., Gavaia, P., Mira, S., Dias, J., 2010. Novel methodologies in marine fish larval nutrition. Fish Physiol. Biochem. 36 (1), 1–16.

Diatchenko, L., Lau, Y.F.C., Campbell, A.P., Chenchik, A., Moqadam, F., Huang, B., Lukyanov, S., Lukyanov, K., Gurskaya, N., Sverdlov, E.D., Siebert, P.D., 1996. Suppression subtractive hybridization: a method for generating differentially regulated or tissue-specific cDNA probes and libraries. Proc. Natl. Acad. Sci. USA 93, 6025–6030.

Drivenes, Ø., Tarranger, G.L., Edvardsen, R.B., 2012. Gene expression profiling of Atlantic cod (*Gadus morhua*) embryogenesis using microarrays. Mar. Biotechnol. 14, 167–176.

Edvardsen, R.B., Malde, K., Mittelholzer, C., Taranger, G.L., Nilsen, F., 2011. EST resources and establishment and validation of a 16 k cDNA microarray from Atlantic cod (*Gadus morhua*). Comp. Biochem. Physiol. Part D 6, 23–30.

Espelid, S., Rødseth, O.M., Jørgensen, T.Ø., 1991. Vaccination experiments and studies of the humoral immune responses in cod, *Gadus morhua* L., to four strains of monoclonal-defined *Vibrio anguillarum*. J. Fish Dis. 14 (2), 185–197.

Feng, C.Y., Johnson, S.C., Hori, T.S., Rise, M., Hall, J.R., Gamperl, A.K., Hubert, S., Kimball, J., Bowman, S., Rise, M.L., 2009. Identification and analysis of differentially expressed genes in the immune tissues of Atlantic cod stimulated with formalin-killed, atypical *Aeromonas salmonicida*. Physiol. Genom. 37, 149–163.

Flajnik, M.F., Kasahara, M., 2001. Comparative genomics of the MHC: glimpses into the evolution of the adaptive immune system. Immunity 15 (3), 351–362.

Flajnik, M.F., Kasahara, M., 2010. Origin and evolution of the adaptive immune system: genetic events and selective pressures. Nat. Rev. Genet. 11 (1), 47–59.

Gemayel, R., Vinces, M.D., Legendre, M., Verstrepen, K.J., 2010. Variable tandem repeats accelerate evolution of coding and regulatory sequences. Annu. Rev. Genet. 44 (1), 445–477.

Gollock, M.J., Currie, S., Petersen, L.H., Gamperl, A.K., 2006. Cardiovascular and hematological responses of Atlantic cod (*Gadus morhua*) to acute temperature increase. J. Exp. Biol. 209, 2961–2970.

Hammock, E.A.D., Young, L.J., 2005. Microsatellite instability generates diversity in brain and sociobehavioral traits. Science 308 (5728), 1630–1634.

Hibino, T., Loza-Coll, M., Messier, C., Majeske, A.J., Cohen, A.H., Terwilliger, D.P., et al., 2006. The immune gene repertoire encoded in the purple sea urchin genome. Dev. Biol. 300 (1), 349–365.

Holen, E., Lie, K.K., Araujo, P., Olsvik, P.A., 2012. Pathogen recognition and mechanisms in Atlantic cod (*Gadus morhua*) head kidney cells—Bacteria (LPS) and virus (poly I:C) signals through different pathways and affect distinct genes. Fish Shellfish Immunol. 33, 267–276.

Holth, T.F., Thorsen, A., Olsvik, P.A., Hylland, K., 2010. Long-term exposure of Atlantic cod (*Gadus morhua*) to components of produced water: condition, gonad maturation, and gene expression". Can. J. Fish. Aquat. Sci. 67 (10), 1685–1698.

Hori, T.S., Gamperl, A.K., Afonso, L.O.B., Johnson, S.C., Hubert, S., Kimball, J., Bowman, S., Rise, M.L., 2010. Heat-shock responsive genes identified and validated in Atlantic cod (*Gadus morhua*) liver, head kidney and skeletal muscle using genomic techniques. BMC Genomics 11, 72.

Hori, T.S., Gamperl, A.K., Booman, M., Nash, G.W., Rise, M.L., 2012. A moderate increase in ambient temperature modulates the Atlantic cod (*Gadus morhua*) spleen transcriptome response to intraperitoneal viral mimic injection. BMC Genomics 13, 431.

Hori, T.S., Gamperl, A.K., Nash, G.W., Booman, M., Barat, A., Rise, M.L., 2013. The impact of a moderate chronic temperature increase on spleen immune-relevant gene transcription depends on whether Atlantic cod are stimulated with bacterial versus viral antigens. Genome 56, 567–576.

Huang, S., Yuan, S., Guo, L., Yu, Y., Li, J., Wu, T., et al., 2008. Genomic analysis of the immune gene repertoire of amphioxus reveals extraordinary innate complexity and diversity. Genome Res. 18 (7), 1112–1126.

Jiang, Q., Li, Q., Yu, H., Kong, L., 2014. Genome-wide analysis of simple sequence repeats in marine animals—a comparative approach. Mar. Biotechnol. 16 (5), 604–619.

Johansen, S.D., Coucheron, D.H., Andreassen, M., Karlsen, B.O., Furmanek, T., Jørgensen, T.E., Emblem, Å., Breines, R., Nordeide, J.T., Moum, T., Nederbragt, A.J., Stenseth, N.C., Jakobsen, K.S., 2009. Large-scale sequence analyses of Atlantic cod. New Biotechnol. 25 (5), 263–271.

Johansen, S.D., Karlsen, B.O., Furmanek, T., Andreassen, M., Jørgensen, T.E., Bizuayehu, T.T., Breines, R., Emblem, Å., Kettunen, P., Luukko, K., Edvardsen, R.B., Nordeide, J.T., Coucheron, D.H., Moum, T., 2011. RNA deep sequencing of the Atlantic cod transcriptome. Comp. Biochem. Physiol. Part D 6, 18–22.

Jørstad, K.E., Van der Meeren, T., Paulsen, O.I., Thomsen, T., Thorsen, A., Svåsand, T, 2008. "Escapes" of eggs from farmed cod spawning in net pens: recruitment to wild stocks. Rev. Fish. Sci. 16 (1–3), 285–295.

Kleppe, L., Karlsen, Ø., Edvardsen, R.B., Norberg, B., Andersson, E., Taranger, G.L., Wargelius, A., 2013. Cortisol treatment of prespawning female cod affects cytogenesis related factors in eggs and embryos. Gen. Comp. Endocrinol. 189, 84–95.

Kleppe, L., Edvardsen, R.B., Furmanek, T., Taranger, G.L., Wargelius, A., 2014. Global transcriptome analysis identifies regulated transcripts and pathways activated during oogenesis and early embryogenesis in Atlantic cod. Mol. Rep. Dev. 81, 619–635.

Kortner, T.M., Rocha, E., Silva, P., Castro, L.F.C., Arukwe, A., 2008. Genomic approach in evaluating the role of androgens on the growth of Atlantic cod (*Gadus morhua*) previtellogenic oocytes. Comp. Biochem. Physiol. Part D 3, 205–218.

Krasnov, A., Kileng, Ø., Skugor, S., Jørgensen, S.M., Afanasyev, S., Timmerhaus, G., Sommer, A., Jensen, I., 2013. Genomic analysis of the host response to nervous necrosis virus in Atlantic cod (*Gadus morhua*) brain. Mol. Immunol. 54, 443–452.

Lanes, C.F.C., Bizuayehu, T.T., Fernandes, J.M.O., Kiron, V., Babiak, I., 2013. Transcriptome of Atlantic cod (*Gadus morhua* L.) early embryos from farmed and wild broodstocks. Mar. Biotechnol. 15, 677–694.

Lie, K.K., Lanzen, A., Breilid, H., Olsvik, P.A., 2009a. Gene expression profiling in Atlantic cod (*Gadus morhua* L.) from two contaminated sites using a custom-made cDNA microarray. Environ. Toxicol. Chem. 28 (8), 1711–1721.

Lie, K.K., Meier, S., Olsvik, P.A., 2009b. Effects of environmental relevant doses of pollutants from offshore oil production on Atlantic cod (*Gadus morhua*). Comp. Biochem. Physiol. Part C 150, 141–149.

Lie, K.K., Hansen, A.C., Eroldogan, O.T., Olsvik, P.A., Rosenlund, G., Hemre, G.I., 2011. Expression of genes regulating protein metabolism in Atlantic cod (*Gadus morhua* L.) was altered when including high diet levels of plant proteins". Aquacult. Nutr. 17 (1), 33–43.

Lilleeng, E., Frøystad, M.K., Vekterud, K., Valen, E.C., Krogdahl, Å., 2007. Comparison of intestinal gene expression in Atlantic cod (*Gadus morhua*) fed standard fish meal or soybean meal by means of suppression subtractive hybridization and real-time PCR. Aquaculture 267 (1–4), 269–283.

Litman, G.W., Rast, J.P., Fugmann, S.D., 2010. The origins of vertebrate adaptive immunity. Nat. Rev. Immunol. 10 (8), 543–553.

Lokesh, J., Fernandes, J.M.O., Korsnes, K., Bergh, O., Brinchmann, M.F., Kiron, V., 2012. Transcriptional regulation of cytokines in the intestine of Atlantic cod fed yeast derived mannan oligosaccharide or beta-Glucan and challenged with *Vibrio anguillarum*. Fish Shellfish Immunol.Immunol. 33 (3), 626–631.

Małachowicz, M., Kijewska, A., Wenne, R., 2015. Transcriptome analysis of gill tissue of Atlantic cod *Gadus morhua* L. from the Baltic Sea. Mar. Genom. 23, 37–40.

Malmstrøm, M., Jentoft, S., Gregers, T.F., Jakobsen, K.S., 2013. Unraveling the evolution of the Atlantic cod's (*Gadus morhua* L.) alternative immune strategy. PLoS One 8 (9), e74004.

Meager, J.J., Skjæraasen, J.E., Ferno, A., Karlsen, Ø., Løkkeborg, S., Michalsen, K., Utskot, S.O., 2009. Vertical dynamics and reproductive behaviour of farmed and wild Atlantic cod *Gadus morhua*. Mar. Ecol. Prog. Ser. 389, 233–243.

Mikkelsen, H., Lund, V., Larsen, R., Seppola, M., 2011. Vibriosis vaccines based on various sero-subgroups of *Vibrio anguillarum* O2 induce specific protection in Atlantic cod (*Gadus morhua* L.) juveniles. Fish Shellfish Immunol. 30 (1), 330–339.

Miller, K.M., Kaukinen, K.H., Schulze, A.D., 2002. Expansion and contraction of major histocompatibility complex genes: a teleostean example. Immunogenetics 53 (10–11), 941–963.

Morais, S., Edvardsen, R.B., Tocher, D.R., Bell, J.G., 2012. Transcriptomic analyses of intestinal gene expression of juvenile Atlantic cod (*Gadus morhua*) fed diets with Camelina oil as replacement for fish oil. Comp. Biochem. Physiol. Part B 161, 283–293.

Olsvik, P.A., Lie, K.K., Sturve, J., Hasselberg, L., Andersen, O.K., 2009. Transcriptional effects of nonylphenol, bisphenol A and PBDE-47 in liver of juvenile Atlantic cod (*Gadus morhua*). Chemosphere 75 (3), 360–367.

Olsvik, P.A., Brattas, M., Lie, K.K., Goksoyr, A., 2011a. Transcriptional responses in juvenile Atlantic cod (*Gadus morhua*) after exposure to mercury-contaminated sediments obtained near the wreck of the German WW2 submarine U-864, and from Bergen Harbor, Western Norway. Chemosphere 83 (4), 552–563.

Olsvik, P.A., Hansen, B.H., Nordtug, T., Moren, M., Holen, E., Lie, K.K., 2011b. Transcriptional evidence for low contribution of oil droplets to acute toxicity from dispersed oil in first feeding Atlantic cod (*Gadus morhua*) larvae". Comp. Biochem. Physiol. Part C Toxicol. Pharmacol. 154 (4), 333–345.

Olsvik, P.A., Lie, K.K., Nordtug, T., Hansen, B.H., 2012. Is chemically dispersed oil more toxic to Atlantic cod (*Gadus morhua*) larvae than mechanically dispersed oil? A transcriptional evaluation. BMC Genomics 13, 702.

O'Malley, K.G., Ford, M.J., Hard, J.J., 2010. Clock polymorphism in Pacific salmon: evidence for variable selection along a latitudinal gradient. Proc. Biol Sci Royal Soc 277 (1701), 3703–3714.

Øverland, H.S., Pettersen, E.F., Rønneseth, A., Wergeland, H.I., 2010. Phagocytosis by B-cells and neutrophils in Atlantic salmon (*Salmo salar* L.) and Atlantic cod (*Gadus morhua* L.). Fish Shellfish Immunol. 28 (1), 193–204.

Penglase, S., Edvardsen, R.B., Furmanek, T., Rønnestad, I., Karlsen, Ø., van der Meeren, T., Hamre, K., 2015. Diet affects the redox system in developing Atlantic cod (*Gadus morhua*) larvae. Redox Biol. 5, 308–318.

Pérez-Casanova, J.C., Hamoutene, D., Samuelson, S., Burt, K., King, T.L., Lee, K., 2010. The immune response of juvenile Atlantic cod (*Gadus morhua* L.) to chronic exposure to produced water. Mar. Environ. Res. 70 (1), 26–34.

Pérez-Casanova, J.C., Rise, M.L., Dixon, B., Afonso, L.O., Hall, J.R., Johnson, S.C., Gamperl, A.K., 2008. The immune and stress response of Atlantic cod to long-term increases in water temperature. Fish Shellfish Immunol. 24, 600–609.

Persson, A.-C., Stet, R.J.M., Pilström, L., 1999. Characterization of MHC class I and *β*2-microglobulin sequences in Atlantic cod reveals an unusually high number of expressed class I genes. Immunogenetics 50 (1–2), 49–59.

Pilström, L., Warr, G.W., Strömberg, S., 2005. Why is the antibody response of Atlantic cod so poor? The search for a genetic explanation. Fish. Sci. 71 (5), 961–971.

Rast, J.P., Smith, L.C., Loza-Coll, M., Hibino, T., Litman, G.W., 2006. Genomic insights into the immune system of the sea urchin. Science 314 (5801), 952–956.

Rise, M.L., Hall, J., Rise, M., Hori, T., Gamperl, A.K., Kimball, J., Hubert, S., Bowman, S., Johnson, S.C., 2008. Functional genomic analysis of the response of Atlantic cod (*Gadus morhua*) spleen to the viral mimic polyriboinosinic polyribocytidylic acid (pIC). Dev. Comp. Immunol. 32, 916–931.

Rise, M.L., Hall, J.R., Rise, M., Hori, T.S., Browne, M., Gamperl, A.K., Hubert, S., Kimball, J., Bowman, S., Johnson, S.C., 2010. Impact of asymptomatic nodavirus carrier state and intraperitoneal viral mimic injection on brain transcript expression in Atlantic cod (*Gadus morhua*). Physiol. Genomics 42, 266–280.

Rise, M.L., Nash, G.W., Hall, J.R., Booman, M., Hori, T.S., Trippel, E.A., Gamperl, A.K., 2014. Variation in embryonic mortality and maternal transcript expression among Atlantic cod (*Gadus morhua*) broodstock: a functional genomics study. Mar. Genomics 18, 3–20.

Rise, M.L., Hall, J.R., Nash, G.W., Xue, X., Booman, M., Katan, T., Gamperl, A.K., 2015. Transcriptome profiling reveals that feeding wild zooplankton to larval Atlantic cod (*Gadus morhua*) influences suites of genes involved in oxidation-reduction, mitosis, and selenium homeostasis. BMC Genomics 16, 1016.

Rønneseth, A., Wergeland, H.I., Pettersen, E.F., 2007. Neutrophils and B-cells in Atlantic cod (*Gadus morhua* L.). Fish Shellfish Immunol. 23 (3), 493–503.

Schrøder, M.B., Ellingsen, T., Mikkelsen, H., Norderhus, E.A., Lund, V., 2009. Comparison of antibody responses in Atlantic cod (*Gadus morhua* L.) to *Vibrio anguillarum, Aeromonas salmonicida* and *Francisella* sp. Fish Shellfish Immunol. 27 (2), 112–119.

Schwender, H., Krause, A., Ickstadt, K., 2006. Identifying interesting genes with siggenes. RNews 6, 45–50.

Skjæraasen, J.E., Salvanes, A.G.V., Karlsen, Ø., Dahle, R., Nilsen, T., Norberg B, 2004. The effect of photoperiod on sexual maturation, appetite and growth in wild Atlantic cod (*Gadus morhua* L.). Fish Physiol. Biochem. 30 (163), 174.

Skjæraasen, J.E., Mayer, I., Meager, J.J., Rudolfsen, G., Karlsen, Ø., Haugland, T., Kleven O, 2009. Sperm characteristics and competitive ability in farmed and wild cod. Mar. Ecol. Prog. Ser. 375, 219–228.

Škugor, A., Krasnov, A., Andersen, Ø., 2014a. Genome-wide microarray analysis of Atlantic cod (*Gadus morhua*) oocyte and embryo. BMC Genomics 15, 594.

Škugor, A., Tveiten, H., Krasnov, A., Andersen, Ø., 2014b. Knockdown of the germ cell factor Dead end induces multiple transcriptional changes in Atlantic cod (*Gadus morhua*) hatchlings. Animal Reprod. Sci. 144, 129–137.

Softeland, L., Holen, E., Olsvik, P.A., 2010. Toxicological application of primary hepatocyte cell cultures of Atlantic cod (*Gadus morhua*) - Effects of BNF, PCDD and Cd. Comp. Biochem. Physiol. C-Toxicol. Pharmacol. 151 (4), 401–411.

Solbakken, M.H., Tørresen, O.K., Nederbragt, A.J., Seppola, M., Gregers, T.F., Jakobsen, K.S., Jentoft, S., 2016. Evolutionary redesign of the Atlantic cod (*Gadus morhua* L.) Toll-like receptor repertoire by gene losses and expansions. Scientific Rep. 6, 25211.

Star, B., Jentoft, S., 2012. Why does the immune system of Atlantic cod lack MHC II? BioEssays 34 (8), 648–651.

Star, B., Nederbragt, A.J., Jentoft, S., Grimholt, U., Malmstrøm, M., Gregers, T.F., Rounge, T.B., Paulsen, J., Solbakken, M.H., Sharma, A., Wetten, O.F., Lanzén, A., Winer, R., Knight, J., Vogel, J.-H., Aken, B., Andersen, Ø., Lagesen, K., Tooming-Klunderud, A., Edvardsen, R.B., Kirubakaran, G.T., Espelund, M., Nepal, C., Previti, C., Karlsen, B.O., Moum, T., Skage, M., Berg, P.R., Gjøen, T., Kuhl, H., Thorsen, J., Malde, K., Reinhardt, R., Du, L., Johansen, S.D., Searle, S., Lien, S., Nilsen, F., Jonassen, I., Omholt, S.W., Stenseth, N.C., Jakobsen, K.S., 2011. The genome sequence of Atlantic cod reveals a unique immune system. Nature 477 (7363), 207–210.

Star, B., Hansen, M.H., Skage, M., Bradbury, I.R., Godiksen, J.A., Kjesbu, O.S., Jentoft, S., 2016. Preferential amplification of repetitive DNA during whole genome sequencing library creation from historic samples. STAR 2 (1), 36–45.

Tingbo, M.G., Pedersen, M.E., Grondahl, F., Kolset, S.O., Veiseth-Kent, E., Enersen, G., Hannesson, K.O., 2012. Type of carbohydrate in feed affects the expression of small leucine-rich proteoglycans (SLRPs), glycosaminoglycans (GAGs) and interleukins in skeletal muscle of Atlantic cod (*Gadus morhua* L.). Fish Shellfish Immunol. 33 (3), 582–589.

Tørresen O.K., Star B., Jentoft S., Reinar W.B., Grove H., Miller J.R., Walenz B., Knight J., Ekholm J.M., Peluso P., Edvardsen R.B., Tooming-Klundrerud A., Skage M., Lien S., Jakobsen K.S., Nederbragt A.J., 2016. An improved genome assembly uncovers a prolific tandem repeat structure in Atlantic cod. submitted.

Tusher, V.G., Tibshirani, R., Chu, G., 2001. Significance analysis of microarrays applied to the ionizing radiation response. Proc. Natl. Acad. Sci. USA 98, 5116–5121.

Xu, P., Zhang, X., Wang, X., Li, J., Liu, G., Kuang, Y., et al., 2014. Genome sequence and genetic diversity of the common carp, *Cyprinus carpio*. Nat. Genet. 46 (11), 1212–1219.

Yadetie, F., Karlsen, O.A., Lanzén, A., Berg, K., Olsvik, P., Hogstrand, C., Goksøyr, A., 2013. Global transcriptome analysis of Atlantic cod (*Gadus morhua*) liver after in vivo methylmercury exposure suggests effects on energy metabolism pathways. Aquat. Toxicol. 126, 314–325.

Yadetie, F., Karlsen, O.A., Eide, M., Hogstrand, C., Goksøyr, A., 2014. Liver transcriptome analysis of Atlantic cod (*Gadus morhua*) exposed to PCB 153 indicates effects on cell cycle regulation and lipid metabolism. BMC Genomics 15, 481.

Catfish genomic studies: progress and perspectives

4

Yulin Jin, Shikai Liu, Zihao Yuan, Yujia Yang, Suxu Tan, Zhanjiang Liu
The Fish Molecular Genetics and Biotechnology Laboratory, Aquatic Genomics Unit, School of Fisheries, Aquaculture and Aquatic Sciences, and Program of Cell and Molecular Biosciences, Auburn University, Auburn, AL, United States

Introduction

Catfish (order Siluriformes) is one of the largest orders of teleosts containing ~4100 species, representing ~12% of all teleosts and ~6.3% of all vertebrates (Eschmeyer and Fong, 2014; Wilson and Reeder, 2005). Most catfish have a cylindrical body with a flattened ventral to allow for benthic feeding (Bruton, 1996). Catfish are so-named because of their whisker-like barbels, which are located on the nose, each side of the mouth, and on the chin. Most catfish possess leading spines in their dorsal and pectoral fins. Catfish are scaleless, a characteristic of catfishes distinguishing them from most other teleost fish. However, some catfish, such as plecos, possess bony dermal plates covering their skin (Arce et al., 2013; Armbruster, 2004; Burgess, 1989; Ferraris and Vari, 2012).

Catfish are highly diverse and distributed worldwide. They are commonly found in inland or coastal waters of all continents, including Antartica where fossils are found (Grande and Eastman, 1986). Catfish are most abundantly distributed in the tropics of South America, Africa and Asia, with one family native to North America and one family in Europe (Lundberg and Friel, 2003). Due to their worldwide distribution and diversity, catfish are interesting models to ecologists and evolutionary biologists, and are important for biogeographical studies (Sullivan et al., 2006).

Over half of all catfish species live in the Americas, of which channel catfish, blue catfish, and flathead catfish are the three primary catfish species in the United States. Populations of at least one of these catfish species are available in large numbers in most rivers, lakes, and reservoirs across the United States. Channel catfish are primarily native to the Mississippi River Valley but are widely introduced throughout the nation (FAO, 2015). Blue catfish are native to the Mississippi, Missouri, and Ohio River basins of central and southern United States, and also occupy south along the Gulf coast to Belize and Guatamala (Graham, 1999). Flathead Catfish (*Pylodictis olivaris*) are a large species in the Ictaluridae family. They spread primarily from the lower Great Lakes through the Mississippi River watershed to the Gulf States.

Catfish are quite hardy such that they are more adaptable for artificial spawning, handling, and culture. They possess all the characteristics necessary for aquaculture including relatively high fecundity, ability for artificial spawning, adaptability to earthen ponds for culture, high tolerance to low dissolved oxygen, relatively high resistance

Genomics in Aquaculture. http://dx.doi.org/10.1016/B978-0-12-801418-9.00004-4

against infectious diseases, and relatively high feed conversion efficiency. It is such characteristics that make catfish one of the most popular groups of fish for aquaculture. In the world, a few major species are widely used for aquaculture, including channel catfish, blue catfish, walking catfish, shark catfish, Thai catfish, and African catfish. Channel catfish (*Ictalurus punctatus*) is the major aquaculture species in the United States. A closely related species, blue catfish (*I. furcatus*), is also important in this industry as the F1 hybrid catfish produced by mating channel catfish females with blue catfish males are now broadly cultured.

Catfish are of considerable economic importance for aquaculture and recreational fisheries. Its global importance is increasing as several countries in Asia, such as China and Vietnam, are now heavily involved in catfish aquaculture (Liu, 2008). Catfish are the primary aquaculture species in the United States, accounting for over 60% of all US aquaculture production (Liu, 2011). In the southeastern United States, particularly in the states of Mississippi, Kansas, and Alabama, catfish is one of the top agricultural commodities. The catfish industry provides employment opportunities for tens of thousands of producers, processors, services providers, marketers, retailers, and restaurant owners.

Catfish production in the United States mainly includes channel catfish, blue catfish, and their F1 hybrid. The market for channel catfish is well developed in the United States. In 2006, the catfish production reached 700 million pounds, becoming one of the fastest growing sectors in American agribusiness. Blue catfish is also important because of its ability to produce hybrid catfish with channel catfish. The F1 hybrid catfish produced by crossing channel catfish female with blue catfish male exhibit a high level of heterosis and is a preferred genotype for aquaculture. Artificial fertilization is required for the production of the interspecific hybrids due to reproduction isolation. Major progress has been made in recent years for efficient production of the hybrid catfish, which account for almost half of catfish fry used in the current catfish industry (Dr Rex Dunham, Auburn University, personal communications).

However, the catfish industry in the United States has recently encountered unprecedented challenges due to devastating diseases, high feed and energy costs, and fierce international competition. As a consequence, the catfish industry has significantly shrunk from its high in 2003 with 662 million pounds down to 334 million pounds in 2013 (Hanson and Sites, 2014). Many catfish producers are not profitable due to the high feed and energy costs, and the economic losses caused by the outbreaks of devastating infectious diseases. Therefore, genetically enhanced brood stocks are much needed to support the catfish industry.

The interspecific hybrid catfish produced by mating channel catfish female with blue catfish male are superior to either parent in most production and performance traits. In comparison with its parents, the hybrid catfish grow faster, have a more aggressive feeding behavior, exhibit a greater feed conversion efficiency, are more resistant to major bacterial diseases, are easier to harvest, and provide a greater fillet yield. These improved production and performance traits make hybridization one of the most effective genetic improvement programs (Liu, 2008).

The hybrid catfish has the potential to revolutionize the catfish industry. However, the bottleneck exists in mass production of the hybrid seeds due to reproductive isolation between channel catfish and blue catfish. Artificial spawning must be conducted to produce the hybrid catfish fingerlings, while the rates of fertilization and hatching of the hybrid catfish are normally lower than those of channel catfish. Fortunately, recent progress in artificial spawning and fertilization has allowed improved fry production. Another important achievement has been made through use of different strains of channel catfish and blue catfish, which has produced genetically distinct Channel catfish X Blue catfish hybrids with even more superior production characteristics (Dunham and Masser, 2012).

Several performance and production traits are important for aquaculture, which include, but are not limited to, growth rate, feed conversion efficiency, disease resistance, processing yield, seinability, stress tolerance, and tolerance to low dissolved oxygen (Liu, 2011). Growth rate is a primary trait for aquaculture because fast growth would allow shorter culture period to reduce the energy and other related costs. Enhanced feed conversion efficiency could reduce feed costs and increase profit margins since feed cost accounts for up to 50% of all variable costs. Disease resistance is one of the most important traits because diseases cause major economic losses, up to one third of the industry every year. Two major devastating diseases include the enteric septicemia of catfish (ESC) caused by *Edwardsiella ictaluri* (Hawke, 1979) and the columnaris disease caused by *Flavobacterium columnare*. In addition, channel catfish virus disease, motile aeromonad septicemia caused by *Aeromonas hydrophila* and the "Ich" disease caused by protozoan parasite *Ichthyophthirius multifiliis*, can also cause significant economic losses. Processing yield is important because most catfish are processed into fillet for marketing. Seinability is an important trait because any fish that is left in the pond after seining can significantly reduce the feed conversion efficiency. Stress tolerance is important because stress of catfish can lead to the so-called red flesh disease, caused by high levels of stress during harvesting and transportation immediately before processing. Tolerance to low dissolved oxygen is important because it directly affects survival rates and the energy cost due to the use of aerators. Genetic factors contribute to controlling all these traits, though most of these traits are affected by the environment and environment × genotype interactions. Efficient methods of selection for improving these production and performance traits within breeding programs must be developed to provide technological advantages to our catfish producers to help them compete in the global marketplace (Liu, 2011).

The genome: structure and organization

Channel and blue catfish both have 29 pairs of chromosomes and genomic sizes of approximately 1 GB per haploid genome (LeGrande et al. 1984; Tiersch and Goudie, 1993). Like most other teleosts, channel catfish is a diploid fish. Recent genome research demonstrated that channel catfish have also undergone the teleost-specific genome duplication (TSGD) during its evolution (Kasahara et al. 2007; Meyer and Van de Peer, 2005). The catfish genome is highly AT-rich, with 60.7% A + T (Xu et al., 2006). It contains one main type of tandem repeats named Xba elements (Liu et al., 1998b) and several types of dispersed repetitive elements, with the predominant dispersed repetitive elements being Tc1/mariner DNA transposons (Nandi et al., 2007). In addition, retrotransposons, LINE, and SINE elements also exist in the catfish genome, with several SINE elements being well characterized such as the Mermaid and Merman SINE elements (Kim et al., 2000). Simple sequence repeats or microsatellites are also highly abundant in the catfish genome with AC and AG microsatellites being the most abundant types of repeats. All these repeats within the catfish genome, especially the dispersed repetitive elements such as the Tc1 DNA transposons, added more difficulty to whole genome sequence assembly and annotation (Jiang et al., 2013a,b; Kim et al., 2000; Liu et al., 2009; Liu, 2011; Xu et al., 2007).

In almost 20 years of catfish genome research, great efforts have been devoted to the understanding of basic genome structure and organization through stages of molecular marker development, genetic and physical mapping, and whole genome sequencing, assembly and annotation.

Development of molecular markers

Development of molecular markers has been one of the major efforts in the first decade of catfish genome research (Liu and Cordes, 2004). In the early days of catfish genome research, most of these efforts were devoted to markers that were readily available without the availability of genome information or resources. This included the development of RAPD and AFLP markers (Liu, 2003; Liu et al., 1998a,b, 1999b,c). These efforts certainly contributed to the development of genome programs at that time, but these dominant markers are not very useful in terms of polymorphic DNA markers in the long term. Since 1997, great effort has been devoted to microsatellite development in catfish. Several approaches were adopted including construction of microsatellite-enriched genomic libraries, targeted sequencing (Serapion et al., 2004b), and identification of microsatellites through data mining using genome resources such as expressed sequence tags (ESTs) or BAC-end sequences (Li et al., 2007; Liu et al., 1999a, 2009; Serapion et al., 2004a; Wang et al., 2010; Xu et al., 2006). Large numbers of useful microsatellites have been obtained through data mining of genome sequence surveys such as BAC end sequences, and through EST data mining. However, it is apparent that such huge efforts can be avoided now with the application of next generation sequencing technologies.

The effort of marker development has shifted to the identification of single nucleotide polymorphisms (SNPs) over time (He et al., 2003; Liu et al., 2011; Sun et al., 2014; Wang et al., 2010). SNPs are now the markers of choice for genetic studies because they are the most abundant genetic variations widely distributed in the genome, and are generally bi-allelic polymorphisms that are amenable to automated genotyping. SNPs are efficient for genome-wide association studies (GWAS) because linkage disequilibrium can be detected with high-density SNPs when dealing with complex traits (Liu et al., 2014). Once again, the huge efforts devoted in SNP discoveries can now be easily achieved through next generation sequencing.

Gene-associated SNPs have been identified using EST resources in the absence of a whole genome sequence (He et al., 2003; Wang et al., 2010). A key issue of

EST-derived SNPs is the possibility of sequencing errors that are subsequently interpreted as potential SNPs by data mining. In order to circumvent this problem, quality standards for identifying EST-derived SNPs have been established (Wang et al., 2008). Two factors are the most important: the contig size and the minor sequence allele frequency. Contigs with at least four sequences with the minor allele sequence being represented at least twice provided a high rate of SNP validation (Wang et al., 2008).

SNPs are able to be identified with high confidence more recently because of much deeper sequencing coverage provided by the next-generation sequencing technologies in comparison with the Sanger sequencing. The large-scale sequence data enabled efficient and effective identification of SNPs from genomes of various organisms. Using Illumina based RNA-Seq technology, we generated genome-scale gene-associated SNP markers from both channel catfish and blue catfish (Liu et al., 2011), yielding large sets of gene-associated SNPs within channel catfish, blue catfish, and between channel catfish and blue catfish. Moreover, we conducted whole genome resequencing of four major channel catfish aquaculture populations and one wild population to identify a total of 8.4 million putative SNPs (Sun et al., 2014). On average, there are one SNP every 116 bp in the catfish genome. The abundant and high-quality SNPs were devoted to the construction of high-density SNP arrays (Liu et al., 2014), enabling large-scale genotyping of genetic markers for GWAS (Geng et al., 2015), high-density linkage mapping (Li et al., 2015), fine QTL mapping, haplotype analysis, and whole genome-based selection.

Linkage mapping

With the accumulation of channel catfish genome resources including sequences and polymorphic markers, high resolution genetic maps are needed to facilitate genetic analysis and increase the efficiency of selection. With genetic mapping, molecular markers correlated with genetic loci controlling economic traits can be identified and used to select superior brood stocks with genomic merits. A genetic map will be more useful to assist selection of fish for traits in which measurement is difficult or expensive (e.g., disease resistance) or lethal to brood stocks (e.g., carcass composition). In addition, a genetic map can provide a guide to selection on multiple traits. For instance, fast growth is often correlated with low reproductive capacity. A genetic map would allow both traits to be mapped such that brood stocks can be obtained to harbor superior traits on both growth and reproduction traits (Liu, 2003, 2011). In the course of the catfish genome research, we have constructed linkage maps with various densities of markers including the very first sets of framework linkage maps (Liu et al., 2003; Waldbieser et al., 2001), higher densities of maps with gene-associated markers (Kucuktas et al., 2009), the second generation of linkage map that included microsatellite sequences within BAC-end sequence, allowing partial integration of the linkage map with the physical map (Ninwichian et al., 2012a). With this second generation linkage map, the average marker density across the 29 linkage groups was 1.4 cM/marker. The increased marker density highlighted variations in recombination rates within and among catfish chromosomes. The integration of this genetic map with

BAC-based physical map enabled comparative studies with teleost model species as well as provided a framework for ordering and assembling whole-genome scaffolds (Ninwichian et al., 2012a; Zhang et al., 2013). Most recently, we constructed a high-density and high-resolution genetic map for channel catfish with three large resource families genotyped using the catfish 250K SNP array (Li et al., 2015). A total of 54,342 SNPs were placed onto the linkage map, which is one of the genetic maps with the highest marker density among aquaculture species. This genetic linkage map was used to validate the whole genome assembly and anchor the scaffolds to chromosomes. In parallel, a high density genetic linkage map was also constructed using the interspecific hybrid catfish resource families (Liu et al., 2015). Comparison of these maps made using channel catfish resource families and the interspecific hybrid catfish resource families allowed identification of incompatible genomic regions between the channel catfish and blue catfish, possibly accounting, at least in part, for the reproductive isolation of the two species.

Physical mapping

Two channel catfish BAC (Bacterial Artificial Chromosome) libraries were constructed and characterized: the CHORI-212 BAC library (Wang et al., 2007) and the CCBL1 BAC library (Quiniou et al., 2003). BAC-based physical maps of the channel catfish genome were generated using high information content fingerprinting (Quiniou et al., 2007; Xu et al., 2007). Fingerprinting of 46,548 BAC clones (6.8 X genome coverage) from the CCBL1 BAC library developed from an inbred meiotic gynogen provided the assembly of 1,782 contigs (Quiniou et al., 2007). Fingerprinting of 34,580 BAC clones (5.6 X genome coverage) from the CHORI-212 BAC library developed from an outbred male resulted in the assembly of 3,307 contigs (Xu et al., 2007). The CCBL1 BAC library was based on HindIII partial digestion of genomic DNA while the CHORI-212 BAC library was based on EcoRI partial digestion, therefore many genomic regions not covered within one library could have been covered in the other. These two physical maps have been integrated to close gaps (unpublished data).

Over 103,000 BAC-end sequences (over 61 Mb, 6% of the channel catfish genome) were generated by sequencing BAC clones using the traditional Sanger sequencing from the two catfish BAC libraries, which provided a valuable resource toward whole-genome sequencing of catfish (Liu et al., 2009; Xu et al., 2006). Furthermore, many genes and markers were assigned to BACs, allowing them to be used for comparative genome analysis. A high-quality integrated map with genetic map already provided a valuable tool for validating and improving the catfish whole-genome assembly. Additionally, it will facilitate fine-scale QTL mapping and positional cloning of genes responsible for economically important traits and large-scale comparative genome analysis.

Whole genome sequencing

A whole genome sequence is essential for understanding evolutionary and biological characteristics of catfish, and for genetic improvement programs. Recent advances in next-generation sequencing technologies have led to both huge increases in throughput and drastic reduction in costs that have made the previously difficult sequencing tasks readily practical in catfish.

Sequencing the channel catfish genome using the next generation sequencing technology was funded by the U S Department of Agriculture in 2010. In this project, a doubled haploid channel catfish (Waldbieser et al., 2010) was used as sequencing template to reduce the complexity caused by whole genome duplication. The initiation of the catfish whole genome sequencing project is historic. The genome sequence assembly will facilitate (1) genome-wide comparative analysis to provide insights into evolutionary biology and genome function; (2) establishment of gene orthologies with other vertebrate species to facilitate functional genomic studies and identify candidate genes for performance traits; and (3) haplotype mapping to provide the basis for genome-based selection. The catfish is positioned basally in the teleost phylogeny, so its genome assembly will be useful in genome alignments with other fish (e.g., rainbow trout, Atlantic salmon, zebrafish, pufferfish, medaka, and stickleback) to inform teleost evolution and function. Comparison of genomes of channel catfish and blue catfish will benefit researchers to investigate mechanisms and consequences of speciation. Additionally, it will facilitate introgression of blue catfish genomic regions into the channel catfish genome for production of self-reproducing hybrid catfish (Liu, 2011).

A primary goal of genome sequencing programs is the annotation of protein-coding genes. The most common annotation pipeline for annotation of the assemblies is performing transcript mapping and putative annotation based on comparative sequence information. Establishment of conserved syntenies and orthologies will permit researchers to identify virtually all coding regions in the catfish genome and enable comprehensive functional genomics research. The development of more complete microarray platforms will benefit from the whole genome sequences. The assemblies will serve as references for quantification of gene expression using RNA-Seq.

Genome sequencing will not only permit efficient identification of sequence polymorphisms, but also allow their in silico mapping. The genome sequence assembly can expedite the identification and location of SNPs and microsatellite markers evenly spaced throughout the genome. It will also enable the identification of insertions/deletions, segmental duplications, multicopy loci, and changes in local DNA topography (Freeman et al., 2006; Parker et al., 2009) to permit researchers to investigate the contribution of these types of genomic variations accounting for phenotypic variations. The whole genome assembly should serve as a reference for the selection of evenly spaced SNPs for the construction of high-density SNP arrays.

The channel catfish reference genome has been assembled with a high quality among genomes sequenced with next generation sequencing technology. The vast majority (97.2%) of the assembled genome sequence was anchored onto the 29 chromosomes, the highest proportion among all sequenced fish genomes. A total of 26,661 protein-coding genes were annotated: these were characterized by the presence of recently duplicated genes involved in immune function and metabolism of xenobiotics (Liu et al., 2016). This great progress provides genomic and transcriptional resources for investigation of biological characteristics of catfish, and genomic hallmarks of teleosts.

Genome evolution and comparative genomics

Comparative genome map is a powerful tool in genomics studies, especially for non-model species, which facilitates the transferring of genomic information from well-studied model species (Liu et al., 2009). It not only allows better understanding of genome arrangement during evolution, but also benefits the discovery of orthologies among species. For aquaculture species, direct analysis of gene functions will prove to be difficult. Orthologies should allow inference of gene functions from model species based on evolutionary conservation of genes and their functions.

Comparative genome analysis is readily applicable in catfish with the large amount of genome resources, including genetic linkage maps, physical maps, BAC end sequences (BES), integrated linkage and physical maps using BES-derived markers, physical map contig-specific sequences, (Jiang et al., 2013b), and reference genome sequences. These genomic resources facilitate comparative analysis of catfish at the genome scale with several other model fish species. In a pilot study, comparative analysis was conducted with other model fish species in an approximately 1-Mb region in the channel catfish genome (Jiang et al., 2011). Comparative genomic analysis of one catfish linkage group (LG8) with several model fish allowed the identification of conserved syntenies in the catfish genome with respective to other sequenced fish genomes (Zhang et al., 2013). Whole genome comparative analysis between channel catfish and four model fish species, including zebrafish, medaka, stickleback, and tetraodon, suggested that the organization of the catfish genome was similar to that of these fishes, allowing homologous chromosomes to be identified (Jiang et al., 2013a). Extended syntenic blocks were evident within each chromosome, but the conserved syntenies at the chromosome level involved extensive inter- and intrachromosomal rearrangements. The whole genome comparative analysis should facilitate whole genome sequence assembly and annotation, as well as provide insights into genome evolution.

The transcriptome: expression and function

Early efforts of EST sequencing

ESTs are unique DNA sequences generated by sequencing cDNA clones from cDNA libraries. In the Sanger sequencing era, EST sequencing served as an important approach in identification and analysis of gene transcripts and expression (Adams et al., 1991). The number of the ESTs sequenced from a gene can serve as an indicator for the expression level of the gene. In other words, the higher one gene is expressed, the more its ESTs should be present in the cDNA library. For instance, the gene expression profile in spleen and brain of channel catfish have been identified via EST sequencing (Ju et al., 2000; Kocabas et al., 2002), which were useful for functional genomics research.

EST sequencing is also applied to the development of molecular markers. For instance, by comparing the ESTs between channel catfish and blue catfish, 1020 putative SNPs among 161 genes that are different between two species were identified (He et al., 2003). Similarly, the simple sequence repeats (SSR) markers were also identified via EST sequencing (Serapion et al., 2004a). By sequencing a total of nearly 500,000 ESTs from both channel catfish and blue catfish, we identified a total of 20,757 microsatellites and over 300,000 putative SNPs, of which approximately 48,000 SNPs are high-quality SNPs identified from contigs with at least four sequences and the minor allele presence of at least two sequences in the contigs (Wang et al., 2010).

RNA-Seq based transcriptome analysis

RNA sequencing (RNA-Seq) is a technique where RNA is sequenced by next generation sequencing (Chu and Corey, 2012). Compared with the traditional microarray technology, which depends on the prior genomic sequence information, RNA-Seq is revolutionizing the analysis of transcriptomes by providing high-throughput expression data (Ekblom and Galindo, 2011). RNA-Seq has been applied not only to assemble and annotate transcriptomes but also to determine differential gene expression under various physiological or environmental conditions in numerous aquatic animals (Palstra et al., 2013; Pauli et al., 2012; Qian et al., 2014), including catfish (Li et al., 2012; Liu et al., 2012, 2013; Liu, 2011; Sun et al., 2012).

RNA-Seq has been used to characterize the catfish transcriptome, reconstruct gene transcripts, annotate full-length transcripts and identify gene-associated markers. A comprehensive and accurate transcriptome assembly was generated by deep sequencing of various tissues of a doubled haploid channel catfish (Liu et al., 2012). A large set containing over 14,000 full-length transcripts was identified, greatly enriching catfish genomic resources for functional analysis and genome annotation. RNA-Seq was also used to efficiently and cost-effectively generate genome-scale gene-associated SNPs for developing high-density SNP arrays for genetic analysis in catfish (Liu et al., 2011). In a study to identify male-biased expressed genes, RNA-Seq was used to construct the first testis transcriptome in catfish, resulting in identification of 5450 genes highly expressed in the testis, with many of which being involved in gonadogenesis, spermatogenesis, testicular determination, gametogenesis, gonad differentiation, and possibly sex determination (Sun et al., 2013).

RNA-Seq has also been used to characterize differentially expressed genes in catfish in responses to disease infection and stresses such as ESC disease (Li et al., 2012; Wang et al., 2013), columnaris disease (Sun et al., 2013), heat stress (Liu et al., 2013), and hypoxia stress. For instance, using RNA-Seq technology, we identified 1255 differentially expressed genes between groups of catfish resistant and susceptible to ESC disease. In addition, 56,419 SNPs residing in 4,304 unique genes were identified as significant SNPs between susceptible and resistant catfish to ESC (Wang et al., 2013). In response to heat stress, RNA-Seq analysis revealed expression signatures of genes involved in oxygen transport, protein synthesis, folding, and degradation in catfish (Liu et al., 2013).

Protein-coding genes

Protein-coding genes are the "major" part of the genetics central dogma, though the importance of noncoding RNAs has increasingly drawn researchers' attention over the past decades (Kapranov et al., 2007). Annotation of protein-coding genes is a major step after the genome is sequenced. Therefore, the delineation of the complete set of protein coding genes is essential to the task of translating the information in the sequence of the genome into biologically relevant knowledge (Harrow et al., 2009).

Determining a full list of the protein-coding genes is not a trivial task. Even in the human genome project, the actual number of protein-coding genes in the human genome is still not finalized many years after releasing the first draft of the human genome sequence (Lander et al., 2001; Venter et al., 2001). The initial sequencing of the human genome suggested that the number of protein-coding genes falls somewhere between 26,000 (Venter et al., 2001) and 30,000 (Lander et al., 2001). With the publication of the final draft of the Human Genome Project (International Human Genome Sequencing Consortium, 2004), the number of protein-coding genes was revised downwards to between 20,000 and 25,000. The latest studies suggested even smaller numbers of protein-coding genes, just 20,500 genes (Clamp et al., 2007) or as few as 19,000 genes (Ezkurdia et al., 2014). The number of genes annotated in the Ensembl database (Cunningham et al., 2015) has been on a downward trend since its inception in 1999. The latest Ensembl results based on build CRch38.p2 and the GENCODE annotation (GENCODE 22) show that there are 20,300 protein coding genes in the human genome.

Gene prediction is a bioinformatics-driven task. Since the early methods developed to distinguish coding from noncoding regions (Fickett, 1982), numerous strategies have been explored and many methods have been developed to elucidate the exonic structure of genes in eukaryotic genomes (Harrow et al., 2009). Given a genome DNA sequence, information on the location of genes can be predicted from different sources: (1) conservation with one or more genomes of model organisms; (2) intrinsic signals involved in gene specification, such as start and stop codons and splice sites; (3) the statistical properties of coding sequences; and, most importantly, (4) known transcript sequences (either full-length cDNAs or partial ESTs) and protein sequences.

The prediction of protein-coding genes is more inconclusive in sequenced genomes of aquaculture species. Gene prediction is more difficult in fish species due to the genome complexities derived from teleost-specific whole genome duplication. Moreover, gene prediction strategies and programs are not as well developed as those applied to human genome annotation. This is further complicated by the completeness and quality of genome assemblies, and the lack of transcriptome resources.

With catfish, a total of 26,661 protein coding genes were predicted from the channel catfish genome, comparable to that of zebrafish (Liu et al., 2016). The vast majority (>99%) of predicted genes were supported by transcript evidence. Further orthology analysis revealed that a large amount of genes were conserved among several vertebrates, while the channel catfish harbored a larger number of single copy genes among all sequenced diploid fish species (Table 4.1).

Table 4.1 Number of protein-coding genes in several fish species with sequenced genomes

Species	Number of coding genes	Sources
Zebrafish	25,642	Ensembl release 80
Fugu	18,523	Ensembl release 80
Medaka	19,699	Ensembl release 80
Tetraodon	19,602	Ensembl release 80
Stickleback	20,787	Ensembl release 80
Amazon molly	23,615	Ensembl release 80
Cave fish	23,042	Ensembl release 80
Platy fish	20,379	Ensembl release 80
Cod	20,095	Ensembl release 80
Tilapia	21,437	Ensembl release 80
Tongue sole	21,516	Chen et al. (2014)
Common carp	52,610	Xu et al. (2014)
Grass carp	27,263	Wang et al. (2015)
Yellow croaker	19,362	Wu et al. (2014)
Sea bass	26,719	Tine et al. (2014)
Catfish	26,661	(Liu et al., 2016)

Long noncoding RNAs

Long noncoding RNAs (lncRNAs) are comprised of diverse noncoding RNAs that are longer than 200 nucleotides, which structurally resemble mRNAs but do not encode proteins (Kung et al., 2013). Recent genome-wide annotation of human and mouse genomes indicated that lncRNAs were expressed in various tissues. The results showed that the lncRNAs were expressed 4 times more than the protein-coding transcripts (Kapranov et al., 2007). Although the importance of lncRNA studies has gradually been emphasized, systematic analysis of lncRNAs in fish has been rare, especially with aquaculture species.

Previous lncRNA analysis has been conducted in zebrafish (Pauli et al., 2012). At least 1133 long noncoding transcripts originating from diverse genomic loci have been identified through transcriptome sequencing of 8 developmental stages of zebrafish. The zebrafish lncRNAs shared many of the characteristics of mammals, including relatively short lengths, small exon numbers, low expression levels, and less conservative compared to that of introns. The lncRNAs were expressed in time-specific and tissue-specific patterns in both zebrafish embryos and adult tissues (Pauli et al., 2012). Large numbers of unique unknown transcripts were also found in previous RNA-Seq studies in catfish, which were likely derived from lncRNAs (Liu et al., 2011, Wang et al., 2013).

The lncRNAs are shown to carry out various functions in living organisms. An extensive identification and analysis of lncRNAs in catfish is ongoing, further investigation would warrant better understanding of their roles in many biological processes including development and responses to bacterial infections and abiotic stresses.

MicroRNAs

MicroRNAs (miRNAs) are endogenous, single-stranded noncoding RNAs with lengths of ~22 nucleotides. They play critical roles in posttranscriptional gene regulation and are commonly found in animals and plants (Bartel, 2004). Mature miRNA can execute function by targeting homologous sequences in mRNAs for degradation or translational repression.

The mature miRNA is derived from a stable stem-loop structure called precursor miRNA (pre-miRNA). A special RNase III-like endonuclease known as Dicer (Bernstein et al., 2001) trims the pre-miRNA and removes the loop, creating a short double-stranded RNA (dsRNA), of which one strand acts as mature miRNA. Then Dicer integrates the mature miRNA into the RNA induced silencing complex (RISC) which hybridizes the mature miRNA to its targeted mRNA (Hammond et al., 2000). Subsequently, this complex could negatively regulates gene expression either by triggering mRNA degradation or by inhibiting translation depending on complete or partial complementarities of miRNA with its target (Barozai, 2012).

Several miRNA studies in channel catfish have been reported, focusing on identification through computational methods and high throughput sequencing. Computational methods (Li et al., 2010) have been used to identify miRNAs based on the phylogenetic conservation of miRNAs (Chen and Rajewsky, 2007). Utilizing computational genomic homology, Barozai (2012) identified and characterized 73 mature miRNA in channel catfish and targeted a total of 341 protein coding genes involved in diverse biological processes such as immune response and signaling transduction. Similarly, Xu et al. (2013) identified 237 conserved miRNAs and 45 novel miRNAs in 10 tissues of channel catfish by applying bioinformatic analysis. Some novel miRNAs were expressed in all examined tissues, while some others appeared to exhibit tissue-specific expression. Furthermore, target genes of miRNAs were found to be potentially involved in immune regulation, transcriptional regulation, metabolism, and many other biological processes (Xu et al., 2013). In a study of yellow catfish, a total of 384 conserved miRNAs and 113 potentially novel miRNAs were discovered from gonad tissues; a number of sex-biased miRNAs were also identified with some of which being potentially involved in testis development and spermatogenesis (Jing et al., 2014). Extensive miRNA identification and analysis through deep small RNA sequencing of various tissues from samples exposed to bacterial infections is underway in channel catfish, which will contribute to a better understanding of the roles that miRNAs play in regulating different biological processes in catfish.

Analysis of gene expression under various stress conditions

Living in aquatic environments, fish are greatly affected by diverse stressors such as pathogen infection, temperature change, and hypoxia challenge. The stress tolerance and mechanisms underlying the molecular response could be revealed by transcriptome profiling of gene expression. Therefore, transcriptome profiling has been essential for the functional analysis of genes. Over the past decades, microarrays have been developed for gene expression analysis. Most recently, gene expression altered by bacterial infection in catfish has been investigated using RNA-Seq.

Numerous studies have been conducted to elucidate molecular mechanisms of the immune response against bacterial infection. Many genes, such as end-binding protein 1, heat shock protein 70, transferrin, hepcidin, leptin-like peptide, tumor suppressor, bactericidal permeability-increasing protein gene, pathogen recognition receptors, and many other genes were shown to be regulated after ESC infection (Bao et al., 2005; Elibol-Flemming et al., 2009; Kobayashi et al., 2011; Liu et al., 2010; Mu et al., 2015; Rajendran et al., 2012a; Rajendran et al., 2012b; Sun et al., 2014; Xu et al., 2005; Zhang et al., 2012). These results added knowledge to the understanding of immune responses and gene functional analysis in teleosts. Recently, RNA-Seq based transcriptome profiling allowed identification of differentially expressed genes in catfish in response to ESC disease (Li et al., 2012; Wang et al., 2013), and columnaris disease (Sun et al., 2013).

Temperature is one of the major environmental factors for catfish. In response to heat stress, genes involved in oxygen transport (hemoglobin, myoglobin, and cytoglobin genes), protein folding and degradation, and metabolic process were found to be dramatically upregulated, while general protein synthesis was highly repressed (Feng et al., 2014, 2015; Liu et al., 2013). After cold-stress challenge, Ju et al. (2002) identified differential gene expression in response to cold acclimation in the brain of channel catfish. Weber & Bosworth (2005) observed that gene expression of myostatin, myosin heavy chain, proliferating cell nucleus antigen, and heat shock protein 40 increased over different time-points.

Hypoxia is also regarded as an important stress factor to aquaculture species. Oxygen availability in water varies significantly temporally and spatially and fish have to adapt rapidly to the changing environment in order to survive. Studying at gene expression level, hypoxia inducible factors alpha and their inhibiting factor of channel catfish were shown to be altered under low oxygen conditions, suggesting their importance in responses to hypoxia (Geng et al., 2014).

Gene pathways, networks, and their regulation

Signal transduction pathway involves the binding of extracellular signaling molecules and ligands to receptors located on the cell surface or inside the cell that trigger events inside the cell, to invoke a response. The response can then alter the cell's metabolism, shape, and gene expression (Krauss, 2006). Signaling pathways in multicellular organisms are triggered by various environmental stimuli. Responsible genes function coordinately as networks responding to stresses. Two examples are provided later based on the results in catfish.

In a genome-wide association study, we identified a number of genes within the significantly associated genomic regions that were involved in the PI3K pathway (Fig. 4.1), suggesting the importance of the PI3K pathway in the resistance to columnaris disease (Geng et al., 2015). In a RNA-Seq study of catfish in response to heat stress, a combination of KEGG pathway analysis, manual annotation based on the several databases, and manual literature searches provided six broad functional categories

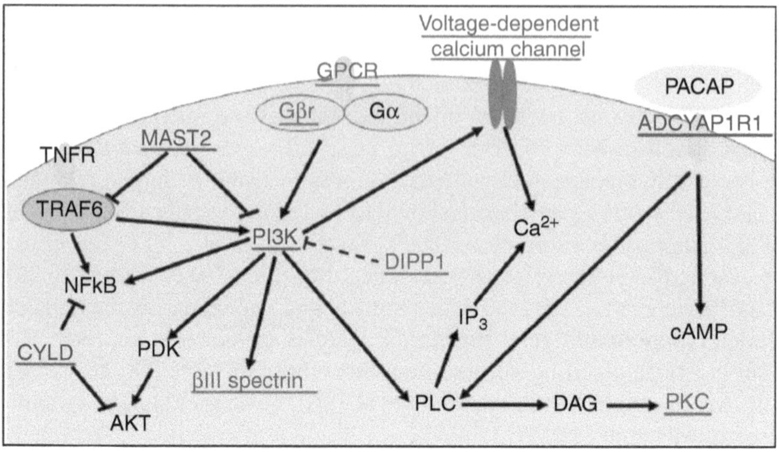

Figure 4.1 Signal transduction pathways involving PI3Ks and other candidate genes. Candidate genes are underlined. The candidate genes in the significant QTL are CYLD, MAST2, PI3K, and ADCYAP1R1. The candidate genes in the suggestive QTLs are βIII spectrin, Gβr, GPCR, DIPP1, voltage-dependent calcium channel, and PKC (Geng et al., 2015).

of the differentially expressed genes (Fig. 4.2) involved in heat stress tolerance. These genes are mainly involved in oxygen transport, protein synthesis, protein folding, and degradation (Liu et al., 2013).

One of the most important concepts is the observation of the presence of functional hubs. As demonstrated in the resistant QTLs against columnaris, many genes involved in the same pathway were found also reside in a similar genomic region, together in the form of a "hub". It is possible that co-presence in the same genomic location would allow rapid responses involving coordinated regulation of a set of genes involved in the same gene pathway (Fig. 4.1).

The genome: single nucleotide polymorphisms and genomic variations

Single nucleotide polymorphisms

Single nucleotide polymorphisms (SNPs) are polymorphisms that are caused by point mutations that give rise to different alleles containing alternative bases at a given position of nucleotide within a locus. Due to their high abundance in the genome, SNPs already serve as the predominant marker type. Many sequencing techniques and bioinformatics methods are applied for SNP development and identification, such as random shotgun sequencing, PCR amplicon targeted sequencing, RNA sequencing, and bioinformatic mining of ESTs databases (Liu and Cordes, 2004). For identification of EST-derived SNPs, quality standards have been established in order to reduce the possibility of sequencing errors (Wang et al., 2008). Contigs with at least four sequences with the minor allele sequence being represented at least twice provided a high rate

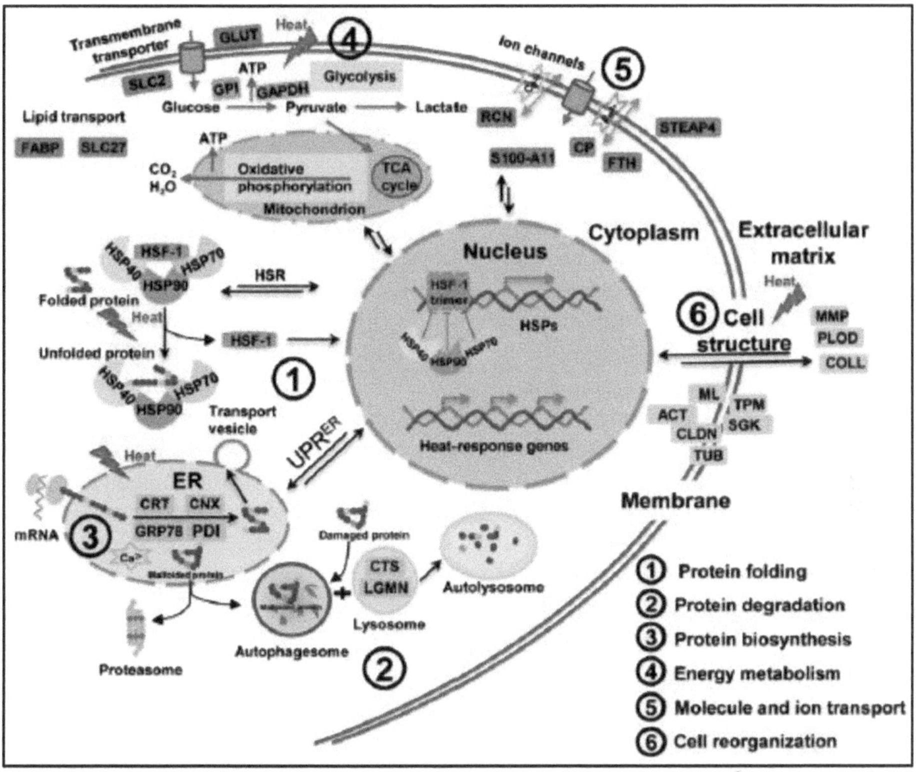

Figure 4.2 Schematic presentation of six gene pathways (as circled numbers 1–6) involved in heat stress responses based on RNA-Seq expression signatures in catfish gill and liver. Abbreviations: *HSP*, heat shock response; *UPR*, unfolding protein response; *ER*, endoplasmic reticulum; *TCA cycle*, tricarboxylic acid cycle (Liu et al., 2013).

of SNP validation. Using EST resources of channel catfish, Wang et al. (2010) have identified over 300,000 putative SNPs. Based on RNA-Seq, over 2 million putative SNPs were identified from channel catfish, while almost 2.5 million putative SNPs were identified from blue catfish using next generation DNA sequencing technology (Liu et al., 2011). Of these putative SNPs, a set of filtered SNPs were identified, which included 342,104 intraspecific SNPs for channel catfish, 366,269 intraspecific SNPs for blue catfish, and 420,727 interspecific SNPs between channel catfish and blue catfish. These filtered SNPs were distributed within 16,562 unique genes in channel catfish and 17,423 unique genes in blue catfish (Liu et al., 2011). Recently, whole genome resequencing of 4 major channel catfish aquaculture populations, as well as one natural population, has provided a total of 8.4 million putative SNPs (Sun et al., 2014). The identified SNPs provide a much needed resource for genetic studies in the catfish scientific community, contributing to the development of high-density, cost-effective genotyping platforms. Validation and testing of these SNPs using SNP arrays will form the material basis for genome association studies and whole genome-based selection in catfish (Liu et al., 2011, 2014).

Strain and population level genomic variations

As a result of environment adaptation, many mutations and other genomic variations have accumulated through the long process of evolution. The fundamental basis of genetic diversity is accumulation of various types of mutations in reproductively isolated populations. The basic principle behind population genetic analysis is to unravel such differences and their inheritance among populations (Liu, 2007).

With DNA markers, genetic variation can be observed across the entire genome. Differences in the relative frequencies of alleles, presence or absence of alleles are used to quantify genetic variation and distinguish genetic units at the levels of populations, species and higher taxonomic designations (Liu and Cordes, 2004).

In a study of catfish intraspecific genetic variations using RAPD, we discovered that intraspecific genetic variations were low among domestic strains of channel catfish. Moreover, genetic analysis of strains of channel catfish and blue catfish suggested the overall low polymorphism among strains within a species for both channel catfish and blue catfish (Liu et al., 1998a). However, considerably higher levels of polymorphisms were detected between channel catfish and blue catfish. Notably, there was no difference between channel catfish × blue catfish F1 hybrids or their reciprocal hybrids (Liu et al., 1999b).

Various domestic strains of channel catfish were also explored using AFLP (Liu et al., 1999c). The study showed that the profiles of various channel catfish domestic strains were quite similar, but variations were prominent. Among 78 total amplified bands, 25 bands were shared by all 10 strains, while the remaining bands showed various levels of polymorphism (Liu et al., 1999c).

Species-specific markers were developed for species identification. Interspecific SNPs for catfish, which were defined as sequence variations between channel catfish and blue catfish, were not identified within channel catfish or within blue catfish (Liu et al., 2011). Similarly, strain-specific SNPs were identified by comparing polymorphisms among strains and populations. In a whole-genome resequencing study in channel catfish, the number of putative strain-specific SNPs varied from 66,487 to 143,126, accounting for 0.9, 2.9, 2.2, 1.3, and 1.7% of SNPs that were identified from the Hatchery strain, USDA103, Thompson strain, Marion strain, and wild population, respectively (Sun et al., 2014).

Historical application of molecular markers for the analysis of populations

The application of molecular markers has allowed rapid progress in investigations of genetic variability and inbreeding, parentage assignments, and species and strain identification in catfish. Popular genetic markers, which include allozymes, mitochondrial DNA, RFLP, RAPD, AFLP, microsatellite, and SNP, have been widely used in catfish population genetics studies (Liu and Cordes, 2004).

Low levels of intraspecific variation were found in RAPD profiles among strains of channel catfish or blue catfish (Liu et al., 1999b). It has been reported that AFLP could be more efficient than RAPD for identifying genotypes within populations

(Chong et al., 2000). Given the power for differentiation, applications of AFLP include stock identification, genetic resource analysis, identification of hybrids, relative contribution estimation in genetic enhancement programs, discrimination of hatchery-reared and wild fish, gender identification have been identified, parentage or forensic analysis, taxonomy and systematics based on cluster relationships of AFLP bands, behavior studies, performance evaluation, and QTL mapping (Liu et al., 1999c).

The AFLP marker system has been used to analyze genetic diversity of domestic channel catfish (Mickett et al., 2003) and wild channel catfish (Simmons et al., 2006). Mickett et al. (2003) analyzed genetic diversity of 16 domestic channel catfish populations within Alabama using AFLP markers which indicated that many channel catfish farm strains in Alabama were genetically similar, but some very distinct differences existed (Mickett et al., 2003). Such information provided implications for future broodstock selection and facilitated further analysis of interactions between domestic and wild populations of channel catfish (Mickett et al., 2003). The analysis of the genetic diversity of wild channel catfish by Simmons et al. (2006) indicated the wild channel catfish populations exhibited higher levels of polymorphisms and heterozygosities than the domestic populations using AFLP markers. Interestingly, strong genetic structures were related with the geographical distribution of catfish samples (Simmons et al., 2006).

Microsatellite markers have been more popularly used for population genetics because of its natures of codominance and high polymorphism. Utilizing microsatellite markers, the genetic variability of populations of channel catfish were detected to be more variable than the blue catfish (Lamkom et al., 2008). It has been summarized that there were five main clusters among channel catfish populations: (1) GK, AS, and GKal, (2) S1, S2, and MR, (3) MS, TA, and ARMK, (4) T, AF, and AR1, and (5) KR, MK, and KS. While, there were two clusters among blue catfish: (1) ARR and DxR, and (2) D, R, and AR2 (Lamkom et al., 2008).

Development of high-density SNP arrays

The large numbers of SNPs provided a critical resource for genetic studies in catfish and contributed to the development of high-density SNP arrays. Using SNP arrays, the SNPs within and among families or populations can be validated and tested for genetic analysis of performance and production traits in catfish, which will provide the basis for genome association studies and whole genome-based selection in catfish (Liu et al., 2011).

A catfish 250K SNP array was developed using Affymetrix Axiom genotyping technology (Liu et al., 2014), which has been used for genome-wide association studies, high-density linkage mapping, fine-scale QTL mapping, haplotype analysis, and whole genome-based selection. The SNPs interrogated on the SNP array were obtained from multiple sources including gene-associated SNPs, anonymous genomic SNPs, and interspecific SNPs between channel catfish and blue catfish. Notably, the performance evaluated by genotyping individuals from wild populations and multiple backcross families suggested the success and good utility of the catfish 250K SNP array (Liu et al., 2014). Currently, a 675K SNP array is being constructed, which will include all the validated SNPs from the 250K SNP array and additional more and well-spaced SNPs selected based on the well-assembled channel catfish genome.

The genome: its relationship with phenome

QTL analysis

Quantitative trait locus (QTL) analysis is a statistical method which links the phenotypic data (trait measurements) to genotypic data (usually molecular markers) to explain the genetic basis of variation underlying complex traits (Miles and Wayne, 2008). For aquaculture species, economically important traits include growth rate, disease resistance, feed conversion efficiency, body conformation and processing yield, meat quality, stress response, tolerance to low dissolved oxygen, and tolerance to low water quality. Aquaculture species, such as catfish, offer unique advantages for QTL analysis because they are able to produce thousands of individuals per spawn. The use of large full-sib families for analysis of quantitative traits should minimize variations due to the use of different families. However, phenotypic evaluations of aquatic animals can be challenging (Kocher and Kole, 2008). Marking fish is often difficult and intrusive, and causes wounding or stress to fish that interferes with phenotypic evaluations and measurements. For instance, labeling fish, no matter whether it is heat branding or PIT tagging, often leads to wounding that may stress the fish and result in infections by bacterial pathogens (Liu, 2003).

Selective genotyping is an effective approach for initial QTL analysis. Because families are large, phenotypic extremes can be selected from each family for genotyping. For certain molecular markers, such as SNPs, selective genotyping coupled with pooling of DNA samples has proven to be very efficient for initial identification of markers linked to the genes or loci which are associated with performance traits. Identification of markers linked to favorable genetic variation is effective for hunting the genes or loci which are associated with performance traits, especially traits that have low heritability, or are difficult, expensive, or lethal to be measured (Liu, 2011). The key step for QTL analysis is to map and position the QTLs on specific regions of chromosomes. Based on this QTL mapping, the phenotypes of the organisms could be analyzed, as in phenomics (Hegele and Oshima, 2007). One of the great advantages of QTL analysis is to provide insights into whether differences of phenotype are primarily affected by a few loci with large effects, or by numerous loci, each with tiny effects. Following QTL mapping, a series of molecular techniques and bioinformatic analyses are required to narrow down the QTL region for the analysis of candidate genes, which could be protein-coding genes or genes with regulatory functions to encode transcription factors and other signaling proteins (Miles and Wayne, 2008).

The ultimate goal of QTL analysis in aquaculture is to conduct marker assisted selection (MAS) or genome selection (Wolters et al., 1981). Breeding values and heritability of production can be predicted based on QTL, which provide useful information for estimating the breeding potential of selection. To date, in the case of catfish, a few performance traits have been identified: three markers have been tentatively linked to feed conversion efficiency; several markers are being evaluated for their linkage with resistance to ESC (Liu, 2008). Most recently, a QTL map was constructed for the seven morphometric traits, which would be a huge leap to initial evaluation of MAS for body weight and morphology in catfish (Hutson et al., 2014). In the near future, with the sufficient genome information, it is promising to fill in the blanks of QTL research in catfish. QTL analyses of sex-determination locus, columnaris disease, and hypoxia tolerance are being conducted in catfish.

Genome-wide association studies

High-throughput genotype data coupled with phenotypic data can be used to identify marker-trait associations via genome-wide association studies (GWAS). Genome-wide association mapping allows detection of markers (SNPs) closely linked to QTLs, which based upon the principle of linkage disequilibrium (LD) between genetic markers and QTL that affect the trait (Geng et al., 2015). Generally, loci that are physically close together exhibit stronger LD than loci that are farther apart on a chromosome (Visscher et al., 2012). Moreover, GWAS can offer the chance or opportunity to systematically analyze the genetic structure of complex traits, and can provide information including high diversity and rapid LD decay in a species (Li et al., 2013a; Yan et al., 2011).

Although QTL mapping is well-suited for family-based samples, association studies, especially GWAS, do not rely on pedigree information and can potentially offer higher mapping resolution by genotyping population samples with high-density SNP arrays. Moreover, the rapid development of next-generation sequencing technologies and recent developments in GWAS methodologies have offered mature software packages for association analysis. The continuing decrease in the genotyping cost makes GWAS a standard tool for detecting natural variation that accounts for complex quantitative phenotypes in organisms (Li et al., 2013a).

GWAS have provided hundreds of common variants whose allele frequencies are statistically correlated with various disease and other traits (McClellan and King, 2010). To date, there is a merging trend for GWAS as the leading method to identify genes underlying human diseases and agriculturally important traits (Gu et al., 2011; Li et al., 2014; Nishimura et al., 2012; Wang et al., 2014). However, GWAS has been rarely performed in aquaculture species. Very recently, we conducted a GWAS study to identify four QTL regions associated with columnaris resistance in catfish, and revealed that the candidate genes identified may function coordinately as "functional hubs" (Geng et al., 2015). Further determination of the causative genes for disease resistance will be followed up in the near future. For instance, GWAS resolution and power will be increased by using larger and/or more families. Ultimately, gene knockout experiments will be performed to demonstrate the candidate genes as the source of disease resistance. This strategy with its discovery potential provides us valuable insight into the mining of other QTLs which are associated with commercial traits of catfish in the near future.

Sex determination and progress toward understanding of sex in catfish

It has been reported that channel catfish exhibit an XY sex-determining system but do not demonstrate sex chromosome difference (Davis et al., 1990; Wolters et al., 1981). Utilizing genomic DNA hybridization-based methods, the linkage of known sex-associated elements in mammals, such as Zfy, Sry, human telomeric sequence (TTAGGG), and the bkm minisatellite sequences, with sex phenotype in catfish has not been identified (Tiersch et al., 1992). Glucosephosphate isomerase-B has been reported to be linked to the sex-determining gene via an isozyme polymorphism between channel catfish and blue catfish (Liu et al., 1996). Genetic and temperature-dependent mechanisms of sex determination were both functional in channel catfish (Patiño et al., 1996). The signs of the ovarian differentiation in channel catfish were detected on day 19, while testicular formation was observed between day 90 and 102, which showed a positive histological signs of male differentiation (Patiño et al., 1996).

Channel catfish sex-determining locus has been mapped and investigated with linkage mapping. Waldbieser et al. (2001) identified seven microsatellite loci that were closely linked with the sex-determining locus (Waldbieser et al., 2001). Although three linkage groups displayed higher recombination rates in males, the sex-specific map distances revealed overall recombination was 3.18-fold more frequent during oogenesis than spermatogenesis in channel catfish. Sex-specific maps have been constructed with gene-associated markers, which revealed that the recombination rate for females was 1.6 times higher than that for males. Among 331 EST-derived microsatellites and SNPs mapped on the combined map, a total number of 25 markers showed evidence of identification of linkage with sex (Kucuktas et al., 2009). Notably, marker AUEST0678 has been tested that it showed a 100% genotyping accuracy in correctly identifying the sex of channel catfish (Ninwichian et al., 2012b).

At the genomic level, the first testis transcriptome was constructed for channel catfish using the RNA-Seq (Sun et al., 2013). Through a meta-analysis of expressed genes in the testis and gynogen transcriptome, 5450 genes were identified to be preferentially expressed in the testis, providing a pool of putative male-biased genes. In gene ontology and annotation analysis, it has been identified that many of these male-biased genes were involved in gonadogenesis, spermatogenesis, testicular determination, gametogenesis, gonad differentiation, and possibly sex determination (Sun et al., 2013). With the recently constructed high-density and high-resolution genetic map (Li et al., 2015), a single sex-determination locus was determined. Additional analyses are ongoing, including performing association analysis by genotyping large numbers of samples from wild populations, conducting fine-scale QTL mapping by genotype more SNPs located sex QTL region with more individuals and families, and functional analysis of putative sex determination genes with genome-editing techniques such as CRISPR/Cas-9.

Gene expression and regulation in control of phenotypes

Through altering expression of a large number of genes, organisms adjust to a wide range of physiological states and environmental changes. When the physiological state of the catfish was experimentally controlled, differential expression of various genes in catfish was observed depending on treatment, challenge, tissue type, physiological stage, and among families.

Differential gene expression has been shown during a reproductive cycle (Kazeto et al., 2005; Kumar et al., 2000) and embryogenesis process (Peterson et al., 2005) in channel catfish. In fast and slow growing families of catfish, differences in expression levels of IGF-II mRNA between fast and slow growing families of fish suggested a role of IGF-II in growth of channel catfish (Peterson et al., 2004).

Differential gene expression was observed in the channel catfish in response to a wide range of environmental changes. Gene expression under various stress conditions have been reported, including cold acclimation (Ju et al., 2002) and copper sulphate (Schlenk et al., 1999), low-water stress (Karsi et al., 2005), heat stress (Liu et al., 2013) and bacterial infections (Bao et al., 2005; Baoprasertkul et al., 2004, 2006; Bilodeau and Waldbieser, 2005; Chen et al., 2005; Elibol-Flemming et al., 2009; Li et al., 2012; Liu et al., 2010; Peatman et al., 2006, 2007, 2008; Takano et al., 2008). For instance, differential gene expression was identified in response to cold acclimation in the brain of channel catfish (Ju et al., 2002). Besides, it has been reported that copper sulphate has an acute effect on sublethal endpoints of stress and tissue damage in channel catfish (Schlenk et al., 1999). During low-water stress, the expression of POMC related-genes were detected in channel catfish (Karsi et al., 2005). A RNA-Seq analysis was conducted to identify heat stress-induced genes in catfish at the transcriptome level, which revealed the expression signatures of genes involved in oxygen transport, protein synthesis, folding, and degradation (Liu et al., 2013). The expression of genes, which were presumed to function in the inducible innate defense, was regulated when channel catfish were challenged with the bacterial pathogens, *E. ictaluri* (Li et al., 2012; Peatman et al., 2006, 2007) and *F. columnare* (Beck et al., 2012; Sun et al., 2012).

Future directions

Functional analysis of genes

The completion of catfish genome sequencing is not the end of catfish genome project rather the beginning of better understanding of catfish biology. The channel catfish reference genome sequence provides essential genomic resources to investigate genomic and transcriptional hallmarks for the biological characteristics of the catfish. Also, it is indispensable for understanding genome evolution as well as for enhancing genetic improvement programs in aquaculture. The catfish genome work will focus on functional analysis of genes in the near future. The main objectives would include: (1) identification and mapping of all functional elements (both coding and noncoding) in the genome; (2) analysis of steady-state RNA and protein expression levels in a given cell type; of comparative levels of gene products in different cell types; and of temporal or induced changes in RNA and protein expression; (3) analysis of naturally occurring or induced mutations that alter RNA and/or protein expression; (4) analysis of cellular localization of proteins and of protein–protein or protein–nucleic acid interactions; (5) comparative analysis of protein sequences; (6) analysis of genome organization and its effect on cellular functions; and (7) analysis of the effects beyond genetic materials-epigenome.

Selecting available or suitable tools is critical for efficiently capturing and analyzing the function of genes and pathways. During the past several years, loss-of-function techniques have emerged to study gene function in diverse experimental systems. As a powerful tool to block the expression of target genes, RNA interference (RNAi) methods already been used to transiently silence or knock-down gene expression in mammalian cells and zebrafish cell lines (Gruber et al., 2005; Silva et al., 2004), opening the door to the execution of such screens in catfish. Notably, bioinformatic enrichment tools with the biological knowledge accumulated in public databases (e.g., Gene Ontology) have played an increasingly important role in contributing to the gene functional analysis of large numbers of genes in various large-scale biological studies (Huang et al., 2009).

Gene knockout systems and their potential use in catfish

Gene knockout is considered to be a major component of the functional genomics toolbox, and is a top priority in revealing and clarifying the function of genes discovered by large-scale sequencing programs (Bouché and Bouchez, 2001). It is accomplished through a combination of techniques. Homologous recombination is a DNA repair mechanism that is employed in gene targeting to insert a designed mutation into the homologous genetic locus (Hall et al., 2009). In such a way, it is enforceable to create a mutation into a selected gene by directly utilizing a potentially important genomic clone. This approach is widely used in yeast genetics to assess or modify gene function, and thousands of knockouts have been obtained in mice (Deutscher et al., 2006; Vogel, 2007). In the aspect of animals, knockout mouse has been viewed as a powerful tool for geneticists to identify the role of a gene in embryonic development and to discern its function in normal physiological homeostasis (Hall et al., 2009). In this regard, gene inactivation by knockout might be the best way to delineate the biological role of a protein.

Knockout requires recognition and replacement of the gene sequence by a defective copy via homologous recombination. However, gene targeting has never been easy for other organisms. In terms of farmed fish, the lack of methodologies for homologous recombination and embryonic stem cell derivation makes it difficult to conduct specific gene targeting technologies to unravel the function of genes (Li et al., 2013b). Up to date, only a few targeted gene knockout have been reported in aquaculture species. The targeted disruption of the mstn gene using ZFNs was conducted in yellow catfish (Dong et al., 2011). The knockout of Dmrt1 and Foxl2 to investigate their effects on sex differentiation was conducted using TALENs in tilapia (Li et al., 2013b). As for catfish, with the completion of the whole genome sequencing and genome annotation, it's readily applicable to perform functional analysis with gene knockout or editing with the state-of-the-art technologies such as TALEN and CRISPR/Cas-9. It is the time to establish an efficient and effective genome editing protocol to study the functional genomics in catfish.

Concluding remarks

Whole genome sequences are the most essential genome resources as demonstrated by decades of animal genomic research. Genome research holds great promise because it deciphers all the genetic material into a road map that shows genes important for disease resistance, growth rate, feed conversion efficiency, stress response, and other important production and performance traits. Through the development of genome technologies, we hope to provide the catfish industry with tools to identify brood stocks based on genetics, and make breeding decisions based on genome selection. Several important production and performance traits such as growth rate, feed conversion efficiency, and disease resistance, must be improved in order to make the catfish industry profitable in the face of stiff global competition. As a historical landmark, the completion of channel catfish reference genome sequence opens the real first step of the long march toward genetic enhancement. The research community needs to focus on aquaculture performance and production traits, taking advantage of these unprecedented genome information and technology to make real progress toward genetic improvements of aquaculture brood stocks.

References

Adams, M.D., Kelley, J.M., Gocayne, J.D., Dubnick, M., Polymeropoulos, M.H., Xiao, H., Merril, C.R., Wu, A., Olde, B., Moreno, R.F., 1991. Complementary DNA sequencing: expressed sequence tags and human genome project. Science 252, 1651–1656.

Arce, M., Reis, R.E., Geneva, A.J., Pérez, M.H.S., 2013. Molecular phylogeny of thorny catfishes (Siluriformes: Doradidae). Mol. Phylogenet. Evol. 67, 560–577.

Armbruster, J.W., 2004. Phylogenetic relationships of the suckermouth armoured catfishes (Loricariidae) with emphasis on the Hypostominae and the Ancistrinae. Zool. J. Linnean Soc. 141, 1–80.

Bao, B., Peatman, E., Li, P., He, C., Liu, Z., 2005. Catfish hepcidin gene is expressed in a wide range of tissues and exhibits tissue-specific upregulation after bacterial infection. Dev. Comp. Immunol. 29, 939–950.

Baoprasertkul, P., Peatman, E., Chen, L., He, C., Kucuktas, H., Li, P., Simmons, M., Liu, Z., 2004. Sequence analysis and expression of a CXC chemokine in resistant and susceptible catfish after infection of *Edwardsiella ictaluri*. Dev. Comp. Immunol. 28, 769–780.

Baoprasertkul, P., Peatman, E., Somridhivej, B., Liu, Z., 2006. Toll-like receptor 3 and TICAM genes in catfish: species-specific expression profiles following infection with *Edwardsiella ictaluri*. Immunogenetics 58, 817–830.

Barozai, M.Y., 2012. The MicroRNAs and their targets in the channel catfish (*Ictalurus punctatus*). Mol. Biol. Rep. 39, 8867–8872.

Bartel, D.P., 2004. MicroRNAs: genomics, biogenesis, mechanism, and function. Cell 116, 281–297.

Beck, B.H., Farmer, B.D., Straus, D.L., Li, C., Peatman, E., 2012. Putative roles for a rhamnose binding lectin in *Flavobacterium columnare* pathogenesis in channel catfish *Ictalurus punctatus*. Fish Shellfish Immunol. 33, 1008–1015.

Bernstein, E., Caudy, A.A., Hammond, S.M., Hannon, G.J., 2001. Role for a bidentate ribonuclease in the initiation step of RNA interference. Nature 409, 363–366.

Bilodeau, A.L., Waldbieser, G.C., 2005. Activation of TLR3 and TLR5 in channel catfish exposed to virulent *Edwardsiella ictaluri*. Dev. Comp. Immunol. 29, 713–721.

Bouché, N., Bouchez, D., 2001. Arabidopsis gene knockout: phenotypes wanted. Curr. Opin. Plant Biol. 4, 111–117.

Bruton, M.N., 1996. Alternative life-history strategies of catfishes. Aquat. Living Res. 9, 35–41.

Burgess W. 1989. An atlas of freshwater and marine catfishes: a preliminary survey of the Siluriformes: tfh publications.

Chen, K., Rajewsky, N., 2007. The evolution of gene regulation by transcription factors and microRNAs. Nat. Rev. Genet. 8, 93–103.

Chen, L., He, C., Baoprasertkul, P., Xu, P., Li, P., Serapion, J., Waldbieser, G., Wolters, W., Liu, Z., 2005. Analysis of a catfish gene resembling interleukin-8: cDNA cloning, gene structure, and expression after infection with *Edwardsiella ictaluri*. Dev. Comp. Immunol. 29, 135–142.

Chen, S., Zhang, G., Shao, C., Huang, Q., Liu, G., Zhang, P., Song, W., An, N., Chalopin, D., Volff, J.-N., 2014. Whole-genome sequence of a flatfish provides insights into ZW sex chromosome evolution and adaptation to a benthic lifestyle. Nat. Genet. 46, 253–260.

Chong, L., Tan, S., Yusoff, K., Siraj, S., 2000. Identification and characterization of Malaysian river catfish, *Mystus nemurus* (C&V): RAPD and AFLP analysis. Biochem. Genet. 38, 63–76.

Chu, Y., Corey, D.R., 2012. RNA sequencing: platform selection, experimental design, and data interpretation. Nucleic Acid Ther. 22, 271–274.

Clamp, M., Fry, B., Kamal, M., Xie, X., Cuff, J., Lin, M.F., Kellis, M., Lindblad-Toh, K., Lander, E.S., 2007. Distinguishing protein-coding and noncoding genes in the human genome. Proc. Natl. Acad. Sci. 104, 19428–19433.

Cunningham, F., Amode, M.R., Barrell, D., Beal, K., Billis, K., Brent, S., Carvalho-Silva, D., Clapham, P., Coates, G., Fitzgerald, S., 2015. Ensembl 2015. Nucleic Acids Res. 43, D662–D669.

Davis, K.B., Simco, B.A., Goudie, C.A., Parker, N.C., Cauldwell, W., Snellgrove, R., 1990. Hormonal sex manipulation and evidence for female homogamety in channel catfish. Gen. Comp. Endocrinol. 78, 218–223.

Deutscher, D., Meilijson, I., Kupiec, M., Ruppin, E., 2006. Multiple knockout analysis of genetic robustness in the yeast metabolic network. Nat. Genet. 38, 993–998.

Dong, Z., Ge, J., Li, K., Xu, Z., Liang, D., Li, J., Li, J., Jia, W., Li, Y., Dong, X., 2011. Heritable targeted inactivation of myostatin gene in yellow catfish (*Pelteobagrus fulvidraco*) using engineered zinc finger nucleases. PLoS One 6, e28897.

Dunham, R.A., Masser, M.P., 2012. Production of Hybrid Catfish. Southern Regional Aquaculture Center Publication, Texas.

Ekblom, R., Galindo, J., 2011. Applications of next generation sequencing in molecular ecology of non-model organisms. Heredity 107, 1–15.

Elibol-Flemming, B., Waldbieser, G.C., Wolters, W.R., Boyle, C.R., Hanson, L.A., 2009. Expression analysis of selected immune-relevant genes in channel catfish during *Edwardsiella ictaluri* infection. J. Aquat. Anim. Health 21, 23–35.

Eschmeyer, W., Fong, J., 2014. Available from: http://research.calacademy.org/research/ichthyology/catalog/SpeciesByFamily.asp

Ezkurdia, I., Juan, D., Rodriguez, J.M., Frankish, A., Diekhans, M., Harrow, J., Vazquez, J., Valencia, A., Tress, M.L., 2014. Multiple evidence strands suggest that there may be as few as 19 000 human protein-coding genes. Hum. Mol. Genet. 23, 5866–5878.

FAO, 2015. *Ictalurus punctatus*. Avalilable from: http://www.fao.org/fishery/culturedspecies/Ictalurus_punctatus/en

Feng, J., Liu, S., Wang, X., Wang, R., Zhang, J., Jiang, Y., Li, C., Kaltenboeck, L., Li, J., Liu, Z., 2014. Channel catfish hemoglobin genes: Identification, phylogenetic and syntenic analysis, and specific induction in response to heat stress. Comp. Biochem. Physiol. Part D Genomics Proteomics 9, 11–22.

Feng, J., Liu, S., Wang, R., Zhang, J., Wang, X., Kaltenboeck, L., Li, J., Liu, Z., 2015. Molecular characterization, phylogenetic analysis and expression profiling of myoglobin and cytoglobin genes in response to heat stress in channel catfish *Ictalurus punctatus*. J. Fish Biol. 86, 592–604.

Ferraris, Jr., C.J., Vari, R.P., 2012. Review of the African catfish genus Andersonia (Teleostei: Siluriformes). Zootaxa 3210, 39–49.

Fickett, J.W., 1982. Recognition of protein coding regions in DNA sequences. Nucleic Acids Res. 10, 5303–5318.

Freeman, J.L., Perry, G.H., Feuk, L., Redon, R., McCarroll, S.A., Altshuler, D.M., Aburatani, H., Jones, K.W., Tyler-Smith, C., Hurles, M.E., 2006. Copy number variation: new insights in genome diversity. Genome Res. 16, 949–961.

Geng, X., Feng, J., Liu, S., Wang, Y., Arias, C., Liu, Z., 2014. Transcriptional regulation of hypoxia inducible factors alpha (HIF-α) and their inhibiting factor (FIH-1) of channel catfish (*Ictalurus punctatus*) under hypoxia. Comp. Biochem. Physiol. Part B Biochem. Mol. Biol. 169, 38–50.

Geng, X., Sha, J., Liu, S., Bao, L., Zhang, J., Wang, R., Yao, J., Li, C., Feng, J., Sun, F., 2015. A genome-wide association study in catfish reveals the presence of functional hubs of related genes within QTLs for columnaris disease resistance. BMC Genomics 16, 196.

Graham, K., 1999. A review of the biology and management of blue catfish. In: Irwin, E., Hubert, W., Rabeni, C., Schramm, Jr., H., Coon, T. (Eds.), Catfish 2000 Proceedings of the international ictalurid symposium. American Fisheries Society, Symposium 24, 37–49.

Grande, L., Eastman, J.T., 1986. A review of Antarctic ichthyofaunas in the light of new fossil discoveries. Palaeontology 29, 113–137.

Gruber, J., Manninga, H., Tuschl, T., Osborn, M., Weber, K., 2005. Specific RNAi mediated gene knockdown in zebrafish cell lines. RNA Biol. 2, 101–105.

Gu, X., Feng, C., Ma, L., Song, C., Wang, Y., Da, Y., Li, H., Chen, K., Ye, S., Ge, C., 2011. Genome-wide association study of body weight in chicken F2 resource population. PLoS One 6, e21872.

Hall, B., Limaye, A., Kulkarni, A.B., 2009. Overview: generation of gene knockout mice. Curr. Protocols Cell Biol. 19, 19.12.1–19.12.17.

Hammond, S.M., Bernstein, E., Beach, D., Hannon, G.J., 2000. An RNA-directed nuclease mediates post-transcriptional gene silencing in *Drosophila* cells. Nature 404, 293–296.

Hanson, T., Sites, D., 2014. 2013 U.S. Catfish Database. USDA National Agriculture USDA National Agricultural Statistics Service (NASS). Mississippi Agricultural Statistics Service (MASS), Mississippi.

Harrow, J., Nagy, A., Reymond, A., Alioto, T., Patthy, L., Antonarakis, S.E., Guigó, R., 2009. Identifying protein-coding genes in genomic sequences. Genome Biol. 10, 201.

Hawke, J.P., 1979. A bacterium associated with disease of pond cultured channel catfish, *Ictalurus punctatus*. J. Fish. Board Can. 36, 1508–1512.

He, C., Chen, L., Simmons, M., Li, P., Kim, S., Liu, Z., 2003. Putative SNP discovery in interspecific hybrids of catfish by comparative EST analysis. Anim. Genet. 34, 445–448.

Hegele, R.A., Oshima, J., 2007. Phenomics and lamins: from disease to therapy. Exp. Cell Res. 313, 2134–2143.

Huang, D.W., Sherman, B.T., Lempicki, R.A., 2009. Bioinformatics enrichment tools: paths toward the comprehensive functional analysis of large gene lists. Nucleic Acids Res. 37, 1–13.

Hutson, A., Liu, Z., Kucuktas, H., Umali-Maceina, G., Su, B., Dunham, R., 2014. Quantitative trait loci map for growth and morphometric traits using a channel catfish× blue catfish interspecific hybrid system. J. Anim. Sci. 92, 1850–1865.

International Human Genome Sequencing Consortium, 2004. Finishing the euchromatic sequence of the human genome. Nature 431, 931–945.

Jiang, Y., Lu, J., Peatman, E., Kucuktas, H., Liu, S., Wang, S., Sun, F., Liu, Z., 2011. A pilot study for channel catfish whole genome sequencing and de novo assembly. BMC Genomics 12, 629.

Jiang, Y., Gao, X., Liu, S., Zhang, Y., Liu, H., Sun, F., Bao, L., Waldbieser, G., Liu, Z., 2013a. Whole genome comparative analysis of channel catfish (*Ictalurus punctatus*) with four model fish species. BMC Genomics 14, 780.

Jiang, Y., Ninwichian, P., Liu, S., Zhang, J., Kucuktas, H., Sun, F., Kaltenboeck, L., Sun, L., Bao, L., Liu, Z., 2013b. Generation of physical map contig-specific sequences useful for whole genome sequence scaffolding. PLoS One 8, e78872.

Jing, J., Wu, J., Liu, W., Xiong, S., Ma, W., Zhang, J., Wang, W., Gui, J.-F., Mei, J., 2014. Sex-biased miRNAs in gonad and their potential roles for testis development in yellow catfish. PLoS One 9, e107946.

Ju, Z., Karsi, A., Kocabas, A., Patterson, A., Li, P., Cao, D., Dunham, R., Liu, Z., 2000. Transcriptome analysis of channel catfish (*Ictalurus punctatus*): genes and expression profile from the brain. Gene 261, 373–382.

Ju, Z., Dunham, R., Liu, Z., 2002. Differential gene expression in the brain of channel catfish (*Ictalurus punctatus*) in response to cold acclimation. Mol. Genet. Genomics 268, 87–95.

Kapranov, P., Cheng, J., Dike, S., Nix, D.A., Duttagupta, R., Willingham, A.T., Stadler, P.F., Hertel, J., Hackermüller, J., Hofacker, I.L., 2007. RNA maps reveal new RNA classes and a possible function for pervasive transcription. Science 316, 1484–1488.

Karsi, A., Waldbieser, G.C., Small, B.C., Wolters, W.R., 2005. Genomic structure of the proopiomelanocortin gene and expression during acute low-water stress in channel catfish. Gen. Comp. Endocrinol. 143, 104–112.

Kasahara, M., Naruse, K., Sasaki, S., Nakatani, Y., Qu, W., Ahsan, B., Yamada, T., Nagayasu, Y., Doi, K., Kasai, Y., 2007. The medaka draft genome and insights into vertebrate genome evolution. Nature 447, 714–719.

Kazeto, Y., Goto-Kazeto, R., Trant, J.M., 2005. Membrane-bound progestin receptors in channel catfish and zebrafish ovary: changes in gene expression associated with the reproductive cycles and hormonal reagents. Gen. Comp. Endocrinol. 142, 204–211.

Kim, S., Karsi, A., Dunham, R.A., Liu, Z., 2000. The skeletal muscle α-actin gene of channel catfish (*Ictalurus punctatus*) and its association with piscine specific SINE elements. Gene 252, 173–181.

Kobayashi, Y., Quiniou, S., Booth, N.J., Peterson, B.C., 2011. Expression of leptin-like peptide (LLP) mRNA in channel catfish (*Ictalurus punctatus*) is induced by exposure to *Edwardsiella ictaluri* but is independent of energy status. Gen. Comp. Endocrinol. 173, 411–418.

Kocabas, A.M., Li, P., Cao, D., Karsi, A., He, C., Patterson, A., Ju, Z., Dunham, R.A., Liu, Z., 2002. Expression profile of the channel catfish spleen: analysis of genes involved in immune functions. Mar. Biotechnol. 4, 526–536.

Kocher, T.D., Kole, C., 2008. Genome Mapping and Genomics in Fishes and Aquatic Animals. Springer Science & Business Media, Berlin.

Krauss, G., 2006. Biochemistry of Signal Transduction and Regulation, third ed. Wiley-VCH, Weinheim.

Kucuktas, H., Wang, S., Li, P., He, C., Xu, P., Sha, Z., Liu, H., Jiang, Y., Baoprasertkul, P., Somridhivej, B., 2009. Construction of genetic linkage maps and comparative genome analysis of catfish using gene-associated markers. Genetics 181, 1649–1660.

Kumar, R.S., Ijiri, S., Trant, J.M., 2000. Changes in the expression of genes encoding steroidogenic enzymes in the channel catfish (*Ictalurus punctatus*) ovary throughout a reproductive cycle. Biol. Reprod. 63, 1676–1682.

Kung, J.T., Colognori, D., Lee, J.T., 2013. Long noncoding RNAs: past, present, and future. Genetics 193, 651–669.

Lamkom, T., Kucuktas, H., Liu, Z., Li, P., Na-Nakorn, U., Klinbunga, S., Hutson, A., Chaimongkol, A., Ballenger, J., Umali, G., Dunham, R.A., 2008. Microsatellite variation among domesticated populations of channel catfish (*Ictalurus punctatus*) and blue catfish (*I. furcatus*). KU. Fish. Res. Bull. 32, 37–47.

Lander, E.S., Linton, L.M., Birren, B., Nusbaum, C., Zody, M.C., Baldwin, J., Devon, K., Dewar, K., Doyle, M., FitzHugh, W., 2001. Initial sequencing and analysis of the human genome. Nature 409, 860–921.

LeGrande, W., DR, A., RO, S., 1984. Comparative karyology of three species of North American catfish (siluriformes: Ictaluridae: Ictalurus) and four of their hybrid combinations. Copeia 1984, 873–878.

Li, P., Peatman, E., Wang, S., Feng, J., He, C., Baoprasertkul, P., Xu, P., Kucuktas, H., Nandi, S., Somridhivej, B., 2007. Towards the ictalurid catfish transcriptome: generation and analysis of 31,215 catfish ESTs. BMC Genomics 8, 177.

Li, L., Xu, J., Yang, D., Tan, X., Wang, H., 2010. Computational approaches for microRNA studies: a review. Mamm. Genome 21, 1–12.

Li, C., Zhang, Y., Wang, R., Lu, J., Nandi, S., Mohanty, S., Terhune, J., Liu, Z., Peatman, E., 2012. RNA-seq analysis of mucosal immune responses reveals signatures of intestinal barrier disruption and pathogen entry following *Edwardsiella ictaluri* infection in channel catfish, *Ictalurus punctatus*. Fish Shellfish Immunol. 32, 816–827.

Li, H., Peng, Z., Yang, X., Wang, W., Fu, J., Wang, J., Han, Y., Chai, Y., Guo, T., Yang, N., 2013a. Genome-wide association study dissects the genetic architecture of oil biosynthesis in maize kernels. Nat. Genet. 45, 43–50.

Li, M.-H., Yang, H.-H., Li, M.-R., Sun, Y.-L., Jiang, X.-L., Xie, Q.-P., Wang, T.-R., Shi, H.-J., Sun, L.-N., Zhou, L.-Y., 2013b. Antagonistic roles of Dmrt1 and Foxl2 in sex differentiation via estrogen production in tilapia as demonstrated by TALENs. Endocrinology 154, 4814–4825.

Li, F., Chen, B., Xu, K., Wu, J., Song, W., Bancroft, I., Harper, A.L., Trick, M., Liu, S., Gao, G., 2014. Genome-wide association study dissects the genetic architecture of seed weight and seed quality in rapeseed (*Brassica napus* L.). DNA Res. 21, 355–367.

Li, Y., Liu, S., Qin, Z., Waldbieser, G., Wang, R., Sun, L., Bao, L., Danzmann, R.G., Dunham, R., Liu, Z., 2015. Construction of a high-density, high-resolution genetic map and its integration with BAC-based physical map in channel catfish. DNA Res. 22, 39–52.

Liu, Z., 2003. A review of catfish genomics: progress and perspectives. Comp. Funct. Genomics 4, 259–265.

Liu, Z., 2007. Fish genomics and analytical genetic technologies, with examples of their potential applications in management of fish genetic resources. America 5, 145–179.

Liu, Z., 2008. Catfish. In: Thomas, K.D., Chittaranjan, K. (Eds.), Genome Mapping and Genomics in Fishes and Aquatic Animals Volume 2. Berlin Heidelberg. p. 85–100.

Liu, Z., 2011. Development of genomic resources in support of sequencing, assembly, and annotation of the catfish genome. Comp. Biochem. Physiol. Part D Genomics Proteomics 6, 11–17.

Liu, Z., Liu, S., Yao, J., Bao, L., Zhang, J., Li, Y., Jiang, C., Sun, L., Wang, R., Zhang, Y., Zhou, T., Zeng, Q., Fu, Q., Gao, S., Li, N., Koren, S., Jiang, Y., Zimin, A., Xu, P., Phillippy, A., Geng, X., Song, L., Sun, F., Li, C., Wang, X., Chen, A., Jin, Y., Yuan, Z., Yang, Y., Tan, S., Peatman, E., Lu, J., Qin, Z., Dunham, R., Li, Z., Sonstegard, T., Feng, J., Danzmann, R., Schroeder, S., Scheffler, B., Duke, M., Ballard, L., Kucuktas, H., Kaltenboeck, L., Liu, H., Armbruster, J., Xie, Y., Kirby, M., Tian, Y., Flanagan, M., Mu, W., Waldbieser, G., 2016. The channel catfish genome sequence provides insights into the evolution of scale formation in teleosts. Nat. Commun. 7, 11757.

Liu, Z., Cordes, J., 2004. DNA marker technologies and their applications in aquaculture genetics. Aquaculture 238, 1–37.

Liu, Q., Goudie, C., Simco, B., Davis, K., 1996. Sex-linkage of glucosephosphate isomerase-B and mapping of the sex-determining gene in channel catfish. Cytogenet. Genome Res. 73, 282–285.

Liu, Z., Li, P., Argue, B., Dunham, R., 1998a. Inheritance of RAPD markers in channel catfish (*Ictalurus punctatus*), blue catfish (*I. furcatus*), and their F1, F2 and backcross hybrids. Anim. Genet. 29, 58–62.

Liu, Z., Li, P., Dunham, R.A., 1998b. Characterization of an A/T-rich family of sequences from channel catfish (*Ictalurus punctatus*). Mol. Mar. Biol. Biotechnol. 7, 232–239.

Liu, Z., Karsi, A., Dunham, R.A., 1999a. Development of polymorphic EST markers suitable for genetic linkage mapping of catfish. Mar. Biotechnol. 1, 437–447.

Liu, Z., Li, P., Argue, B., Dunham, R., 1999b. Random amplified polymorphic DNA markers: usefulness for gene mapping and analysis of genetic variation of catfish. Aquaculture 174, 59–68.

Liu, Z., Li, P., Kucuktas, H., Nichols, A., Tan, G., Zheng, X., Argue, B.J., Dunham, R.A., Yant, D.R., 1999c. Development of amplified fragment length polymorphism (AFLP) markers suitable for genetic linkage mapping of catfish. Trans.Am. Fish. Soc. 128, 317–327.

Liu, Z., Karsi, A., Li, P., Cao, D., Dunham, R., 2003. An AFLP-based genetic linkage map of channel catfish (*Ictalurus punctatus*) constructed by using an interspecific hybrid resource family. Genetics 165, 687–694.

Liu, H., Jiang, Y., Wang, S., Ninwichian, P., Somridhivej, B., Xu, P., Abernathy, J., Kucuktas, H., Liu, Z., 2009. Comparative analysis of catfish BAC end sequences with the zebrafish genome. BMC Genomics 10, 592.

Liu, H., Takano, T., Peatman, E., Abernathy, J., Wang, S., Sha, Z., Kucuktas, H., Xu, Dh, Klesius, P., Liu, Z., 2010. Molecular characterization and gene expression of the channel catfish ferritin H subunit after bacterial infection and iron treatment. J. Exp. Zool. Part A Ecol. Genet. Physiol. 313, 359–368.

Liu, S., Zhou, Z., Lu, J., Sun, F., Wang, S., Liu, H., Jiang, Y., Kucuktas, H., Kaltenboeck, L., Peatman, E., 2011. Generation of genome-scale gene-associated SNPs in catfish for the construction of a high-density SNP array. BMC Genomics 12, 53.

Liu, S., Zhang, Y., Zhou, Z., Waldbieser, G., Sun, F., Lu, J., Zhang, J., Jiang, Y., Zhang, H., Wang, X., 2012. Efficient assembly and annotation of the transcriptome of catfish by RNA-Seq analysis of a doubled haploid homozygote. BMC Genomics 13, 595.

Liu, S., Wang, X., Sun, F., Zhang, J., Feng, J., Liu, H., Rajendran, K., Sun, L., Zhang, Y., Jiang, Y., 2013. RNA-Seq reveals expression signatures of genes involved in oxygen transport, protein synthesis, folding, and degradation in response to heat stress in catfish. Physiol. Genomics 45, 462–476.

Liu, S., Sun, L., Li, Y., Sun, F., Jiang, Y., Zhang, Y., Zhang, J., Feng, J., Kaltenboeck, L., Kucuktas, H., 2014. Development of the catfish 250K SNP array for genome-wide association studies. BMC Res. Notes 7, 135.

Liu, S., Li, Y., Qin, Z., Geng, X., Bao, L., Kaltenboeck, L., Kucuktas, H., Dunham, R., Liu, Z., 2015. High-density interspecific genetic linkage mapping provides insights into genomic incompatibility between channel catfish and blue catfish. Anim. Genet. 47 (1), .

Lundberg, J.G., Friel, J.P., 2003. Siluriformes—Catfishes. Version 20. The Tree of Life Web Project. Available from: http://tolweb.org/Siluriformes/15065/2003.01.20.

McClellan, J., King, M.-C., 2010. Genetic heterogeneity in human disease. Cell 141, 210–217.

Meyer, A., Van de Peer, Y., 2005. From 2R to 3R: evidence for a fish-specific genome duplication (FSGD). Bioessays 27, 937–945.

Mickett, K., Morton, C., Feng, J., Li, P., Simmons, M., Cao, D., Dunham, R., Liu, Z., 2003. Assessing genetic diversity of domestic populations of channel catfish (*Ictalurus punctatus*) in Alabama using AFLP markers. Aquaculture 228, 91–105.

Miles, C., Wayne, M., 2008. Quantitative trait locus (QTL) analysis. Nat. Educ. 1, 208.

Mu, W., Yao, J., Zhang, J., Liu, S., Wen, H., Feng, J., Liu, Z., 2015. Expression of tumor suppressor genes in channel catfish after bacterial infections. Dev. Comp. Immunol. 48, 171–177.

Nandi, S., Peatman, E., Xu, P., Wang, S., Li, P., Liu, Z., 2007. Repeat structure of the catfish genome: a genomic and transcriptomic assessment of Tc1-like transposon elements in channel catfish (*Ictalurus punctatus*). Genetica 131, 81–90.

Ninwichian, P., Peatman, E., Liu, H., Kucuktas, H., Somridhivej, B., Liu, S., Li, P., Jiang, Y., Sha, Z., Kaltenboeck, L., 2012a. Second-generation genetic linkage map of catfish and its integration with the BAC-based physical map. G3 2, 1233–1241.

Ninwichian, P., Peatman, E., Perera, D., Liu, S., Kucuktas, H., Dunham, R., Liu, Z., 2012b. Identification of a sex-linked marker for channel catfish. Anim. Genet. 43, 476–477.

Nishimura, S., Watanabe, T., Mizoshita, K., Tatsuda, K., Fujita, T., Watanabe, N., Sugimoto, Y., Takasuga, A., 2012. Genome-wide association study identified three major QTL for carcass weight including the PLAG1-CHCHD7 QTN for stature in Japanese black cattle. BMC Genetics 13, 40.

Palstra, A.P., Beltran, S., Burgerhout, E., Brittijn, S.A., Magnoni, L.J., Henkel, C.V., Jansen, H.J., Van Den Thillart, G., Spaink, H.P., Planas, J.V., 2013. Deep RNA sequencing of the skeletal muscle transcriptome in swimming fish. PLoS One 8, e53171.

Parker, S.C., Hansen, L., Abaan, H.O., Tullius, T.D., Margulies, E.H., 2009. Local DNA topography correlates with functional noncoding regions of the human genome. Science 324, 389–392.

Patiño, R., Davis, K.B., Schoore, J.E., Uguz, C., Strüssmann, C.A., Parker, N.C., Simco, B.A., Goudie, C.A., 1996. Sex differentiation of channel catfish gonads: normal development and effects of temperature. J. Exp. Zool. 276, 209–218.

Pauli, A., Valen, E., Lin, M.F., Garber, M., Vastenhouw, N.L., Levin, J.Z., Fan, L., Sandelin, A., Rinn, J.L., Regev, A., 2012. Systematic identification of long noncoding RNAs expressed during zebrafish embryogenesis. Genome Res. 22, 577–591.

Peatman, E., Bao, B., Peng, X., Baoprasertkul, P., Brady, Y., Liu, Z., Catfish, C.C., 2006,. chemokines: genomic clustering, duplications, and expression after bacterial infection with *Edwardsiella ictaluri*. Mol. Genet. Genomics 275, 297–309.

Peatman, E., Baoprasertkul, P., Terhune, J., Xu, P., Nandi, S., Kucuktas, H., Li, P., Wang, S., Somridhivej, B., Dunham, R., 2007. Expression analysis of the acute phase response in channel catfish (*Ictalurus punctatus*) after infection with a Gram-negative bacterium. Dev. Comp. Immunol. 31, 1183–1196.

Peatman, E., Terhune, J., Baoprasertkul, P., Xu, P., Nandi, S., Wang, S., Somridhivej, B., Kucuktas, H., Li, P., Dunham, R., 2008. Microarray analysis of gene expression in the blue catfish liver reveals early activation of the MHC class I pathway after infection with *Edwardsiella ictaluri*. Mol. Immunol. 45, 553–566.

Peterson, B.C., Waldbieser, G.C., Bilodeau, L., 2004. IGF-I and IGF-II mRNA expression in slow and fast growing families of USDA103 channel catfish (*Ictalurus punctatus*). Comp. Biochem. Physiol. Part A Mol. Integr. Physiol. 139, 317–323.

Peterson, B.C., Bosworth, B.G., Bilodeau, A.L., 2005. Differential gene expression of IGF-I, IGF-II, and toll-like receptors 3 and 5 during embryogenesis in hybrid (channel × blue) and channel catfish. Comp. Biochem. Physiol. Part A Mol. Integr. Physiol. 141, 42–47.

Qian, X., Ba, Y., Zhuang, Q., Zhong, G., 2014. RNA-Seq technology and its application in fish transcriptomics. OMICS 18, 98–110.

Quiniou, S.M., Katagiri, T., Miller, N.W., Wilson, M., Wolters, W.R., Waldbieser, G.C., 2003. Construction and characterization of a BAC library from a gynogenetic channel catfish *Ictalurus punctatus*. Genet. Sel. Evol. 35, 673–684.

Quiniou, S.M., Waldbieser, G.C., Duke, M.V., 2007. A first generation BAC-based physical map of the channel catfish genome. BMC Genomics 8, 40.

Rajendran, K., Zhang, J., Liu, S., Kucuktas, H., Wang, X., Liu, H., Sha, Z., Terhune, J., Peatman, E., Liu, Z., 2012a. Pathogen recognition receptors in channel catfish: I. Identification, phylogeny and expression of NOD-like receptors. Dev. Comp. Immunol. 37, 77–86.

Rajendran, K., Zhang, J., Liu, S., Peatman, E., Kucuktas, H., Wang, X., Liu, H., Wood, T., Terhune, J., Liu, Z., 2012b. Pathogen recognition receptors in channel catfish: II. Identification, phylogeny and expression of retinoic acid-inducible gene I (RIG-I)-like receptors (RLRs). Dev. Comp. Immunol. 37, 381–389.

Schlenk, D., Davis, K.B., Griffin, B.R., 1999. Relationship between expression of hepatic metallothionein and sublethal stress in channel catfish following acute exposure to copper sulphate. Aquaculture 177, 367–379.

Serapion, J., Kucuktas, H., Feng, J., Liu, Z., 2004a. Bioinformatic mining of type I microsatellites from expressed sequence tags of channel catfish (*Ictalurus punctatus*). Mar. Biotechnol. 6, 364–377.

Serapion, J., Waldbieser, G., Wolters, W., Liu, Z., 2004b. Development of type I markers in channel catfish through intron sequencing. Anim. Genet. 35, 463–466.

Silva, J.M., Mizuno, H., Brady, A., Lucito, R., Hannon, G.J., 2004. RNA interference microarrays: high-throughput loss-of-function genetics in mammalian cells. Proc. Natl. Acad. Sci. USA 101, 6548–6552.

Simmons, M., Mickett, K., Kucuktas, H., Li, P., Dunham, R., Liu, Z., 2006. Comparison of domestic and wild channel catfish (*Ictalurus punctatus*) populations provides no evidence for genetic impact. Aquaculture 252, 133–146.

Sullivan, J.P., Lundberg, J.G., Hardman, M., 2006. A phylogenetic analysis of the major groups of catfishes (Teleostei: Siluriformes) using rag1 and rag2 nuclear gene sequences. Mol. Phylogenet. Evol. 41, 636–662.

Sun, F., Peatman, E., Li, C., Liu, S., Jiang, Y., Zhou, Z., Liu, Z., 2012. Transcriptomic signatures of attachment, NF-κB suppression and IFN stimulation in the catfish gill following columnaris bacterial infection. Dev. Comp. Immunol. 38, 169–180.

Sun, F., Liu, S., Gao, X., Jiang, Y., Perera, D., Wang, X., Li, C., Sun, L., Zhang, J., Kaltenboeck, L., 2013. Male-biased genes in catfish as revealed by RNA-Seq analysis of the testis transcriptome. PLoS One 8, e68452.

Sun, L., Liu, S., Wang, R., Jiang, Y., Zhang, Y., Zhang, J., Bao, L., Kaltenboeck, L., Dunham, R., Waldbieser, G., 2014. Identification and analysis of genome-wide SNPs provide insight into signatures of selection and domestication in channel catfish (*Ictalurus punctatus*). PLoS One 9, e109666.

Takano, T., Sha, Z., Peatman, E., Terhune, J., Liu, H., Kucuktas, H., Li, P., Edholm, E.-S., Wilson, M., Liu, Z., 2008. The two channel catfish intelectin genes exhibit highly differential patterns of tissue expression and regulation after infection with *Edwardsiella ictaluri*. Dev. Comp. Immunol. 32, 693–705.

Tiersch, T., Goudie, C., 1993. Inheritance and variation of genome size in half-sib families of hybrid catfishes. J. Hered. 84, 122–125.

Tiersch, T., Simco, B., Davis, K., Wachtel, S., 1992. Molecular genetics of sex determination in channel catfish: studies on SRY, ZFY, Bkm, and human telomeric repeats. Biol. Reprod. 47, 185–192.

Tine, M., Kuhl, H., Gagnaire, P.-A., Louro, B., Desmarais, E., Martins, R.S., Hecht, J., Knaust, F., Belkhir, K., Klages, S., 2014. European sea bass genome and its variation provide insights into adaptation to euryhalinity and speciation. Nature communications 5, 5770.

Venter, J.C., Adams, M.D., Myers, E.W., Li, P.W., Mural, R.J., Sutton, G.G., Smith, H.O., Yandell, M., Evans, C.A., Holt, R.A., 2001. The sequence of the human genome. Science 291, 1304–1351.

Visscher, P.M., Brown, M.A., McCarthy, M.I., Yang, J., 2012. Five years of GWAS discovery. Am. J. Hum. Genet. 90, 7–24.

Vogel, G., 2007. A knockout award in medicine. Science 318, 178–179.

Waldbieser, G.C., Bosworth, B.G., Nonneman, D.J., Wolters, W.R., 2001. A microsatellite-based genetic linkage map for channel catfish, *Ictalurus punctatus*. Genetics 158, 727–734.

Waldbieser, G.C., Bosworth, B.G., Quiniou, S.M., 2010. Production of viable homozygous, doubled haploid channel catfish (*Ictalurus punctatus*). Mar. Biotechnol. 12, 380–385.

Wang, S., Xu, P., Thorsen, J., Zhu, B., de Jong, P.J., Waldbieser, G., Kucuktas, H., Liu, Z., 2007. Characterization of a BAC library from channel catfish *Ictalurus punctatus*: indications of high levels of chromosomal reshuffling among teleost genomes. Mar. Biotechnol. 9, 701–711.

Wang, S., Sha, Z., Sonstegard, T.S., Liu, H., Xu, P., Somridhivej, B., Peatman, E., Kucuktas, H., Liu, Z., 2008. Quality assessment parameters for EST-derived SNPs from catfish. BMC Genomics 9, 450.

Wang, S., Peatman, E., Abernathy, J., Waldbieser, G., Lindquist, E., Richardson, P., Lucas, S., Wang, M., Li, P., Thimmapuram, J., 2010. Assembly of 500,000 inter-specific catfish expressed sequence tags and large scale gene-associated marker development for whole genome association studies. Genome Biol. 11, R8.

Wang, R., Sun, L., Bao, L., Zhang, J., Jiang, Y., Yao, J., Song, L., Feng, J., Liu, S., Liu, Z., 2013. Bulk segregant RNA-seq reveals expression and positional candidate genes and allele-specific expression for disease resistance against enteric septicemia of catfish. BMC Genomics 14, 929.

Wang, Q., Tian, F., Pan, Y., Buckler, E.S., Zhang, Z., 2014. A SUPER Powerful Method for Genome Wide Association Study. PLoS One 9, e107684.

Wang, Y., Lu, Y., Zhang, Y., Ning, Z., Li, Y., Zhao, Q., Lu, H., Huang, R., Xia, X., Feng, Q., 2015. The draft genome of the grass carp (*Ctenopharyngodon idellus*) provides insights into its evolution and vegetarian adaptation. Nat. Genet. 47, 625–631.

Weber, T.E., Bosworth, B.G., 2005. Effects of 28 day exposure to cold temperature or feed restriction on growth, body composition, and expression of genes related to muscle growth and metabolism in channel catfish. Aquaculture 246, 483–492.

Wilson, D.E., Reeder, D.M., 2005. Mammal Species of the World: a Taxonomic and Geographic Reference, third ed. Johns Hopkins University Press, Baltimore, Maryland.

Wolters, W., Chrisman, C., Libey, G., 1981. Lymphocyte culture for chromosomal analyses of channel catfish, *Ictalurus punctatus*. Copeia 1981, 503–504.

Wu, C., Zhang, D., Kan, M., Lv, Z., Zhu, A., Su, Y., Zhou, D., Zhang, J., Zhang, Z., Xu, M., Jiang, L., Guo, B., Wang, T., Chi, C., Mao, Y., Zhou, J., Yu, X., Wang, H., Weng, X., Jin,

J.G., Ye, J., He, L., Liu, Y., 2014. The draft genome of the large yellow croaker reveals well-developed innate immunity. Nature Communications 5, 5227.

Xu, P., Bao, B., He, Q., Peatman, E., He, C., Liu, Z., 2005. Characterization and expression analysis of bactericidal permeability-increasing protein (BPI) antimicrobial peptide gene from channel catfish *Ictalurus punctatus*. Dev. Comp. Immunol. 29, 865–878.

Xu, P., Wang, S., Liu, L., Peatman, E., Somridhivej, B., Thimmapuram, J., Gong, G., Liu, Z., 2006. Channel catfish BAC-end sequences for marker development and assessment of syntenic conservation with other fish species. Anim. Genet. 37, 321–326.

Xu, P., Wang, S., Liu, L., Thorsen, J., Kucuktas, H., Liu, Z., 2007. A BAC-based physical map of the channel catfish genome. Genomics 90, 380–388.

Xu, Z., Chen, J., Li, X., Ge, J., Pan, J., Xu, X., 2013. Identification and characterization of microRNAs in channel catfish (*Ictalurus punctatus*) by using Solexa sequencing technology. PLoS One 8, e54174.

Xu, P., Zhang, X., Wang, X., Li, J., Liu, G., Kuang, Y., Xu, J., Zheng, X., Ren, L., Wang, G., 2014. Genome sequence and genetic diversity of the common carp, *Cyprinus carpio*. Nat. Genet. 46, 1212–1219.

Yan, J., Warburton, M., Crouch, J., 2011. Association mapping for enhancing maize (L.) genetic improvement. Crop Sci. 51, 433–449.

Zhang, H., Peatman, E., Liu, H., Feng, T., Chen, L., Liu, Z., 2012. Molecular characterization of three L-type lectin genes from channel catfish, *Ictalurus punctatus* and their responses to *Edwardsiella ictaluri* challenge. Fish Shellfish Immunol. 32, 598–608.

Zhang, Y., Liu, S., Lu, J., Jiang, Y., Gao, X., Ninwichian, P., Li, C., Waldbieser, G., Liu, Z., 2013. Comparative genomic analysis of catfish linkage group 8 reveals two homologous chromosomes in zebrafish and other teleosts with extensive inter-chromosomal rearrangements. BMC Genomics 14, 387.

Tilapia genomic studies

5

Brendan J. McAndrew, David J. Penman, Michaël Bekaert, Stefanie Wehner
The Genetics and Reproduction Research Group, Institute of Aquaculture, University of Stirling, Stirling, Scotland, United Kingdom

Introduction

Tilapia belong to the family Cichlidae, in the order Perciformes. This is one of the most species-rich families of fish, particularly in the large African rift valley lakes such as Lake Malawi and Lake Tanganyika, where there has been extensive speciation. The Cichlidae have a widespread tropical distribution and are divided into four subfamilies. The African species belong to the Psuedocrenilabrinae, which can be subdivided into three informal groups, hemichromine, haplochromine, and tilapiine. There is still some doubt about the present taxonomy of the tilapiini because many species have still to be studied in detail and in those that have, the work has often been restricted to a small number of preserved museum specimens.

Tilapia is used as the common name for over 70 species subdivided into as many as 10 genera. Phylogenetic studies using different molecular marker systems clearly suggest that the tilapiines are a monophyletic group with clear dichotomy between the ancestral substrate spawning and nest-guarding species in the genus *Tilapia* and the mouthbrooding *Sarotherodon* and *Oreochromis* genera (McAndrew, 2000). In terms of importance for aquaculture and fisheries production, the majority of the species fall into these three genera.

The species in the genus *Tilapia* are characterized by their substrate spawning and nest guarding behaviour, and are mainly found in the lakes and rivers of western and central Africa. The *Sarotherodon* species are characterized as either paternal or biparental mouthbrooders and are generally restricted to West Africa. The *Oreochromis* species are all characterized as maternal mouthbrooders and are mostly found in central and east African rivers and lakes. Four tilapia species (*T. zillii*, *S. galilaeus*, *O. aureus,* and *O. niloticus*) have a wider Sudo-Nilotic-Sahel distribution because interconnections that existed between the Nile, L. Chad, and Volta basins during a wetter period in the Pleistocene enabled these species to spread over this whole region (Trewavas, 1983).

Four *Oreochromis* species, *O. niloticus*, *O. aureus, O. mossambicus,* and *O. hornorum,* are the most significant species in terms of aquaculture production. *O. niloticus* is by far the most important tilapia species globally with over 3.4 million tonnes being produced in 2013 (FAO, 2014) making it the most important farmed noncyprinid species. Despite the many attempts to develop tilapia as farmed fish in various African countries, today the vast majority of the production comes from outside the species natural range, mostly from Asia. The development of the global

Genomics in Aquaculture. http://dx.doi.org/10.1016/B978-0-12-801418-9.00005-6

industry was achieved by the widespread transfer of farmed stocks, and a limited number of wild populations, by governments and commercial companies. These often uncontrolled introductions caused considerable worries about the genetic status of many farmed stocks and the potential loss of biodiversity of natural populations within Africa. These issues and lists of the various species and introductions can be found in the reviews by Pullin and Lowe-McConnell (1982) and Pullin (1988).

The genetic status of farmed tilapia today can still be described as complex and it is important to know the origins and history of how a given stock has been managed to understand its potential production performance. The main complication has been the widespread adoption of hybridization between species in hatcheries, primarily to produce all-male or nearly all-male fry to reduce fry production during the grow-out in pond-based culture systems (but also in some cases to combine traits of different species, such as good growth, cold tolerance, or salinity tolerance).

Research in Malaysia showed that crosses between *O. mossambicus* and a species introduced from Zanzibar that was eventually identified as *O. hornorum,* generated all-male fry in crosses between female *O. mossambicus* and male O. *hornorum.* This hybrid production system was subsequently adopted in many countries because it overcame the problems of precocious maturation and excessive fry production in pond based production systems (Hickling, 1960). A similar approach was developed in Israel using crosses between female *O. niloticus* and male *O. aureus* and subsequently widely used elsewhere to generate all or nearly all-male fry for production in several countries (Mires, 1977; FAO, 2014). Difficulties in species identification caused by the hybrids and poor broodstock management resulted in these production systems becoming much less reliable as hybrid fish entered and bred with the pure species broodstock pools resulting in a reduction in performance and value of the fish. Additionally, the percentage of males in such hybrid crosses varied depending on the individuals and strains being used as parents.

In an attempt to overcome some of the problems of the genetic deterioration of farmed tilapia stocks (including introgression of feral *O. mossambicus* into farmed *O. niloticus* stocks in Asia), the genetically improved farmed tilapia (GIFT) program was setup in the Philippines to develop a new improved *O. niloticus* strain (Eknath et al., 1993; Bentsen et al., 1998). The program compared the performance of four existing commercial *O. niloticus* based strains with four new collections of wild *O. niloticus* from across Africa (Bentsen et al., 2012) together with all possible inter-stock crosses across a wide range of culture conditions. In general, the program showed that Nile and Nile rift valley populations of *O. niloticus* outperformed the commercial stocks and West African populations from the Volta and Senegal Rivers. The GIFT base population was constructed from the best of the pure and hybrid crosses and has now undergone many generations of selective improvement. Fish from the original GIFT project have been dispersed to a private company GenoMar and the GIFT foundation in the Philippines. The Worldfish Center (WFC, formerly the International Center for Living Aquatic Resource Management, ICLARM) have maintained the GIFT breeding program. Isolates from the GIFT breeding program have been distributed in a controlled way to many different countries and now make a significant contribution to global tilapia production (Asian Development Bank, 2005).

The distribution of the GIFT strain in Africa has been very restricted because of the risks of introgression of the farmed strain into wild *O. niloticus* and other endemic tilapia populations. However, the popularity of *O. niloticus* as a farmed species has resulted in it being introduced into a range of African countries. Escapees from these introductions mean that the species is spreading into new drainages in central and southern Africa and introgressing with endemic species (Deines et al., 2014).

Today *O. niloticus*-based strains dominate tilapia farming, but other tilapia species can perform better under particular environmental conditions. *Oreochromis aureus* has been used in tilapia farming in Israel from the earliest days. It is sympatric with *O. niloticus* in the Levant and lakes in the Nile delta. The species is more cold- and saline-tolerant than *O. niloticus* and has been widely used to generate all or nearly all-male hybrid fry, which are more resistant to colder winter conditions than pure *O. niloticus*. The increased tolerance of the hybrid to cold is beneficial in production areas at altitude or higher latitudes, such as Israel and parts of China, which enables the fish to be overwintered without the losses that would occur with pure *O. niloticus*. Today the production of the hybrid often needs to be supplemented with hormonal sex reversal to ensure all-male fry because of genetic introgression of hybrids into the pure species (Penman and McAndrew, 2000).

The contribution of *O. mossambicus* (and to a lesser extent *O. hornorum*) today relates to their euryhaline tolerance, that allows commercial exploitation brackish and saltwater conditions. A number of red variants that have occurred independently in both Asian and the USA stocks of *O. mossambicus* and *O. mossambicus* × *O. hornorum* hybrids have been developed into a variety of red coloured saline tolerant strains. In order to improve the performance of these strains they have been crossed to *O. aureus* and *O. niloticus* or their hybrids, which has resulted in a wide range of different strains with variable species compositions and colour variations (Penman and McAndrew, 2000).

Life history

Tilapias are naturally found in a wide range of tropical and subtropical African riverine and lacustrine ecosystems with some of the species also entering brackish water estuaries and lagoons. Tilapia are primarily herbivorous and detritivorous feeders: *Tilapia spp.* have coarse pharyngeal teeth and few gill rakers and tend to eat higher plants whereas the *Sarotherodon* and *Oreochromis* have finer teeth and gill rakers and are predominantly microphagous, grazing on epiphythic and benthic algae, detritus and filter feeding on plankton in more open environments (Trewavas, 1983). However, most have a flexible and opportunistic feeding behaviour, which has made them ideal species for aquaculture.

All tilapias are lek-spawners, in that the males dig circular nests in areas with a suitable substrate and females are attracted into the area and will enter and mate with the male. In *Tilapia spp.* the female lays eggs that stick directly to the bottom of the nest and the male fertilizes the eggs in situ. Both parents will then guard the eggs and young. In the mouthbrooding *Sarotherodon* and *Oreochromis* species, small batches of eggs are laid by the female and fertilized by the male. In the case of *Sarotherodon*, which are paternal or biparental mouthbrooders, either the male or both the male and female will pick up and incubate the eggs in their mouths. In *Oreochromis*, which are maternal mouthbrooders, it is the female alone that picks up the eggs and incubates them in her mouth, moving to more protected areas away from the lake until the fry are ready to feed, roughly 10–12 days postfertilization. The female will then move into shallow water to release the fry to minimize predation. However, if the fry are threatened they will return to the female's mouth. This behaviour can last for several weeks in the wild (Turner and Robinson, 2000).

The breeding behavior of tilapia has a major impact on the hatchery techniques used to produce the large numbers of fry needed by ongrowers. Most largescale production systems try to produce single sex male fry to minimize problems associated with breeding and fry production during grow out, particularly in pond-based systems. Single sex populations also reduce size differences in harvested fish related to the sexual dimorphism in growth performance between male and female fish. The difficulties described in trying to generate all-male fry through the use of hybridization means that most hatchery managers have turned to methods of direct sex-reversal of fry, using 17α-methyltestosterone in the feed. To be able to consistently produce high proportions of males (over 99%), hatchery managers need to begin the sex-reversal treatments on young fry, at first feeding. This requires hatcheries to be set up in such a way that brooding females can be easily captured and their eggs/fry "robbed" on a regular basis. Batches of eggs/fry are pooled into similar developmental stages for incubation so that large, synchronized batches of first feeding fry are produced for hormone treatment (Little and Hulata, 2000). Genetic methodologies based on the production of YY supermales that can be crossed to normal XX females have been used on a small scale commercially (Beardmore et al., 2001) but complexities with the sex-determination system in tilapia mean that the sex-ratio of males often falls below that obtained by well managed hormonal sex-reversal.

The genome: structure and organization

Only a limited number (13) of the 70 or so tilapia species have been karyotyped (Majumdar and McAndrew, 1986; Klinkhardt et al., 1995) and the evidence is that the karyotype has been highly conserved. Cytogenetic mapping using repetitive sequences (Ferreira et al., 2010) has shown that there have been at least two chromosome fusion events in the evolution of typical cichlid karyotypes. The first occurred before the divergence of the East African and South American cichlids and a second occurred independently in the African lineage, resulting in a characteristic single large pair of chromosomes in Tilapines and two large pairs in the nontilapiine species. Additional chromosome fusions have also occurred later within the tilapines, giving reduced chromosome numbers in four species showing deviation from the modal $2n = 44$ pattern [*T. mariae* $2n = 40$ (Thompson, 1981); *T. sparmanii* $2n = 42$ (Thompson, 1981;

Vervoort, 1980); *O. alcalicus* $2n = 48$ (Feldberg et al., 2003; Poletto et al., 2010), and *O. karongae* $2n = 38$ (Harvey et al., 2002; Mota-Velasco et al., 2010)]. The genome size of *O. niloticus* was estimated as 1.06 gigabases (Majumdar and McAndrew, 1986).

Development of molecular markers

Molecular markers have been critical for the management of farmed tilapia stocks. Some of the earliest applications were for species identification to ensure broodstock purity of the species used to produce hybrid all-male crosses. The range of markers used matched the development of the technology, initially general proteins and haemoglobin (Illes and Howlett, 1967) and easily stained enzyme loci, such as esterases (Chen and Tsuyuki, 1970; Avtalion and Wojdani, 1971; Basasibwaki, 1975; Avtalion et al., 1975; Avtalion, 1982). The development of allozyme techniques increased the number of loci and the level of species and hybrid discrimination (McAndrew and Majumdar 1983; Sodsuk and McAndrew, 1991). Seyoum and Kornfield (1992) applied mtDNA RFLP to discriminate all seven sub-species of *O. niloticus*. With the development of early PCR-based techniques such as RAPD markers Bardakci and Skibinski (1994, 1999), Dinesh et al. (1996), and Naish et al. (1995) could discriminate between some farmed species and subspecies (Penman and McAndrew, 2000).

Microsatellite markers have not proved very useful as species-diagnostic loci but have proved useful in the population genetic analysis of natural populations of *O. niloticus* (Bezault et al., 2011), the development of genetic maps (Lee et al., 2005) and pedigree assignment within breeding programs (Trong et al., 2013). However, recent developments in next generation sequencing (NGS) have enabled large numbers of single nucleotide polymorphisms (SNPs) to be accurately identified. Syaifudin (2015) used double-digest restriction-site associated DNA sequencing (ddRADseq) (Peterson et al., 2012) to identify >1000 shared RAD loci from 10 species of tilapia. Phylogenetic analysis of these markers confirmed previous molecular-based phylogenies. Analysis of the SNP loci identified many species-diagnostic loci that could potentially be used to discriminate between species pairs. Xia et al. (2014, 2015) identified much larger numbers of SNPs from RNA-seq and genome resequencing data (over a million from the latter).

Linkage mapping

The first linkage map in *O. niloticus* (Kocher et al., 1998) was based on the segregation of 62 microsatellite and 112 AFLP markers in 41 haploid *O. niloticus* embryos derived from a single female, in which resolved 30 LGs spanning 704 cM, with an estimated total map length of approximately 1200 cM. A second-generation map (Lee et al., 2005), based on F_2 progeny from a cross between an *O. niloticus* (red) male × *O. aureus* female and the segregation of 525 microsatellite loci and 21 gene-based markers, spanned 1311 cM in 24 linkage groups (resolved to 22 from subsequent studies by Shirak et al., 2006 and Guyon et al., 2012). The two maps were generally in good agreement with each other and with less dense maps generated by Agresti et al. (2000) and McConnell et al. (2000). Palaiokostas et al. (2013) produced a sire-based

Table 5.1 **Summary of tilapia genetic maps**

Family type	Number of markers	Marker type	References
F_1	3802	SNP	Palaiokostas et al. (2013)
F_2	550	Microsatellite/SNP	Lee et al. (2005)
F_2	229	Microsatellite/AFLP	Agresti et al. (2000)
BC	37	Microsatellite	McConnell et al. (2000)
haploid	174	Microsatellite/AFLP	Kocher et al. (1998)

BC, backcross.

map for the Stirling *O. niloticus* (Manzala) strain containing 3802 SNP markers, which grouped into 23 linkage groups, with an estimated map distance of 1,176 cM. The linkage groups were named in accordance with the Orenil1.1 genome assembly (NCBI assembly GCA_00188235.2). By reference to the genome assembly, two of these linkage groups were coalesced to form LG3, which contained a broad region of recombination suppression. This gave 22 LGs corresponding to the expected $n = 22$ chromosome number. Table 5.1 compares these maps.

Repetitive sequences in the tilapia genome

Repetitive elements can account for a large proportion of the genome in some species and can cause serious problems in the assembly of genome sequences. Shirak et al. (2010) estimated that 14% of the tilapia genome is composed of a variety of different repeat sequences, based on a search of tilapia sequence databases and analysis of a BAC library end-sequencing project. A tilapia-specific repeat library was constructed to identify the occurrence of different repeats and showed that the tilapia contained many species-specific repeats not identified in other model fish species. Long interspersed nuclear elements (LINEs) and short interspersed repetitive elements (SINEs) made up a significant proportion of the repetitive DNA as did SATB elements, with between 1000–10,000 copies of this 1,900 bp repeat in a single chromosome (Franck and Wright, 1993).

The SATA satellite DNA family of tandem repeat sequences, comprising three size variants (type 1, 237 bp; type 2, 230 bp, and type 3, 209 bp), is conserved in the tilapiine and haplochromiine genomes (Franck et al., 1992, 1994). SATA DNA has been mapped to the centromeric regions of most African cichilids (Ferreira et al., 2010) and appears to play an important role in the organisation and function of the centromeres in these species. Analysis of the distribution of SATA elements has also helped to map the chromosome fusions that have occurred independently in the *Oreochromis* and *Tilapia* genera that led to the reduced chromosome number detected in *O. karongae* and *T. mariae*. The position of SATA elements in the centromeres of *O. karongae*, but also in interstitial sites in three medium sized chromosome pairs, has been interpreted as evidence of relics of centromeres involved in past chromosome fusions (Mota-Velasco et al., 2010). Martins et al. (2004) were able to distinguish several of the chromosomes in the tilapia karyotype using different repetitive DNAs, including SATA, as FISH probes.

Although the main sex determining locus has been mapped to LG1 in the Stirling-Manzala *O. niloticus* population (and in LG23 in some others: see Section Reproduction—Sex Determination and Differentiation), analysis of synaptonemal complexes during meiosis in this species showed delayed pairing in LG3 (the largest chromosome pair) in the XY genotype but not in XX or YY genotypes (Foresti et al., 1993; Carrasco et al., 1999), which led to the early hypothesis that this was the sex chromosome pair. Harvey et al. (2002, 2003a) developed FISH chromosome paints by microdissection of LG3, and showed that these contained large amounts of repetitive DNA elements, in particular CiLINE2 and ROn-1. It has been suggested that LG3 is the ancestral sex chromosome pair in tilapias, and accumulated repeat sequences, which restricted recombination during sex chromosome evolution (Cnaani et al., 2008). However, this does not solve the puzzle of delayed pairing in meiosis in the heterogametic sex of LG3 chromosomes, which are not the sex chromosomes in *O. niloticus*.

More recently, Valente et al. (2016) provided insights in the genomic organization of transposable elements under an integrated view based on cytogenetics and genomics in *O. niloticus*. The study showed that some elements are not randomly distributed. They found extensive overlap between genomics and cytogenetics data and that tandem duplication may be the major mechanism responsible for the genomic dynamics of some transposable elements.

Physical mapping

The first BAC libraries were generated from sperm from a single male sourced from the Stirling Manzala strain (Katagiri et al., 2001). Four libraries with different average insert sizes (TBL1 with 65 kb, TBL2 with 105 kb, TBL3 with 145 kb and TBL4 with 194 kb), confirmed by pulsed field gel electrophoresis (PFGE), were produced. The two larger insert libraries were used by Katagiri et al. (2005) to physically map the *O. niloticus* genome. They restriction fingerprinted 35,245 clones from the T3 and T4 library and used the data to assemble contigs from the BACs. The average contig contained 9 clones and had an average length of 389.9 kb, and the map had a total length of 1.752 Gb or 1.65× the presumed genome length. This allowed the position of the regulatory regions of opsin genes and the red body color mutation to be identified. Chromosome-specific probes have had to be developed to identify individual chromosomes because of the similarity in size and lack of banding in the 20 small chromosomes in tilapia (Majumdar and McAndrew, 1986). The BAC library has been used to develop probes for in-situ hybridization. High density filters containing all clones were hybridized with DNA from genes of interest. Harvey et al. (2003b) used two such probes to show that the two aromatase genes identified in *O. niloticus* (Stirling Manzala) were located on separate chromosomes. Mota-Velasco et al. (2010) used BAC probes from known LGs to look at the chromosome fusions that resulted in the reduced chromosome number in *O. karongae*. Guyon et al. (2012) developed a series of BAC probes from the Katagiri et al. (2005) libraries that covered every LG in the tilapia karyotype, and integrated this with mapping information from a high-resolution radiation hybrid panel for the Nile tilapia genome. The RH panel was developed using irradiated splenocytes derived from an isogenic all-female line, again sourced from the Stirling Lake Manzala *O. niloticus* strain.

Whole genome sequencing

The first article based on a whole genome sequence for a tilapia (*O. niloticus*) and four species representative of other African cichlid lineages was published in 2014 (Brawand et al., 2014) [https://www.broadinstitute.org/models/tilapia]. The other species represented a broad cross-section of the East African Lake lineages: *Astatotilapia burtoni*, *Malawi* (*Maylandia*) *zebra*, *Pundamila nyererei*, and *Neolamprologus brichardi*. Unlike the tilapiini, these lineages were chosen because they have undergone recent rapid adaptive radiation. The *O. niloticus* DNA was derived from an isogenic clonal female line of the Stirling Lake Manzala strain (the same line used to develop the RH panel by Guyon et al., 2012). The main online resources for working with the tilapia genome assembly can be accessed through the Cichlid Genome Consortium at http://cichlid.umd.edu.

Xia et al. (2015) recently reported whole genome resequencing of 47 tilapia individuals using next generation techniques in order to identify selection signatures. The fish originated from five populations collected in South Africa, China, and Singapore and included GIFT individuals, *O. niloticus* of Egyptian origin, *O. mossambicus*, and a red hybrid tilapia strain. The study identified 1.43 million high-quality SNPs with linkage disequilibrium (LD) block size ranging from 10–100 kb.

The complete nucleotide sequence of the mitochondrial genome of several species of tilapia, have been published. The sequences are all around 16 Kbp in size, containing the same order and an identical number of genes and regions with the other reported cichlid fishes, which consists of 13 protein-coding genes, 22 transfer RNA (tRNA) genes, and ribosomal RNA (rRNA) genes (*O. niloticus* and *O. aureus*, He et al., 2011; *O. esculentus*, Kinaro et al., 2015).

Genome evolution and comparative genomics

Soler et al. (2010) constructed a BAC library (VMRC-44) from the same isogenic line used for the whole genome sequence, consisting of 73,728 clones with an average insert size of 150 bp. This new library and the Katagiri et al. (2001) TBL3 and TBL4 libraries were end sequenced. Soler et al. (2010) used these data and the sequence assemblies available for Stickleback, Medaka, and Fugu to develop a comparative physical map with tilapia. The sequencing identified 7,230 microsatellite motifs and BLAST hits for 16,636 genes. The BAC end sequences were aligned against the genome assemblies. The highest BLAST hit rate was against the Stickleback genome (17%). Four types of BLAST hit were classified: T1 clones with a single hit in the target genome; T2 where both ends of the clone hit in the appropriate opposing direction within 300 kb in the target genome; T3 where both ends hit within the target genome but outside the 300 kb range; and T4 where the two sequences hit different chromosomes in the target genome. By studying the ratio between the T2, T3, and T4 hits across the different species, they estimated that 15–20 rearrangements had occurred in each lineage since they diverged from a common ancestor. The combination of the end sequencing and the BAC fingerprinting map, which contains much larger multiple contigs, proved to be important tools in the genome assembly (see Brawand et al., 2014). Guyon et al. (2012) identified the position of a large number of important genes enabling synteny analysis with Stickleback, Medaka, and Fugu, which indicated that multiple inter-chromosomal rearrangements have occurred between Nile tilapia and these model species.

Transcriptome: expression and function

Expressed sequence tags (ESTs)

Lee et al. (2010) described a major project and the techniques they used to generate 116,899 ESTs from 17 normalized and 2 nonnormalized cDNA libraries from 16 different tissues extracted from *O. niloticus*. The individual ESTs have been deposited with GenBank (Acc# GR588780-705678). The normalization of the libraries greatly reduced the level of redundancy in sequencing output and resulted in a high level of gene discovery with 17,505 transcripts being mapped to Gene Ontology (GO) terms. This resource was used to help in the annotation of the cichlid genome sequences in the genome project.

RNA-seq

Tao et al. (2013) characterized gonadal transcription sequences from four pairs of XX and XY gonads at four developmental stages (5, 30, 90, and 180 days post hatch, dph) using RNAseq. These fishes were derived from an *O. niloticus* strain maintained at the National Institute for Basic Biology in Japan (originally from Egypt). This strain is known to have the main sex-determination locus in LG23 (Li et al., 2015a,b). Single sex XX and XY progeny groups were produced by crossing neomale XX and YY males with normal XX females. They identified 21,334 genes of which 259 were specific to the XY gonads and 69 specific to the XX. This enabled them to study global gene expression in the development of both sexes. The results also showed that the XX fry expressed oestrogen receptors at 5 dph (about 10 days postfertilization), while in contrast the XY fry showed no androgen receptor expression at this timepoint.

Xia et al. (2014) used Nile tilapia RNA-seq data in order to detect and map SNPs. They were able to position 17,699 SNPs onto the 22 linkage groups with an average of 805 SNPs per LG. The genotyping of a selected subset of SNPs validated polymorphisms for 93% of the loci. The authors emphasized the requirement of a SNP genotyping system, for example, an oligonucleotide SNP array, for economic genotyping larger numbers of SNPs in large sample sets in the future. From analysis of four populations in this study (*O. niloticus*, *O. mossambicus*, GIFT, and a red tilapia stock) they were able to demonstrate mixed ancestry in the red tilapia (mostly *O. mossambicus*, but with contribution from other species).

Xu et al. (2015) conducted a transcriptome profiling study of salinity adaptation in *O. niloticus* to identify the key molecular pathways involved in the process. Fish were exposed to a range of salinities (0, 8, or 16 practical salinity units, psu) with gene expression being measured in hepatocytes by RNAseq analysis. A total of 296,000 "genes" or variants were reported. A total of 934, 1087, and 734 transcripts were differently expressed in the 0 versus 8 psu set, 0 versus16 psu set, and 8 versus 16 psu set, respectively. This study revealed that several pathways were involved in salinity tolerance and that steroid hormones, osmoregulation, lipid metabolism, and cell-connected components are critical measures for salinity domestication in aquatic animals.

microRNAs

MicroRNAs (miRNAs) are small noncoding RNAs, generally with a length of 21 nucleotides. They suppress translation by base-pairing with target mRNAs and so regulate gene expression and play an important role in the regulation of cellular functions including differentiation and proliferation (Bartel, 2005).

Huang et al. (2012) analyzed miRNA expression profiles in the skeletal muscle of Nile tilapia in order to find growth-related miRNAs resulting from, or as a possible explanation for, phenotypic differences in growth rate and weight of two tilapia strains. They found 184 known, conserved miRNAs, of which miR-1, miR-206, and miR-133a were highly expressed, and 16 potentially growth-related miRNAs that were differentially expressed.

Yan et al. (2012a) identified 25 conserved miRNAs expressed in tilapia skeletal muscle using small RNA cloning. Among these miR-1, miR-27a, miR133a, and mirR-206 were differentially expressed during different development stages. Suppression of miRNA function using Antagomir (synthetic oligonucleotides designed to silence endogenous miRNAs) resulted in a significant increase in the expression of a number of muscle target genes. Yan et al. (2013) identified IGF-1 as a target gene of miR-206, which interacts with the 3′ UTR of IGF-1 and influences its gene expression, regulating growth of tilapia. The expression of miR-30c, a kidney-enriched miRNA, and miR-429 were found to be influenced by salinity (Yan et al., 2012b) and blood plasma osmolality (Yan et al., 2012c) respectively.

Tang et al. (2013) investigated the effects of vitamin E (VE) on expression of eight microRNAs in liver tissue by feeding juvenile tilapia with semi-purified diets containing three different levels of VE (0, 50, and 2500 mg/kg). VE is a known antioxidant. The VE deficient diet resulted in decreased superoxide dismutase (SOD) activity and downregulation of four of the eight miRNAs, whereas all eight miRNAs were upregulated in the fish receiving 2500 mg/kg, while the SOD activity was reduced. The study confirms that dietary VE levels affects hepatic miRNA expression and points to possible molecular mechanisms underpinning VE activity.

In order to detect genes and miRNAs that were differentially expressed by gender in early embryonic development, Eshel et al. (2014) analyzed embryos from single sex male and female batches of eggs at 2, 5, and 9 days post fertilization (dpf) and found sexually differentially expressed genes and miRNAs from as early as 2 dpf. miRNA expression in ovary and testis was investigated by Xiao et al. (2014) to better understand the role of miRNA mediated post-transcriptional regulation in *O. niloticus*. They detected in total 763 mature miRNAs, 209 miRNA-5p, and 202 miRNA-3p. miRNAs showed a different length distribution in ovary and testis (21 nt in ovary and 27 nt in testis). Furthermore, differential expression was observed in ovary and testis for 430 known miRNAs. miR-727, miR-120, and miR-29 were highly expressed in ovary and could be important for follicular growth or ovulation. In the testis miR-132, miR-212, miR-33a, and miR-135 were the most highly expressed and could play crucial roles during spermatogenesis. miRNAs seem to have multiple regulatory functions during sex differentiation as different expression patterns were present during gonadal development.

Li et al. (2015a) applied an in silico approach using known fish miRNAs to look for homologous sequences in tilapia and were able to predict 21 novel miRNAs belonging to 17 families.

Gene expression under different conditions

Liver transcriptome response to variation in dietary fat level

The impact of low and high fat diets on lipid metabolism in Nile tilapia was analyzed by He et al. (2015) using qPCR. Three groups of fish were fed diets containing different lipid levels (1%, 7%, and 13%) for a period of 10 weeks. The study showed that the low-fat diet resulted in elevated glycolysis and accelerated biosynthesis of fatty acids in the liver, which is the primary responding organ, whereas the high-fat diet led to an increased capability for fatty acid uptake and triglyceride synthesis and an increased number of adipocytes.

Response to pathogen infection

In recent years, streptococcal diseases have become a serious threat in tilapia aquaculture, while effective prevention and control methods have not yet been established. In order to understand the immunological response of tilapia to infection by *Streptococcus agalactiae*, Zhang et al. (2013) employed RNA-seq and digital gene expression technology to investigate changes in the tilapia transcriptome before and after *S. agalactiae* infection. They identified differentially-expressed immune-related genes and signaling pathways.

Functional and applied aspects

Reproduction—sex determination and differentiation

Phenotypic sex is an important commercial trait in tilapias. In many farming situations, tilapias tend to mature sexually before the desired harvest size. This often leads to reproduction in the ongrowing environment and competition between the resultant progeny and the stocked fish for resources, reducing growth rates, and value at harvest. Though not a significant problem in some circumstances (e.g., less likely to be a problem in cage farming due to the mechanical difficulties of spawning and collecting eggs by the females for mouthbrooding), and reproduction in the ongrowing environment may sometimes even be desirable (e.g., in small-scale pond farming where reproduction in the pond may avoid the need to purchase fingerlings), in many tilapia farming situations, it is advantageous to be able to minimize or eliminate reproduction before harvest.

A number of techniques have been investigated to achieve this. While rearing of triploids would potentially be an effective strategy (Bramick et al., 1995), production of triploids is not feasible in commercial hatcheries. The reproductive biology of tilapias (frequent and asynchronous spawning of small batches of eggs by many females) presents major difficulties in obtaining large quantities of unfertilized eggs necessary for inducing triploidy: in vitro fertilization is required to allow accurate timing of pressure or temperature shocks, which need to be applied a few minutes after fertilization. It has been suggested that improved growth rates as a result of selective breeding would allow Nile tilapia to reach harvest size before reproduction became significant (Komen and Benzie, 2015), but this is not yet proven and probably dependant on optimal farming environments that would allow such growth rates from improved stocks.

Currently, monosex male production via treatment of fry with 17α-methyltestosterone (MT) is the most widely used commercial method to control reproduction during grow-out; when correctly applied, this technique can produce >99% male fingerlings on a large scale. Alternative ways of producing all-male fingerlings using genetic techniques have been investigated. YY males have been developed and used in crosses to XX females to produce all-male, or nearly all-male progeny (Beardmore et al., 2001). This has been used commercially on a relatively small scale, but the complexity of producing YY males and inconsistency in progeny sex ratios (generally below those obtained by the best MT treatment results) have limited uptake. Sex ratios in Nile tilapia show variable responses to elevated temperature during sexual differentiation, and it has been demonstrated experimentally that the percentage of males can be increased to over 90% by selection for this trait (Wessels and Hörstgen-Schwark, 2011).

Genomic analysis has made a crucial contribution to our understanding of the determination and differentiation of phenotypic sex in tilapias. Prior to the development of linkage maps and identification of sex-linked genetic markers, analysis of sex determination was largely carried out on phenotypic sex ratios alone. Departures from expected sex ratios (e.g., the presence of male progeny in crosses between XX females and hormonally masculinised XX neomales, or female progeny in crosses between YY males and XX females) were frequently observed but difficult to analyze in depth. The second major tilapia linkage map and the markers therein (mostly microsatellites) were key in mapping the major sex determination locus in the Stirling population of Nile tilapia, originating from Lake Manzala in Egypt (Lee et al., 2003: LG8, later renamed as LG1 in Lee et al., 2005).

Sex-linked AFLP markers were also isolated from this region (Ezaz et al., 2004; Lee et al., 2011). The region was more finely mapped using SNP markers derived from a genotyping-by-sequencing technique, restriction-associated DNA sequencing (RADseq) by Palaiokostas et al. (2013). Gammerdinger et al. (2014) have suggested that the difference between the X and Y alleles lies in an inverted region, but the functional polymorphism(s) have not been identified yet.

Lühmann et al. (2012) mapped temperature-dependant sex reversal to LG1, 3, and 23 in families from a population derived from the Stirling-Manzala stock, using microsatellites from the Lee et al. (2005) map. They later refined the analysis of LG23, showing that SNPs in the *amh* gene were associated with thermosensitivity (Wessels et al., 2014). Palaiokostas et al. (2015) used double-digest RADseq (ddRADseq) to map loci associated with departures from 1:1 sex ratios in the Stirling-Manzala population at both normal (28°C) and elevated (36°C) temperatures. This study identified a QTL in LG20 at both temperatures.

Mapping of sex determining loci in other populations of Nile tilapia has revealed, surprisingly, an XX-XY locus in LG23 to be primarily responsible for controlling sex determination. In a population in Israel derived from FishGen, Swansea, Wales, it was found that the X haplotype carried one copy of the *amh* (anti-Mullerian hormone) gene, while the Y haplotype carried one copy of *amh* and a duplicated copy of *amh* with a 233 bp deletion in exon VII (Eshel et al., 2014; Orly Eshel, personal communication), which was named *amhy* and proposed to be the master sex determining gene. Li et al. (2015a,b), working on a population from Japan that originated from Egypt, found one copy of the *amh* gene on the X haplotype and two duplicated copies on the Y haplotype. One of the Y-specific copies was very similar to the Y-specific *amhy* described by Eshel et al. (2014), but was found not to be determining sex, and was named *amhΔ-y*. The other copy of this gene on the Y haplotype, which had a missense SNP in the N-terminal region and a large deletion in the promoter region, was demonstrated to be determining sex in this population (using overexpression in XX fry and CRISPR/Cas9 knockout to suppress expression in XY embryos), and was named *amhy* (rather confusingly, given that the same name was used by Eshel et al., 2014).

It is evident that copy number variation (CNV) in the *amh* gene is important in sex determination in at least some populations of Nile tilapia, and that further study is required to understand variation in sex determination across the species, which has a wide natural distribution in Africa. To date most studies have been on captive rather than wild populations, and it is not known to what extent founder effects, small population sizes, etc., may have affected sex determination in captive populations. Intraspecific variation in sex determination has been detected in other fish species. For example, *sdY*, the sex-determining gene in Atlantic salmon, is found in a genomic region bordered by repeat sequences associated with mobile elements, and this gene has been found in three different chromosomes in a single aquaculture population, strongly suggesting that it can "jump" to different genomic locations (Lubieniecki et al., 2015). Recent studies on sex determination in zebra fish (*Danio rerio*) have shown that sex determination is polygenic in captive populations, with large fluctuations in sex ratio, but the wild populations had a WZ/ZZ sex determination mechanism: it appears that the W allele was lost from these captive populations (Wilson et al., 2014).

Linkage mapping of sex determination loci in five other tilapia species (Lee et al., 2004; Cnaani et al., 2008), using markers from LG1 and LG3 in the Lee et al. (2005) linkage map, revealed that markers from LG1 were associated with sex in *Tilapia zillii*, markers from LG3 were associated with sex in *O. karongae* and *T. mariae*, and markers from both LG1 and LG3 were associated with sex in *O. aureus* and *O. mossambicus*. Lee et al. (2004) studied the epistatic interactions of LG1 and LG3 loci in determining sex in a family of *O. aureus*. It was concluded that LG3 carried the primary (WZ/ZZ) sex determining locus, but while all fish that were heterozygous for the most tightly linked marker (presumed WZ) were female, some fish with the presumed ZZ genotype (expected to be male under a single locus model) were female if they did not carry a specific allele at the most tightly linked LG1 marker. This was interpreted as ZZ fish (LG3) needing to also be XY (LG1) to ensure being male.

Many genes potentially involved in sexual differentiation in tilapias have been identified (Li et al., 2014a and references therein). Recent molecular innovations should allow a more detailed assessment of the influence of these genes. For example, Li et al. (2014a) report the use of CRISPR/Cas9 to disrupt selected genes involved in sexual differentiation and gonad development, leading to germ-cell deficient gonads or masculinization of somatic cells. This study highlights the potential of gene editing for basic research on sexual differentiation in tilapias (as well as many other genes and species) and possibly for manipulation to control reproduction in tilapia aquaculture in the future.

Salinity

Attempts to evaluate the potential of tilapia culture in saline waters have been undertaken by several groups, primarily due to the interest in brackishwater tilapia culture (Hui et al., 2014; Likongwe et al., 1996; Suresh and Lin, 1992; Watanabe et al., 1985a,b). More recently RNA-seq and gene expression profiling have been used to explore the differential expression of genes and gene pathways between fish reared in different salinities. Avarre et al. (2014) studied the black-chinned tilapia *Sarotherodon melanotheron heudelotii,* focusing on transcripts potentially involved in the plasticity of male reproductive success in response to salinity variations. They built a large transcriptomic resource for the community and also identified and validated several genes associated with control of spermatogenesis (e.g., MORC family CW-type zinc finger 2 protein) or an active ion excretion and uptake for maintaining the intracellular ionic balance (e.g., Na^+/K^+ -transporting ATPase subunit alpha-1). They also reported several overlooked pitfalls in the RNA-seq methodology (including issue in the gene annotations due to ambiguities and poor reproducibility) and advocated the use of multiple reference genes for qPCR.

As mentioned earlier, Xu et al. (2015) exposed *O. niloticus* to 0, 8, or 16 practical salinity units (psu) and classified the gene expressions from freshwater to saline water, using RNA-seq. Despite the low number of sequenced samples (one pool per condition), the authors identified several pathways perturbed by salinity changes, including immune-related pathways, lipid metabolism-related pathways, osmoregulation-related pathways, steroid metabolism-related pathways, signaling pathways, protein metabolism-related pathways, and cell activity-related pathways.

Future direction and aspects

Whole genome sequence

The work on analysis and interpretation of the Nile tilapia genome data is still in its initial stages. The current genome assembly is still incomplete, the current draft still containing many gaps and ambiguities. The use of multiple sequencing techniques, and the association of genetic data and transcriptomic information will prove to be essential to advance the genome assembly. As Avarre et al. (2014) reported for the transcriptomic annotation, the genomic annotation is very versatile, and rigorous annotation and validation is needed.

Genetic improvement based on genomics

Large-scale SNP-based genetic mapping has already been carried out in tilapias, providing denser maps and enabling more insightful interpretations than earlier mapping based on genotyping of much fewer individual markers. Genotyping by sequencing, using RADseq (Palaiokostas et al., 2013) and ddRADseq (Palaiokostas et al., 2015) have been applied to analyze sex determination loci in the Nile tilapia at the level of one to a few thousand SNPs across the genome. To date, no SNP chips are available for tilapia, although large numbers of SNPs have been identified, for example, Xia et al. (2014) identified over 23,000 potential SNPs from RNA-seq data, while Xia et al. (2015) identified 1.43 million SNP variants from resequencing of 47 individual tilapia. It seems likely that SNP chips will be developed for tilapias in the near future (e.g., http://www.genomar.no/), but the scale (number of SNPs), cost, targeting to include markers linked to specific traits, etc. of such initiative have yet to be established. Sequencing costs are still dropping rapidly and new techniques are being developed. Resequencing of individual fish may become an alternative way of genotyping very large numbers of SNPs in the not so distant future.

The application of genomics in aquaculture will be closely linked to the state of selective breeding. Selective breeding in tilapias has been largely based on "traditional" phenotypic selection, mostly for harvest weight, as exemplified by the genetic improvement of farmed tilapia (GIFT) program and several other programs based on GIFT fish and/or similar selection models (Komen and Trong, 2014; Komen and Benzie, 2015). Potentially correlated traits such as body shape, fillet yield and age/size at maturation have been studied in these breeding programs, although seldom directly selected for. Analysis of the quantitative genetic basis of fatty acid composition in GIFT tilapia suggested that this could be altered by selective breeding (Nguyen et al., 2010). Thodesen et al. (2013) described selection for survival and colour traits as well as growth in a red tilapia population. One breeding program (Genomar) based on GIFT fish has used DNA markers in selective breeding, for pedigree data and parental selection, but we are not aware of genomic-based selection (QTL/MAS, genomic selection) having been used in any large-scale commercial breeding program to date.

QTL analysis has been carried out for several traits, including sex determination (primarily in the Nile tilapia) and a range of traits in an F2 hybrid population—cold tolerance, size, stress response, and immune response (Cnaani et al., 2003, 2004). Cnaani and Hulata (2011) reviewed studies on salinity tolerance in tilapias, including suggestions for a QTL/MAS approach to this trait. QTL analysis has progressed furthest with sex determination, with loci having been identified in LG 1, 3, 20, and 23 (Cnaani et al., 2008; Lühmann et al., 2012; Eshel et al., 2014; Wessels et al., 2014; Palaiokostas et al., 2015; Li et al., 2015a,b), and MAS developed to assist in monosex male production (Khan, 2011; Sun et al., 2014). The successes to date should encourage further development and commercial application.

Large-scale replacement of hormone treatment for monosex male production by genetic techniques (using YY males) is still difficult to achieve due to the complexity of sex determination in *O. niloticus*. To date, there is little sign of practical alternative approaches. It has been suggested that selective breeding will result in fish that reach harvest weight before maturation and reproduction become problematic (Komen and Benzie, 2015), but this faster growth may only be realized by a small proportion of farmers that are able to provide ideal growth conditions. There may be a role for gene editing techniques (e.g., CRISPR/Cas9) in manipulating sex determination and/or differentiation (Li et al., 2014a) but it is too early to predict if such an approach will become more acceptable in fish and livestock production than transgenic techniques.

It seems likely that incorporation of traits such as cold tolerance, salinity tolerance, and disease resistance into commercial breeding programs would benefit from genomic-based selection, particularly as traits such as these can often only be tackled otherwise by between-family selection where mean survival of a sample from each family is the basis of phenotypic evaluation.

De Verdal et al. (2015) described a breeding program to develop a saline tolerant tilapia strain in the Philippines, from a base population created by hybridizing *O. niloticus* (better growth) and *O. mossambicus* (better salinity tolerance), subsequently backcrossed to *O. mossambicus* and selected for growth performance in brackishwater. The development of such hybrids in which different key traits from the parent species are to be combined offers strong potential for a genomic-based approach.

There are many articles describing the impact of various bacterial pathogens on tilapia culture in a range of countries, for example, *Aeromonas hydrophila* (Tipmongkolsilp et al., 2012; Fu et al., 2014a; Aly et al., 2015), *Streptococcus agalactiae*, and *S. iniae* (Huang et al., 2013; Cheng et al., 2014; Firdau-Nawi et al., 2014; Fu et al., 2014b; Gaikowski et al., 2014; Jantrakajorn et al., 2014; Li et al., 2014b; Lusiastuti et al., 2014), and *Francisella* spp. (Colquhoun and Duodu, 2011; Soto et al., 2013). There is little published research on genetic variation for disease resistance in tilapias, however, the available literature does suggest the presence of genetic variation for such traits. Sarder et al. (2001) and Sayeed (2003) demonstrated variation in resistance to *A. hydrophila* among isogenic lines of Nile tilapia. Fu et al. (2014a) showed that SNP variation in the *lipopolysaccharide-binding protein* gene was associated with variation in resistance to *A. hydrophila*, while Fu et al. (2014b) and Shen et al. (2015) reported that SNP haplotype combinations in the *mast cell protease 8* gene and the *duodenase-1* gene respectively were associated with resistance to *Streptococcus agalactiae* in tilapia (unnamed tilapia or hybrid *Oreochromis* spp.). The impact of bacteria and other pathogens is likely to become more important as tilapia culture increases and intensifies, and a genomic-based approach seems likely to be beneficial in developing selective breeding for disease resistance.

Tilapia culture systems and markets are very diverse, including extensive to intensive culture, and low to moderately high prices (the latter generally for larger fish, sometimes for live markets, sometimes for fillets). There is evidence that only the more intensive, better managed production systems can be used to realize the genetic gain for growth potential from current genetic improvement—this trend likely to intensify with further genetic improvement through the application of genomics. It is also to be hoped that genetic improvement for traits such as disease resistance, salinity tolerance, etc. will be developed and translated into improved performance for tilapia farmers.

Conclusions

A moderately good assembly of the tilapia genome is publicly available in a format that can be used by researchers. Relatively dense linkage maps, a radiation hybrid panel, BAC libraries, RNA-seq data, and other genomics resources have also been developed; some of these have already been used in informing the current genome assembly while others will undoubtedly be exploited to improve future versions of the assembly. With the exception of studies on sex determination, there has been relatively little application of genomics in tilapia genetic improvement to date. This is probably related to the relatively low value of individual tilapia and the focus on very few traits to date in selective breeding programs. We cautiously predict a fairly rapid expansion as decreasing costs and improving technologies make this more economic, and as breeding programmes develop into selection for a wider range of traits, including disease resistance.

Acknowledgments

The authors thank John Taggart for helpful comments on the manuscript.

References

Agresti, J.J., Seki, S., Cnaani, A., Poompuang, S., Hallerman, E.M., Umiel, N., May, B., 2000. Breeding new strains of tilapia: development of an artificial center of origin and linkage map based on AFLP and microsatellite loci. Aquaculture 185 (1–2), 43–56.

Aly, S.M., Albutti, A.S., Rahmani, A.H., Atti, N.M.A., 2015. The response of new-season Nile tilapia to *Aeromonas hydrophila* vaccine. Int. J. Clin. Exp. Med. 8 (3), 4508–4514.

Asian Development Bank, 2005. An impact evaluation of the development of genetically improved tilapia and their dissemination in selected countries. http://www.adb.org/publications.

Avarre, J.C., Dugué, R., Alonso, P., Diombokho, A., Joffrois, C., Faivre, N., … Durand, J.D., 2014. Analysis of the black-chinned tilapia *Sarotherodon melanotheron heudelotii* reproducing under a wide range of salinities: from RNA-seq to candidate genes. Mol. Ecol. Resour., 14(1), 139–149.

Avtalion, R.R., 1982. Genetic markers in *Sarotherodon* and their use for sex and species identification. The Biology and Culture of Tilapias. ICLARM Conference Proceedings 7, pp. 269–277. International Center for Living Aquatic Resources Management, Manila, Philippines.

Avtalion, R.R., Wojdani, A., 1971. Electrophoresis and immunoelectrophoresis of sera from known F1 hybrids of *Tilapia*. Bamidgeh 23, 117–124.

Avtalion, R.R., Pruginin, Y., Rothbard, S., 1975. Determination of allogenic and xenogenic markers in the genus Tilapia:1 identification of sex and hybrids in tilapia by electrophoresis analysis of serum proteins. Bamidgeh 27, 8–13.

Bardakci, F., Skibinski, D.O.F., 1994. Application of RAPD techniques in tilapiafish species and subspecies identification. Heredity 73, 117–123.

Bardakci, F., Skibinski, D.O.F., 1999. A polymorphic SCAR-RAPD marker between species of tilapia (Pisces: Cichlidae). Anim. Genet. 30, 78–79.

Bartel, B., 2005. MicroRNAs directing siRNA biogenesis. Nat. Struct. Mol. Biol. 12 (7), 569–571.

Basasibwaki, P., 1975. Comparative electrophoresis patterns of lactate dehydrogenase and malate dehydrogenase in five Lake Victoria cichlid species. Afr. J. Trop. Hydrobiol. Fish. 4, 21–26.

Beardmore, J.A., Mair, G.C., Lewis, R.I., 2001. Monosex male production in finfish as exemplified by tilapia: applications, problems, and prospects. Aquaculture 197, 283–301.

Bentsen, H.B., Eknath, A.E., Palada-de Vera, M.S., Danting, J.C., Bolivar, H.L., Reyes, R.A., Dionisio, E.E., Longalong, F.M., Circa, A.V., Tayamen, M.M., Gjerde, B., 1998. Genetic improvement of farmed tilapias: Growth performance in a complete diallel cross experiment with eight strains of *Oreochromis niloticus*. Aquaculture 160 (1–2), 145–173.

Bentsen, H.B, Gjerde, B., Nguyen, H.N., Rye, M. Ponzoni, R.W., Palada de Vera, M.S., Bolivar, H.L., Velasco, R.R., Danting, J.C., Dionisio E. E., Longalong, F.M., Reyes, R.A., Abella, T.A., Tayamen, M.M., & Eknath, A.E., 2012. Genetic improvement of farmed tilapias: genetic parameters for body weight at harvest in Nile tilapia (*Oreochromis niloticus*) during five generations of testing in multiple environments. Aquaculture, 338–341, 56–65.

Bezault, E., Balaresque, P., Toguyeni, A., Fermon, Y., Araki, H., Baroiller, J.-F., Rognon, X., 2011. Spatial and temporal variation in population genetic structure of wild Nile tilapia (*Oreochromis niloticus*) across Africa. BMC Genet. 12, 102.

Bramick, U., Puckhaber, B., Langholz, H.J., Hörstgen-Schwark, G., 1995. Testing of triploid tilapia (*Oreochromis niloticus*) under tropical pond conditions. Aquaculture 137, 343–353.

Brawand, D., Wagner, C.E., Li, Y.I., Malinsky, M., Keller, I., Fan, S., Simakov, O., Ng, A.Y., Lim, Z.W., Bezault, E., Turner-Maier, J., Johnson, J., Alcazar, R., Noh, H.J., Russell, P., Aken, B., Alföldi, J., Amemiya, C., Azzouzi, N., Baroiller, J.-F., Barloy-Hubler, F., Berlin, A., Bloomquist, R., Carleton, K.L., Conte, M.A., D'Cotta, H., Eshel, O., Gaffney, L., Galibert, F., Gante, H.F., Gnerre, S., Greuter, L., Guyon, R., Haddad, N.S., Haerty, W., Harris, R.M., Hofmann, H.A., Hourlier, T., Hulata, G., Jaffe, D.B., Lara, M., Lee, A.P., MacCallum, I., Mwaiko, S., Nikaido, M., Nishihara, H., Ozouf-Costaz, C., Penman, D.J., Przybylski, D., Rakotomanga, M., Renn, S.C.P., Riberio, F.J., Ron, M., Salzburger, W., Sanchez-Pulido, L., Santos, M.E., Searle, S., Sharpe, T., Swofford, R., Tan, F.J., Williams, L., Young, S., Yin, S., Okada, N., Kocher, T.D., Miska, E.A., Lander, E.S., Venkatesh, B., Fernald, R.D., Meyer, A., Ponting, C.P., Streelman, J.T., Lindblad-Toh, K., Alfoldi, J., Noh, H.J., Seehausen, O., Di Palma, F., 2014. The genomic substrate for adaptive radiation in African cichlid fish. Nature 513, 375–381.

Carrasco, L.A.P., Penman, D.J., Bromage, N., 1999. Evidence for the presence of sex chromosomes in the Nile tilapia (*Oreochromis niloticus*) from synaptonemal complex analysis of XX, XY and YY genotypes. Aquaculture 173, 207–218.

Chen, F.-Y., Tsuyuki, M., 1970. Zone electrophoresis studies on the proteins of *Tilapia mossambica* and *T. hornorum* and their F1 hybrids, *T. zillii* and *T. melanopleura*. J. Fish. Res. Bd. Can. 27, 2167–2177.

Cheng, Z.X., Ma, Y.M., Li, H., Peng, X.X., 2014. N-acetylglucosamine enhances survival ability of tilapias infected by *Streptococcus iniae*. Fish Shellfish Immunol. 40, 524–530.

Cnaani, A., Hulata, H., 2011. Improving salinity tolerance in tilapias: past experience and future prospects. Bamidgeh 63, 1–21.

Cnaani, A., Hallerman, E.M., Ron, M., Weller, J.I., Indelman, M., Kashi, Y., ... Hulata, G., 2003. Detection of a chromosomal region with two quantitative trait loci, affecting cold tolerance and fish size, in an F2 tilapia hybrid. Aquaculture, 223(1–4), 117–128.

Cnaani, A., Zilberman, N., Tinman, S., Hulata, G., Ron, M., 2004. Genome-scan analysis for quantitative trait loci in an F2 tilapia hybrid. Mol. Genet. Genomics 272 (2), 162–172.

Cnaani, A., Lee, B.-Y., Zilberman, N., Ozouf-Costaz, C., Hulata, G., Ron, M., D'Hont, A., Baroiller, J.-F., D'Cotta, H., Penman, D.J., Tomasino, E., Coutanceau, J.-P., Pepey, E., Shirak, A., Kocher, T.D., 2008. Genetics of sex determination in tilapiine species. Sex. Dev. 2, 43–54.

Colquhoun, D.J., Duodu, S., 2011. *Francisella* infections in farmed and wild aquatic organisms. Vet. Res. 42, 47.

De Verdal, H., Rosario, W., Vandeputte, M., Muyalde, N., Morissens, P., Baroiller, J.-F., Chevassus, B., 2015. Response to selection for growth in an interspecific hybrid between *Oreochromis mossambicus* and *O. niloticus* in two distinct environments. Aquaculture 430, 159–165.

Deines, A., Bbole, I., Katongo, C., Feder, J., Lodge, D., 2014. Hybridisation between native *Oreochromis* species and introduced Nile tilapia *O. niloticus* in the Kafue River, Zambia. Afr. J. Aquat. Sci. 39 (1), 23–34.

Dinesh, K.R., Lim, T.M., Chan, W.K., Phang, V.P.E., 1996. Genetic inforformion inferred from RAPD fingerprinting in three species of tilapia. Aquacult. Int. 4, 19–30.

Eknath, A.E., Tayamen, M.M., Palada-de Vera, M.S., Danting, J.C., Reyes, R.A., Dionisio, E.E., Capili, J.B., Bolivar, H.L., Abella, T.A., Circa, A.V., Bentsen, H.B., Gjedre, B., Gjedrem, T., Pullin, R.S.V., 1993. Genetic improvement of farmed tilapias: the growth performance of eight strains of *Oreochromis niloticus* tested in different farm environments. Aquaculture 111, 171–188.

Eshel, O., Shirak, A., Dor, L., Band, M., Zak, T., Markovich-Gordon, M., Chalifa-Caspi, V., Feldmesser, E., Weller, J.I., Seroussi, E., Hulata, G., Ron, M., 2014. Identification of male-specific *amh* duplication, sexually differentially expressed genes and microRNAs at early embryonic development of Nile tilapia (*Oreochromis niloticus*). BMC Genomics 15, 774.

Ezaz, M.T., Harvey, S.C., Boonphakdee, C., Teale, A.J., McAndrew, B.J., Penman, D.J., 2004. Isolation and physical mapping of sex-linked AFLP markers in Nile tilapia (*Oreochromis niloticus* L.). Mar. Biotechnol. 6, 435–445.

FAO, 2014. FAO Fisheries and Aquaculture Department, 2014. FishStat database. Available from: http://faostat.fao.org/site/629/default.aspx.[ftp://ftp.fao.org/FI/STAT/summary/b-1.pdf.

Feldberg, E., Porto, J.I.R., Bertollo, L.A.C., 2003. Chromosomal changes and adaptation of cichlid fishes during evolution. In: Val, A.L., Kapoor, B.G. (Eds.), Fish Adaptations. Science Publishers, Enfield, pp. 285–308.

Ferreira, I.A., Poletto, A.B., Kocher, T.D., Moto -Velasco, J.C., Penman, D.J., Martins, C., 2010. Chromosome evolution in African cichlid fish: contribution from the physical mapping of repeated DNA. Cytogenet. Genome Res. 129, 314–322.

Firdau-Nawi, M., Yusoff, S.M., Yusof, H., Abdullah, S.Z., Zamri-Saad, M., 2014. Efficacy of feed-based adjuvant vaccine against *Streptococcus agalactiae* in *Oreochromis* spp. in Malaysia. Aquacult. Res. 45, 87–96.

Foresti, F., Oliveira, C., Galetti Jnr, P.M., de Almeida-Toledo, L.F., 1993. Synaptonemal complex analysis in spermatocytes of tilapia, *Oreochromis niloticus* (Pisces, Cichlidae). Genome 36 (6), 1124–1128.

Franck, J.P.C., Wright, J.M., 1993. Conservation of a Satellite DNA-Sequence (Satb) in the tilapiine and haplochromine genome (Pisces, Cichlidae). Genome 36 (1), 187–194.

Franck, J., Wright, J., McAndrew, B., 1992. Genetic variability in a family of satellite DNAs from tilapia (Pisces: Cichlidae). Genome 35 (5), 719–725.

Franck, J.P., Kornfield, I., Wright, J.M., 1994. The utility of SATA satellite DNA sequences for inferring phylogenetic relationships among the three major genera of tilapiine cichlid fishes. Mol. Phylogenet. Evol. 3 (1), 10–16.

Fu, G.H., Liu, F., Xia, J.H., Yue, G.H., 2014a. The *LBP* gene and its association with resistance to *Aeromonas hydrophila* in tilapia. Int. J. Mol. Sci. 15, 22028–22041.

Fu, G.H., Wan, Z.Y., Xia, J.H., Liu, F., Liu, X.J., Yue, G.H., 2014b. The *MCP-8* gene and its possible association with resistance to *Streptococcus agalactiae* in tilapia. Fish Shellfish Immunol. 40, 331–336.

Gaikowski, M.P., Schleis, S.M., Leis, E., Lasee, B.A., Endris, R.G., 2014. Effectiveness of AquaFlor (50% Florfenicol) administered in feed to control mortality associated with *Streptococcus iniae* in tilapia at a commercial tilapia production facility. N. Am. J. Aquacult. 76, 375–382.

Gammerdinger, W.J., Conte, M.A., Acquah, E.A., Roberts, R.B., Kocher, T.D., 2014. Structure and decay of a proto-Y region in tilapia, *Oreochromis niloticus*. BMC Genomics 15, 975.

Guyon, R., Rakotomanga, M., Azzouzi, N., Coutanceau, J.-P., Bonillo, C., D'Cotta, H., Pepey, E., Soler, L., Rodier-Goud, M., D'Hont, A., Conte, M.A., van Bers, N.E.M., Penman, D.J., Hitte, C., Crooijmans, R.P.M.A., Kocher, T.D., Ozouf-Costaz, C., Baroiller, J.-F., Galibert, F., 2012. A high resolution map of the Nile tilapia genome: a resource for studying cichlids and other percomorphs. BMC Genomics 13, 222.

Harvey, S.C., Masabanda, J., Carrasco, L.A.P., Bromage, N.R., Penman, D.J., Griffin, D.K., 2002. Molecular cytogenetic analysis reveals sequence differences between the sex chromosomes of *Oreochromis niloticus*: evidence for an early stage of sex-chromosome differentiation. Cytogenet. Genome Res. 97, 76–80.

Harvey, S.C., Boonphakdee, C., Campos-Ramos, R., Ezaz, M.T., Griffin, D.K., Bromage, N.R., Penman, D.J., 2003a. Analysis of repetitive DNA sequences in the sex chromosomes of *Oreochromis niloticus*. Cytogenet. Genome Res. 101, 314–319.

Harvey, S.C., Kwon, J.Y., Penman, D.J., 2003b. Physical mapping of the brain and ovarian aromatase genes in the Nile tilapia, Oreochromis niloticus, by fluorescence in situ hybridisation. Anim. Genet. 34, 62–64.

He, A., Luo, Y., Yang, H., Liu, L., Li, S., Wang, C., 2011. Complete mitochondrial DNA sequences of the Nile tilapia (*Oreochromis niloticus*) and Blue tilapia (*Oreochromis aureus*): genome characterization and phylogeny applications. Mol. Biol. Rep. 38 (3), 2015–2021.

He, A., Ning, L., Chen, L., Chen, Y., Xing, Q., Li, J., … Du, Z., 2015. Systemic adaptation of lipid metabolism in response to low- and high-fat diet in Nile tilapia (*Oreochromis niloticus*). Physiol. Rep., 3(8), e12485.

Hickling, C.F., 1960. The Malacca tilapia hybrids. J. Genet. 57, 1–10.

Huang, C.W., Li, Y.H., Hu, S.Y., Chi, J.R., Lin, G.H., Lin, C.C., … Wu, J.L., 2012. Differential expression patterns of growth-related microRNAs in the skeletal muscle of Nile tilapia (*Oreochromis niloticus*). J. Anim. Sci., 90(12), 4266–4279.

Huang, B.F., Zou, L.L., Xie, J.G., Huang, Z.C., Li, Y.W., Li, A.X., 2013. Immune responses of different species of tilapia infected with *Streptococcus agalactiae*. J. Fish Dis. 36, 747–752.

Hui, W., Guodong, L., Jiahui, L., Hongshuai, Y., Jun, Q., Pao, X., 2014. Combined effects of temperature and salinity on yolk utilization in Nile tilapia (*Oreochromis niloticus*). Aquacult. Res. 46 (10), 2418–2425.

Illes, T.D., Howlett, C.J., 1967. Electrophoretic analysis of blood of *Tilapia leucostica* Trewavas and Tilapia zilliii (Gervais) from Lake Victoria. East African Freshwater Fisheries Research Organisation Annual Report for 1967, pp. 64–72.

Jantrakajorn, S., Maisak, H., Wongtavatchai, J., 2014. Comprehensive investigation of *Streptococcus* outbreaks in cultured Nile tilapia, *Oreochromis niloticus*, and red tilapia, *Oreochromis* spp., of Thailand. J. World Aquacult. Soc. 45 (4), 392–402.

Katagiri, T., Asakawa, S., Minagawa, S., Shimizu, N., Hirono, I., Aoki, T., 2001. Construction and characterization of BAC libraries for three fish species; rainbow trout, carp and tilapia. Anim. Genet. 32 (4), 200–204.

Katagiri, T., Kidd, C., Tomasino, E., Davis, J.T., Wishon, C., Stern, J.E., Carleton, K.L., Howe, A.E., Kocher, T.D., 2005. A BAC-based physical map of the Nile tilapia genome. BMC Genomics 6, 89.

Khan, M.G.Q., 2011. Marker-assisted selection in enhancing genetically male tilapia (*Oreochromis niloticus* L.) production. PhD Thesis, University of Stirling. 204 p.

Kinaro, Z.O., Xue, L., Nyaundi, K.J., Shen, J., 2015. The mitochondrial genome of an endangered native Singidia tilapia, *Oreochromis esculentus*: genome organization and control region polymorphism. Mitochondrial DNA, 14, 1–3.

Klinkhardt M, Tesche M, Greven H., 1995. Database of fish chromosomes. Westarp Wissenschaften, Magdeburg.

Kocher, T.D., Lee, W.J., Sobolewska, H., Penman, D., McAndrew, B., 1998. A genetic linkage map of a cichlid fish, the tilapia (*Oreochromis niloticus*). Genetics 148 (3), 1225–1232.

Komen, J., Benzie, J., 2015. Tilapia genetic improvement: achievements and future directions. Presentation at the Twelfth International Symposium on Genetics in Aquaculture, 21–27 June, Santiago de Compostela, Spain.

Komen, H., Trong, T.Q., 2014. Nile tilapia genetic improvement: achievements and future directions. Bamidgeh Special Issue: International Symposium on Tilapia in Aquaculture 10. Available from: http://www.siamb.org.il/Content_siamb/editor/ista10.2013.1057.Komen.pdf.

Lee, B.-Y., Penman, D.J., Kocher, T.D., 2003. Identification of a sex-determining region in Nile tilapia (*Oreochromis niloticus*) using bulked segregant analysis. Anim. Genet. 34, 379–383.

Lee, B.-Y., Hulata, G., Kocher, T.D., 2004. Two unlinked loci controlling the sex of blue tilapia (*Oreochromis aureus*). Heredity 92, 543–549.

Lee, B.Y., Lee, W.J., Streelman, J.T., Carleton, K.L., Howe, A.E., Hulata, G., … Kocher, T.D., 2005. A second-generation genetic linkage map of tilapia (*Oreochromis* spp.). Genetics, 170(1), 237–244.

Lee, B.-Y., Howe, A.E., Conte, M.A., D'Cotta, H., Pepey, E., Baroiller, J.-F., Di Palma, F., Carleton, K.L., Kocher, T.D., 2010. An EST resource for tilapia based on 17 normalized libraries and assembly of 116,899 sequence tags. BMC Genomics 11, 278.

Lee, B.-Y., Coutanceau, J.-P., Ozouf-Costaz, C., D'Cotta, H., Baroiller, J.-F., Kocher, T.D., 2011. Genetic and physical mapping of sex-linked AFLP markers in Nile tilapia (*Oreochromis niloticus*). Mar. Biotechnol. 13, 557–562.

Li, M., Yang, H., Zhao, J., Fang, L., Shi, H., Li, M., Sun, Y., Zhang, X., Jiang, D., Zhou, L., Wang, D., 2014a. Efficient and heritable gene targeting in tilapia by CRISPR/Cas9. Genetics 197, 591–599.

Li, Y.W., Liu, L., Huang, P.R., Fang, W., Luo, Z.P., Peng, H.L., Wang, Y.X., Li, A.X., 2014b. Chronic streptococcus in Nile tilapia, *Oreochromis niloticus* (L.), caused by *Streptococcus agalactiae*. J. Fish Dis. 37, 757–763.

Li, X.H., Wu, J.S., Tang, L.H., Hu, D., 2015a. Identification of conserved microRNAs and their target genes in Nile tilapia (*Oreochromis niloticus*) by bioinformatic analysis. Genet. Mol. Res. 14 (1), 2785–2792.

Li, M., Sun, Y., Zhao, J., Shi, H., Zeng, S., Ye, K., Jiang, D., Zhou, L., Sun, L., Tao, W., Nagahama, Y., Kocher, T.D., Wang, D., 2015b. A tandem duplication of *anti-müllerian hormone* with a missense SNP on the Y chromosome is essential for male sex determination in Nile tilapia, *Oreochromis niloticus*. PLoS Genet. 11 (11), e1005678.

Likongwe, J.S., Stecko, T.D., Stauffer, J.R., Carline, R.F., 1996. Combined effects of water temperature and salinity on growth and feed utilization of juvenile Nile tilapia *Oreochromis niloticus* (Linneaus). Aquaculture 146 (1–2), 37–46.

Little, D.C., Hulata, G., 2000. Strategies for tilapia seed production. In: Beveridge, M., McAndrew, B.J. (Eds.), Tilapias: Biology and Exploitation. Kluwer Academic Publishers, Netherlands, pp. 267–326.

Lubieniecki, K.P., Lin, S., Cabana, E.I., Li, J., Lai, Y.Y.Y., Davidson, W.S., 2015. Genomic instability of the sex-determining locus in Atlantic salmon (*Salmo salar*). G3 5, 2513–2522.

Lühmann, L.M., Knorr, C., Hörstgen-Schwark, G., Wessels, S., 2012. First evidence for family-specific QTL for temperature-dependent sex reversal in Nile tilapia (*Oreochromis niloticus*). Sex. Dev. 6, 247–256.

Lusiastuti, A.M., Textor, M., Seeger, H., Akineden, ö., Zschöck, M., 2014. The occurrence of *Streptococcus agalactiae* sequence type 261 from fish disease outbreaks of tilapia *Oroechromis niloticus* in Indonesia. Aquacult. Res. 45, 1260–1263.

Majumdar, K.C., McAndrew, B.J., 1986. Relative DNA content of somatic nuclei and chromosomal studies in three genera, *Tilapia*, *Sarotherodon*, and *Oreochromis* of the tribe Tilapiini (Pisces, Cichlidae). Genetica 68 (3), 175–188.

Martins, C., Oliveira, C., Wasko, A.P., Wright, J.M., 2004. Physical mapping of the Nile tilapia (*Oreochromis niloticus*) genome by fluorescent in situ hybridization of repetitive DNAs to metaphase chromosomes—a review. Aquaculture 231, 37–49.

McAndrew, B.J., 2000. Evolution, phylogenetic relationships and biogeography. In: Beveridge, M., McAndrew, B.J. (Eds.), Tilapias: Biology and Exploitation. Kluwer Academic Publishers, Netherlands, pp. 1–32.

McAndrew, B.J., Majumdar, K.C., 1983. Tilapia stock identification using electrophoretic markers. Aquaculture 30, 249–261.

McConnell, S.K.J., Beynon, C., Leamon, J., Skibinski, D.O.F., 2000. Microsatellite marker based genetic linkage maps of *Oreochromis aureus* and *O. niloticus* (Cichlidae): extensive linkage group segment homologies revealed. Anim. Genet. 31 (3), 214–218.

Mires, D., 1977. Theoretical and practical aspects of the production of all-male tilapia hybrids. Bamidgeh 29, 94–101.

Mota-Velasco, J.C., Ferreira, I.A., Cioffi, M.B., Ocalewicz, K., Campos-Ramos, R., Shirak, A., Lee, B.Y., Martins, C., Penman, D.J., 2010. Characterization of the chromosome fusions in *Oreochromis karongae*. Chromosome Res. 18 (5), 575–586.

Naish, K.A., Warren, M., Bardacki, F., Skibinski, D.O.F., Carvalho, G.R., Mair, G.C., 1995. Multilocus DNA-fingerprinting and RAPD reveal similar genetic relationsips between strains of *Oreochromis niloticus* (Pisces Cichlidae). Mol. Ecol. 4, 271–274.

Nguyen, N.H., Ponzoni, R.W., Yee, H.Y., Abu-Bakar, K.R., Hamzah, A., Khaw, H.L., 2010. Quantitative genetic basis of fatty acid composition in the GIFT strain of Nile tilapia (*Oreochromis niloticus*) selected for high growth. Aquaculture 309, 66–74.

Palaiokostas, C., Bekaert, M., Khan, M.G.Q., Taggart, J.B., Gharbi, K., McAndrew, B.J., Penman, D.J., 2013. Mapping and validation of the major sex-determining region in Nile Tilapia (*Oreochromis niloticus* L.) using RAD sequencing. PLoS ONE 8 (7), e68389.

Palaiokostas, C., Bekaert, M., Khan, M.G.Q., Taggart, J.B., Gharbi, K., McAndrew, B.J., Penman, D.J., 2015. A novel sex-determining QTL in Nile tilapia (*Oreochromis niloticus*). BMC Genomics 16, 171.

Penman, D.J., McAndrew, B.J., 2000. Gentics for the mangement and improvement of cultured tilapias. In: Beveridge, M., McAndrew, B.J. (Eds.), Tilapias: Biology and Exploitation. Kluwer Academic Publishers, Netherlands, pp. 227–266.

Peterson, B.K., Weber, J.N., Kay, E.H., Fisher, H.S., Hoekstra, H.E., 2012. Double digest RADseq: an inexpensive method for de novo SNP discovery and genotyping in model and non-model species. PLoS ONE 7 (5), e37135.

Poletto, A.B., Ferreira, I.A., Cabral-de-Mello, D.C., Nakajima, R., Mazzuchelli, J., Ribeiro, H.B., Venere, P.C., Nirchio, M., Kocher, T.D., Martins, C., 2010. Chromosome differentiation patterns during cichlid fish evolution. BMC Genet. 11, 50.

Pullin, R.S.V., 1988. Tilapia genetic resources for aquaculture. In: Pullin, R.S.V. (Ed.), Proceedings of the Workshop on Tilpaia Resources for Aquaculture 23–24 March 1987. Bangkok, Thailand. ICLARM Conference Proceedings 16.

Pullin R.S.V., Lowe-McConnell, R.H., 1982. The biology and culture of tilapias. ICLARM Conference Proceedings 7, 423 p., International Center for Living Aquatic Resources Management, Manila, Philippines.

Sarder, M.R.I., Thompson, K.D., Penman, D.J., McAndrew, B.J., 2001. Immune responses of Nile tilapia (*Oreochromis niloticus* L.) clones: I. Non-specific responses. Dev. Comp. Immunol. 25, 37–46.

Sayeed, S., 2003. A comparison of immune responses and disease resistance in clonal lines of Nile tilapia *Oreochromis niloticus* L. PhD Thesis, University of Stirling. 245 p.

Seyoum, S., Kornfield, I., 1992. Taxonomic notes on the *Oreochromis niloticus* subspecies complex (Pisces: Cichlidae), with a description of a new subspecies. Can. J. Zool. 70, 2161–2165.

Shen, Y., Fu, G.H., Liu, F., You, G.H., 2015. Characterization of the duodenase-1 gene and its associations with resistance to *Streptococcus agalactiae* in hybrid tilapia (*Oreochromis* spp.). Fish Shellfish Immunol. 45, 717–724.

Shirak, A., Seroussi, E., Cnaani, A., Howe, A.E., Domokhovsky, R., Zilberman, N., Kocher, T.D., Hulata, G., Ron, M., 2006. Amh and Dmrta2 genes map to tilapia (*Oreochromis* spp.) linkage group 23 within quantitative trait locus regions for sex determination. Genetics 174 (3), 1573–1581.

Shirak, A., Grabherr, M., Di Palma, F., Lindblad-Toh, K., Hulata, G., Ron, M., Kocher, T.D., Seroussi, E., 2010. Identification of repetitive elements in the genome of *Oreochromis niloticus*: tilapia repeat masker. Mar. Biotechnol. 12, 121–125.

Sodsuk, P., McAndrew, B.J., 1991. Molecular systematics of three Tilapiine genera *Tilapia, Sarotherodon* and *Oreochromis* using allozyme data. J. Fish Biol. 39 (Suppl. A), 301–308.

Soler, L., Conte, M.A., Katagiri, T., Howe, A.E., Lee, B.-Y., Amemiya, C., Stuart, A., Dossat, C., Poulain, J., Johnson, J., Di Palma, F., Lindblad-Toh, K., Baroiller, J.-F., D'Cotta, H., Ozouf-Costaz, Kocher, T.D., 2010. Comparative physical maps derived from BAC end sequences of tilapia (*Oreochromis niloticus*). BMC Genomics 11, 636.

Soto, E., Kidd, S., Mendez, S., Marancik, D., Revan, F., Hiltchie, D., Camus, A., 2013. *Francisella noatunensis* subsp. *orientalis* pathogenesis analysed by experimental immersion challenge in Nile tilapia, *Oreochromis niloticus* (L.). Vet. Microbiol. 164, 77–84.

Sun, Y.-L., Jiang, D.-N., Zeng, S., Hu, C.-J., Ye, K., Yang, C., Yang, S.-J., Li, M.-H., Wang, D.-S., 2014. Screening and characterization of sex-linked DNA markers and marker-assisted selection in the Nile tilapia (*Oreochromis niloticus*). Aquaculture 433, 19–27.

Suresh, A.V., Lin, C.K., 1992. Tilapia culture in saline waters: a review. Aquaculture 106 (3–4), 201–226.

Syaifudin, M., 2015. Species-specific DNA markers for improving the genetic management of tilapia. PhD Thesis, University of Stirling. 241 p.

Tang, X.-L., Xu, M.-J., Li, Z.-H., Pan, Q., Fu, J.-H., 2013. Effects of vitamin E on expressions of eight microRNAs in the liver of Nile tilapia (*Oreochromis niloticus*). Fish Shellfish Immunol. 34 (6), 1–6.

Tao, W., Yuan, J., Zhou, L., Sun, L., Sun, Y., Yang, S., … Wang, D., 2013. Characterization of gonadal transcriptomes from Nile Tilapia (*Oreochromis niloticus*) reveals differentially expressed genes. PLoS ONE, 8(5), e63604.

Thodesen, J., Rye, M., Wang, Y.-X., Li, S.-J., Bentsen, H.B., Yazdi, M.H., Gjedrem, T., 2013. Genetic improvement of tilapias in China: genetic parameters and selection responses in growth survival and external colour traits of red tilapia (*Oreochromis* spp.) after four generations of multi-trait selection. Aquaculture 416-417, 354–366.

Thompson, K.W., 1981. Karyotypes of six species of African Cichlidae (Pisces, Perciformes). Experientia 37, 351–352.

Tipmongkolsilp, N., del Castilo, C.S., Hikima, J., Jung, T.-S., Kondo, H., Hirono, I., Aoki, T., 2012. Multiple drug resistant strains of *Aeromonas hydrophila* isolated from tilapia farms in Thailand. Fish Pathol. 47 (2), 56–63.

Trewavas E., 1983. Tilapiine fishes of the Genera *Sarotherodon*, *Oreochromis* and *Danakilia*. British Museum (Natural History), 583 p. London, Publication Number 878.

Trong, Q.T., van Bers, N., Crooijmans, R., Dibbits, B., Komen, H., 2013. A comparison of microsatellites and SNPs in parental assignment in the GIFT strain of Nile tilapia (*Oreochromis niloticus*): the power of exclusion. Aquaculture 388–391, 14–23.

Turner, G.F., Robinson, R.L., 2000. Reproductive biology, mating systems and parental care. In Tilapias: Biology and Exploitation, Beveridge,M., McAndrew, B.J., (Eds.) Kluwer Academic Publishers, Netherlands. Fish and Fisheries Series 25.

Valente, G., Kocher, T., Eickbush, T., Simões, R.P., Martins, C., 2016. Integrated cytogenetics and genomics analysis of transposable elements in the Nile tilapia, *Oreochromis niloticus*. Mol. Genet. Genomics, 291(3), 1219–1225.

Vervoort, A., 1980. The karyotypes of seven *Tilapia* (Teleostei:Cichlidae). Cytologia 45, 651–656.

Watanabe, W.O., Kuo, C.M., Huang, M.C., 1985a. Salinity tolerance of Nile tilapia fry (*Oreochromis niloticus*), spawned and hatched at various salinities. Aquaculture 48 (2), 159–176.

Watanabe, W.O., Kuo, C.M., Huang, M.C., 1985b. The ontogeny of salinity tolerance in the tilapias *Oreochromis aureus*, *O. niloticus* and an *O. mossambicus* x *O. niloticus* hybrid, spawned and reared in freshwater. Aquaculture 47 (4), 353–367.

Wessels, S., Hörstgen-Schwark, G., 2011. Temeperature dependent sex ratios in selected lines and crosses with a YY-male in Nile tilapia (*Oreochromis niloticus*). Aquaculture 318, 79–84.

Wessels, S., Sharifi, R.A., Luehmann, L.M., Rueangsri, S., Krause, I., Pach, S., Hörstgen-Schwark, G., Knorr, C., 2014. Allelic variant in the Anti-Müllerian Hormone gene leads to autosomal and temperature-dependent sex reversal in a selected tilapia line. PLOS ONE 9 (8), e104795.

Wilson, C.A., High, S.K., McCluskey, B.M., Amores, A., Yan, Y.-L., Titus, T.A., Anderson, J.L., Batzel, P., Carvan, III, M.J., Schartl, M., Postlethwait, J.H., 2014. Wild sex in zebrafish: loss of the natural sex determinant in domesticated strains. Genetics 198, 1291–1308.

Xia, J.H., Wan, Z.Y., Ng, Z.L., Wang, L., Fu, G.H., Lin, G., Liu, F., Yoe, G.H., 2014. Genome-wide discovery and in silico mapping of gene-associated SNPs in Nile tilapia. Aquaculture 432, 67–73.

Xia, J.H., Bai, Z., Mang, Z., Zhang, Y., Wang, Le, Liu, F., Jing, W., Wan, Z.Y., Li, J., Lin, H., Yue, G.H., 2015. Signatures of selection in tilapia revealed by whole genome resequencing. Sci. Rep. 5, 14168.

Xiao, J., Zhong, H., Zhou, Y., Yu, F., Gao, Y., Luo, Y., … Zhang, Y., 2014. Identification and characterization of MicroRNAs in ovary and testis of Nile tilapia (*Oreochromis niloticus*) by using Solexa sequencing technology. PLoS ONE, 9(1), e86821.

Xu, Z., Gan, L., Li, T., Xu, C., Chen, K., Wang, X., … Li, E., 2015. Transcriptome profiling and molecular pathway analysis of genes in association with salinity adaptation in Nile tilapia *Oreochromis niloticus*. PLoS ONE, 10(8), e0136506.

Yan, B., Guo, J.T., Zhao, L.H., Zhao, J.L., 2012a. MiR-30c: a novel regulator of salt tolerance in tilapia. Biochem. Biophys. Res. Commun. 425 (2), 315–320.

Yan, B., Guo, J.-T., Zhao, L.-H., Zhao, J.-L., 2012b. microRNA expression signature in skeletal muscle of Nile tilapia. Aquaculture 364-365, 240–246.

Yan, B., Zhao, L.H., Guo, J.T., Zhao, J.L., 2012c. MiR-429 regulation of osmotic stress transcription factor 1 (OSTF1) in tilapia during osmotic stress. Biochem. Biophys. Res. Commun. 426 (3), 294–298.

Yan, B., Zhu, C.-D., Guo, J.-T., Zhao, L.-H., Zhao, J.-L., 2013. miR-206 regulates the growth of the teleost tilapia (*Oreochromis niloticus*) through the modulation of IGF-1 gene expression. J. Exp. Biol. 216 (Pt. 7), 1265–1269.

Zhang, R., Zhang, L., Ye, X., Tian, Y., Sun, C., Lu, M., Bai, J., 2013. Transcriptome profiling and digital gene expression analysis of Nile tilapia (*Oreochromis niloticus*) infected by *Streptococcus agalactiae*. Mol. Biol. Rep. 40 (10), 5657–5668.

Turbot (*Scophthalmus maximus*) genomic resources: application for boosting aquaculture production

6

Paulino Martínez, Diego Robledo**, Silvia T. Rodríguez-Ramilo†,¶, Miguel Hermida*, Xoana Taboada**, Patricia Pereiro‡, Juan A. Rubiolo*, Laia Ribas§, Antonio Gómez-Tato**,††, José Antonio Álvarez-Dios**,‡‡, Francesc Piferrer§, Beatriz Novoa‡, Antonio Figueras‡, Belén G. Pardo*, Jesús Fernández†, Ana Viñas*, Carmen Bouza**

*Department of Genetics, Faculty of Veterinary, University of Santiago de Compostela, Lugo, Spain; **Department of Genetics, Faculty of Biology (CIBUS), University of Santiago de Compostela, Santiago de Compostela, Spain; †Departament of Animal Breeding, National Institute for Agricultural and Food Research and Technology, Madrid, Spain; ‡Institute of Marine Research (IIM), Spanish National Research Council (CSIC), Vigo, Spain; §Institute of Marine Sciences, Spanish National Research Council (CSIC), Barcelona, Spain; ¶GenPhySE, National Institute for Agricultural Research, Castanet-Tolosan, France; ††Department of Geometry and Topology, Faculty of Mathematics, University of Santiago de Compostela, Santiago de Compostela, Spain; ‡‡Department of Applied Mathematics, Faculty of Mathematics, University of Santiago de Compostela, Santiago de Compostela, Spain

Biology and aquaculture

Turbot biology

Turbot (*Scophthalmus maximus*; Fig. 6.1) is a flatfish naturally distributed around the European coast, from the Baltic Sea across the Atlantic Ocean and the Mediterranean Sea up to the Black Sea. Its main fisheries are in the Atlantic area and especially in the North Sea. This species exhibits one of the most important growth rates and sexual growth dimorphism observed in flatfish, females largely outgrowing males. Turbot is a demersal species ranging from shallow water down to 150 m. It has a cryptic body color imitating substrate and a carnivorous narrow prey-spectrum diet (Bouza et al., 2014). The spawning season occurs between May and Aug. in the Atlantic region. Females reach maturity at 3 years of age and males at 2 years. Each spawning season more than 5 million of spherical and small eggs (1.1 mm diameter) are produced per female. Larvae are initially symmetric, but by the end of metamorphosis (40 days post hatching—dph), the right eye moves to the left side determining the asymmetrical body plan typical of flatfish (Rodríguez Villanueva and Fernández Souto, 2009).

Genomics in Aquaculture. http://dx.doi.org/10.1016/B978-0-12-801418-9.00006-8

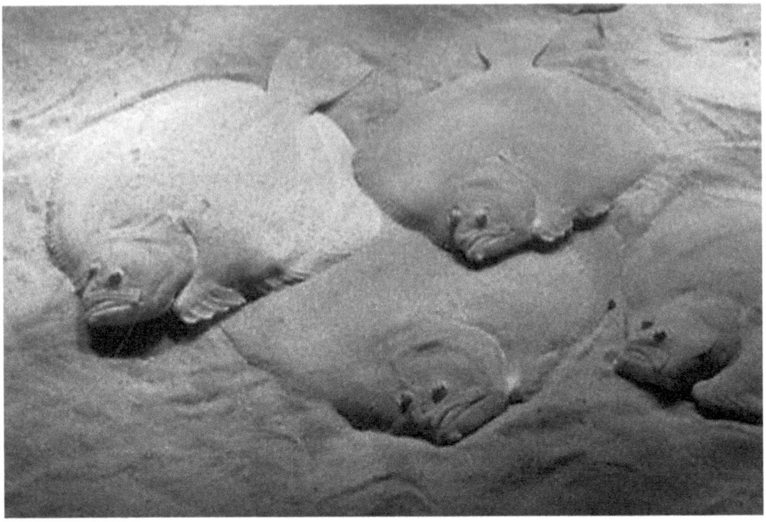

Figure 6.1 The turbot (*S. maximus*). [Aquarium Gijón, Spain (https://acuario.gijon.es/)].

Turbot production: aquaculture

Turbot is a highly appreciated species for human consumption and its aquaculture production has experienced an important increase in the last decade. European production has been estimated in 12,842 tons in 2012, being Galicia (NW Spain) the main producer (> 80% of the EU) due to the optimal growth temperature conditions for growing (http://www.eurofishmagazine.com/news/). Recently, production in PR China has achieved up to 64,000 tons. In Europe, there are three main companies with active breeding programs (Stolt Sea Farm S.A., Insuiña S.L. and Franceturbot). Most turbot production is consumed in the producing countries and the product is generally sold fresh and whole (Bouza et al., 2014).

Broodstock are maintained in square tanks, reproductive males are 2–3 years old (~2 kg weight) and females 4–8 years old (3–6 kg). Males show poor sperm figures as compared to other marine teleosts, but females can produce 5–10 million eggs. Spawning is obtained by hand-stripping all year round and modifying rearing temperatures and day-night rhythms. Embryo development takes 6–7 days at an incubation temperature not exceeding 15°C and turbot larvae measure ~3 mm at hatching. Larval rearing is usually intensive (15–20 larvae/L) in tanks with open-circuit and their survival is generally lower than in other cultured marine fish (20–30%). At the beginning of the exogenous feeding stage (about 3–7 dph), larvae are fed with rotifers, *Artemia* and phytoplankton. Metamorphosis starts 15–25 dph depending on the rearing temperature. After metamorphosis, fish are weaned with dry pellets until they reach 5 g (2–3 months) at a temperature of 16–18°C. At this stage, grading is performed in order to avoid size dispersion and eliminate smaller, deformed, and poorly pigmented individuals. Before transfer to the feeding area, fry are vaccinated against the most common bacterial diseases by immersion (Rodríguez Villanueva and Fernández Souto, 2009; Bouza et al., 2014).

Grow-out can be divided into two stages: pregrowth that ranges from 2–5 g to 60–100 g and growth up to commercial size (1.5–2 kg). Turbot are commonly reared in circular tanks with open-circuit pumped seawater at high stocking density (up to 100 kg/m^2). Turbot culture productivity mainly relies on fry quality, rearing temperatures and on the control of the main pathologies. Another key factor is oxygen concentration, usually between 14 and 16 ppm, which ensures that the saturation level in the outflow water reaches nearly 100%.

Main problems of turbot production

As in most aquaculture species, growth is the main target of turbot breeding programs, since decreasing time to reach commercial size at farm facilities lowers production costs (Gjedrem, 2005). The high sexual growth dimorphism shown by this species has promoted an active investigation on sex determination, ultimately devoted to identify the master sex determinant gene to be applied for producing all-female populations (Martínez et al., 2014). Disease outbreaks, as in most farmed fish, represent the other big challenge for turbot farming. There are some effective vaccines available to control bacterial diseases and a few for parasites, although in most cases, permanent immunity cannot be achieved. Finally, morphological abnormalities, including skin pigmentation, represent a minor but not negligible problem; hence specific routines are implemented at farm facilities to discard abnormal individuals (Bouza et al., 2014).

Breeding programs

Turbot culture began in the 1970s in Scotland and France and has been well established since the 1990s when breeding programs began for this species. Classical breeding programs supported by molecular parentage tools are being carried out by the three main turbot companies, currently in the fifth generation of selection. Different families are usually reared communally in the same tanks at sizes where tagging is not possible, so molecular tools are necessary for parentage assignment (Castro et al., 2004). In addition, sperm cryopreservation is commonly used to facilitate the design of crosses in breeding programs. Growth-related traits have demonstrated medium-high heritability in turbot (Gjerde et al., 1997) and in accordance, a good response to selection has been achieved (growth rate increase per generation: 10–20%).

Genomic approaches have been developed to understand the genetic architecture of the main productive traits in order to provide molecular tools to be applied in marker assisted selection (MAS) programs. Genomic screening with ~100 markers has been applied to identify quantitative trait loci (QTL) related to growth, disease resistance, and sex determination providing a set of genomic regions and molecular markers for MAS (Martínez et al., 2009; Rodríguez-Ramilo et al., 2011, 2013, 2014; Sánchez-Molano et al., 2011). Additionally, the recent assemblage and annotation of the turbot genome has allowed a refined mining of these regions to identify candidate genes and associated markers (Figueras et al., 2016). A set of markers associated with growth rate and sex determination has been evaluated in a large number of families (Martínez et al., 2014; Sciara et al., 2015). This has led to an active MAS program for obtaining all-female populations and to the identification of a set of suitable markers for a pilot MAS program to increase growth. Finally, genomic selection is being evaluated for increasing resistance to *Philasterides dicentrarchi*, currently the main disease threat for turbot industry, in the framework of the FISHBOOST EU project (http://www.fishboost.eu/).

Chromosome set manipulation techniques have been successfully developed in turbot to obtain different products that could be useful for turbot industry (Piferrer et al., 2000, 2004). Triploid turbot are mostly sterile and show a sex bias in favour of females, thus representing an interesting option for turbot companies. Gynogenesis renders both male and female offspring and therefore, they are not useful for obtaining all-female progenies.

Impact of turbot aquaculture on wild populations

The potential escapees of farmed fish to the wild are a matter of increasing concern because natural resources may be affected by genetic introgression from farms. Evaluation of genetic impact is usually a noneasy task because genetic differentiation between domestic and wild populations is not high, especially in those species with young breeding programs. Therefore, genomic screening with thousands of single nucleotide polymorphism (SNP) markers is being applied to identify informative markers capable of detecting escapees from farms and evaluating their genetic impact in the wild within the AQUATRACE EU project (https://aquatrace.eu/). Although not frequent in turbot because it is reared at inland facilities, escapees and even intentional releases have been reported (Bouza et al., 2014).

Genomic resources

Introduction

The application of genomic strategies has increased dramatically as a result of the progressive reduction of sequencing costs and the development of new methodologies for large-scale genetic analysis. Although with some delay with respect to livestock species, aquaculture has also entered into the genomics era. Compared to other vertebrates and many fish species, the turbot genome is small (~600 Mb), one of the smallest genomes among farmed fish (Figueras et al., 2016), and organized in $2n = 44$ chromosomes (Bouza et al., 1994). In turbot, a wide range of genomic resources and tools have been developed (Table 6.1) and applied following different strategies (Table 6.2): (1) genomic and transcriptomic databases essential for identification, annotation, and functional classification of genes; for developing anonymous or gene-linked markers; and for designing oligomicroarrays; (2) genetic maps for tracking key genomic regions associated with productive traits eventually used for MAS strategies; (3) microarrays and large-scale cDNA sequencing (RNA-seq) for gene expression analysis related to the main productive traits; (4) comparative genomics methodologies, very useful for the integration of genomic resources of different species (Bouza et al., 2012); (5) a

Table 6.1 Turbot (*S. maximus*) genomic resources

Genomic resources	Source	Description	References
Whole genome	DNA from a single female	Paired-end (PE), mate pair (MP) and fosmid libraries sequenced using HiSeq2000 and assembled using a hierarchical protocol (219× coverage)	Figueras et al. (2016)
BAC library	DNA of a single highly inbred female	5x genomic BAC library of 46,080 clones with inserts of at least 100 kb	Taboada et al. (2014a)
EST database	Liver, muscle spleen, head kidney, thymus, pyloric caeca, gonads, brain, larvae, gills	ESTs were generated by Sanger and 454 and Illumina sequencing.	Pardo et al. (2008); Ribas et al. (2013); Robledo et al. (2014)
Transcriptome	EST databases and whole genome	Real and in silico information combined to achieve a final transcriptome of 22,751 protein- and 5,808 RNA-coding genes	Figueras et al. (2016)
Microsatellite markers	EST databases and anonymous enriched libraries	433 microsatellites from several projects	Pardo et al. (2007); Bouza et al. (2008); Navajas-Pérez et al. (2012)
AFLP markers	Four gynogenetic families	120 AFLPs; construction of a preliminary linkage map.	Fortes et al. (2008)
SNP markers	EST databases and RAD sequencing	Single nucleotide variation at coding genes and anonymous regions genotyped with Sequenom or Genotyping By Sequencing (GBS) technologies.	Vera et al. (2011, 2013); Robledo et al. (2016); Wang et al. (2015)
Oligomicroarray design	EST databases	Custom 8 × 15; 4 × 44 and 2 × 105 K Agilent Technologies oligomicroarrays, including ~3,000, ~40,000, and ~95,000 oligos.	Millán et al. (2010); Ribas et al. (2013)

(*Continued*)

Table 6.1 **Turbot (*S. maximus*) genomic resources (*cont.*)**

Genomic resources	Source	Description	References
Genetic maps	242 anonymous markers	Consensus map of one F2 and one haploid gynogenetic family. Centromeres located using half-tetrad analysis.	Bouza et al. (2007); Martínez et al. (2008)
	158 anonymous markers	Male and female linkage maps.	Ruan et al. (2010)
	258 anonymous and 180 gene-related markers	Consensus map of an haploid gynogenetic family and eight F2 families with known linkage phase.	Bouza et al. (2008, 2012)
	297 anonymous and 190 gene-related markers	Integration of genetic markers (anonymous and gene-related microsatellites, RAPDs, and SNPs)	Hermida et al. (2013)
	6,647 SNPs	High density consensus map using RAD-seq technologies.	Wang et al. (2015)
	BAC probes	Consolidation of the genetic and cytogenetic maps using FISH with BAC clones.	Taboada et al. (2014a)

Pardo, B.G., Fernández, C., Millán, A., Bouza, C., Vázquez-López, A., Vera, M., Alvarez-Dios, J.A., Calaza, M., Gómez-Tato, A., Vázquez, M., Cabaleiro, S., Magariños, B., Lemos, M.L., Leiro, J.M., Martínez, P., 2008. Expressed sequence tags (ESTs) from immune tissues of turbot (*Scophthalmus maximus*) challenged with pathogens. BMC Vet. Res. 4, 37;
Bouza, C., Hermida, M., Millán, A., Vilas, R., Vera, M., Fernández, C., Pardo, B.G., Martínez, P., 2008. Characterization of EST-derived microsatellites for gene mapping and evolutionary genomics in turbot. Anim. Genet. 39, 666–670;
Fortes, G.G., Nonnis Marzano, F., Bouza, C., Martínez, P., Ajmone-Marsan, P., Gandolfi, G., 2008. Application of amplified fragment length polymorphism markers to assess molecular polymorphims in gynogenetic haploid embryos of turbot (*Scophthalmus maximus*). Aquacult. Res. 39, 41–49;
Millán, A., Gómez-Tato, A., Fernández, C., Pardo, B.G., Álvarez-Dios, J.A., Calaza, M., Bouza, C., Vázquez, M., Cabaleiro, S., Martínez, P., 2010. Design and performance of a turbot (*Scophthalmus maximus*) oligo-microarray based on ESTs from immune tissues. Mar. Biotechnol. 12, 452–465.

Table 6.2 Applications of genomic tools for functional, association, and population studies in turbot (*S. maximus*)

Genomic application	Main features	References
Population genomics	Variation in anonymous and gene linked (GL)-microsatellites and (GL)-SNPs to look for adaptive variation in turbot.	Vilas et al. (2010, 2015)
Microarray gene expression	First high-density immune-enriched turbot oligo-microarray (2,716 oligo-probes).	Millán et al. (2010)
	Expression profiles in spleen, liver, and head kidney after *Aeromonas salmonicida* infection.	Millán et al. (2011)
	Expression profiles in head kidney in resistant and sensitive families to VHS.	Díaz-Rosales et al. (2012)
	Expression profiles in spleen, liver, and head kidney after *Philasterides dicentrarchi* infection.	Pardo et al. (2012)
	Analysis of the inflammatory and immune responses in head kidney turbot leucocytes treated with resveratrol.	Domínguez et al. (2013)
	Analysis of gonad development at several developmental stages	Ribas et al. (2016)
QTL	Sex Determinant (SD)-related QTL screening. Major SD region of turbot at LG 5.	Martínez et al. (2009); Hermida et al. (2013); Wang et al. (2015)
	Growth-related QTL screening for growth traits (body weight and length, condition factor).	Ruan et al. (2010); Sánchez-Molano et al. (2011)
	Resistance and survival QTL screening to *Aeromonas salmonicida.*	Rodríguez-Ramilo et al. (2011)
	Resistance and survival QTL screening to *Philasterides dicentrarchi.*	Rodríguez-Ramilo et al. (2013)
	Resistance and survival QTL to VHSV. Comparative analysis of QTL to the three main bacterial, parasite and viral turbot diseases.	Rodríguez-Ramilo et al. (2014)

Millán, A., Gómez-Tato, A., Fernández, C., Pardo, B.G., Álvarez-Dios, J.A., Calaza, M., Bouza, C., Vázquez, M., Cabaleiro, S., Martínez, P., 2010. Design and performance of a turbot (*Scophthalmus maximus*) oligo-microarray based on ESTs from immune tissues. Mar. Biotechnol. 12, 452–465;
Domínguez, B., Pardo, P.G., Noia, M., Millán, A., Gómez-Tato, A., Martínez, P., Leiro, J., Lamas, J., 2013. Microarray analysis of the inflammatory and immune responses in head kidney turbot leucocytes treated with resveratrol. Int. Immunopharmacol. 15, 588–596.

BAC library of ~45,000 clones of >100 kb inserts, where most turbot genome is physically available for any desired application; and (6) the sequencing and assembly of the whole genome, essential for the integration of all resources and an additional source of information. Each of these resources represents itself a very useful genomic tool, but their integration provides extra information that has been exploited in various applications in turbot aquaculture (Table 6.2).

Transcriptome

Genes are the main functional elements of a genome, and they are responsible, together with environmental factors, for the phenotypic appearance and every single characteristic of an organism. Gene transcription produces RNA transcripts, and the full set of transcripts of an organism is known as "transcriptome." A notable effort has been put forward to sequence the transcriptome of many aquaculture species because it is the basis for any functional research. The transcriptome lets us know the genes of a species and their sequences, which in turn can be used to infer their function if similar genes have been described in other species. Furthermore, the sequence of a gene or transcript is essential to study its regulation and expression. In fact, the tools developed to accurately study gene expression profiles, like microarrays or RNA-seq, rely on the transcriptome sequence and annotation quality. On top of that, for those genes which code protein products, the nucleotide sequence also reveals the protein sequence (Table 6.2).

There are three main types of RNA, messenger RNA (mRNA), ribosomal RNA (rRNA), and transfer RNA (tRNA). Additionally, many other small and long RNAs work in the cells, and their relevance for key cellular functions and gene regulation have been recently disclosed (Morris and Mattick, 2014). However, the mRNA set is the most thoroughly studied. Translation of mRNAs in the ribosome produces proteins, which are critical for every biological process. The turbot mRNA transcriptome contains 22,751 genes, discovered both through transcriptome sequencing and in silico approaches using the genome, which code for 26,823 different protein products (Figueras et al., 2016). These numbers do not match due to alternative splicing, a modification of the primary mRNA transcript, which allows to obtain different mRNAs and the correspondent proteins from combining different exons of a single gene. Through homology searches looking for similar proteins in other species (mainly fish), a total of 23,460 turbot proteins (87.5%) showed some kind of annotation offering clues about their function. Out of these, 19,445 were assigned to a gene description or gene name, clearly identifying them as a putative known gene from other species.

As outlined above, the other components of the transcriptome are the noncoding RNA genes (ncRNA), which include all other RNA types like rRNA, tRNA, microRNA (miRNA), small nuclear RNA (snRNA), small nucleolar RNA (snoRNA), small cytoplasmic RNA (snRNA), and long noncoding RNA (lncRNA). Their roles are diverse but many of them are essential for the correct function of an organism (Morris and Mattick, 2014), such as ribosomal biogenesis, mRNA splicing or gene regulation. In turbot, a total of 5,808 ncRNA genes have been identified through in silico approaches using the genome, including all of the above mentioned types. The number

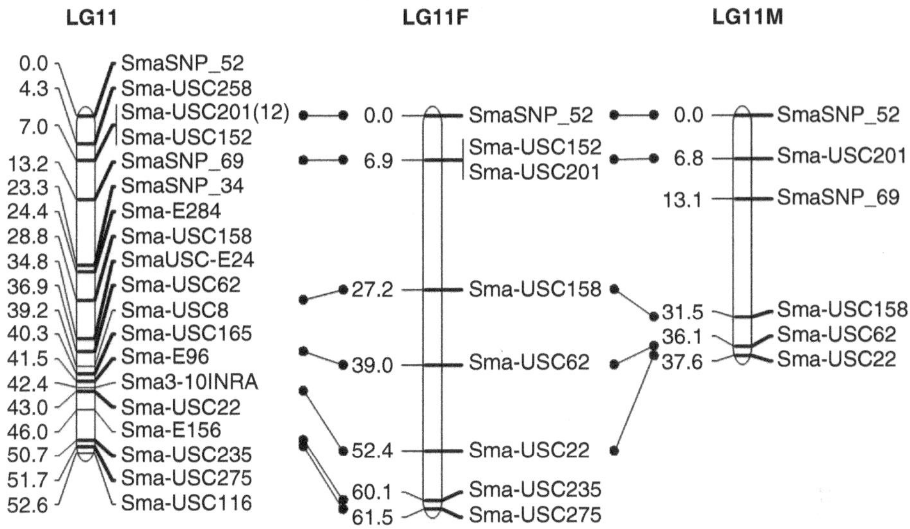

Figure 6.2 Genetic map of LG11 of turbot (*S. maximus*). The maps constructed using male (M) and female (F) segregation, and the consensus of both, are shown. Distances in centimorgans are shown on the left and genetic markers on the right. Notice the smaller size of the male map related to a lower recombination frequency (RF) in males; RF differences between sexes is a common feature of fish.

of ncRNA genes is a matter of dispute even in thoroughly studied species such as both humans and mice. Since the function of many of these ncRNAs is not known, it is also possible that many of them may be nonfunctional. Obviously, ncRNAs are much less characterized in nonmodel species. The latest human estimates for long noncoding RNAs, considered as conservative, suggested the existence of around 10,000 lncRNA genes (Derrien et al., 2012), while in turbot only 12 could be identified (Figueras et al., 2016). There is still a huge knowledge gap on ncRNAs in fish, therefore, discovering new ncRNAs and new kinds of RNA is an interesting research field not only for practical purposes but also to understand the evolution and functionality of many of these RNAs.

Genetic map

The continuous development of genetic markers during the last decades has been essential to foster applications of genetics and genomics to aquaculture. One of the main applications of genetic markers is to provide landmarks for constructing genetic maps (or linkage maps), a roadmap of the genome where the position of markers is established based on the recombination frequencies resulting from crossing-over of homologous chromosomes in the 1st meiotic division (Fig. 6.2). Markers are grouped in linkage groups (LG), whose number would be equal or approach the number of chromosomes of the karyotype depending on the marker density of the map. The markers originally used for mapping were morphological or biochemical polymorphisms (mostly allozymes), but with the advent of DNA recombination technology, DNA-based markers had a dramatic impact on this ground. Microsatellite markers have been extensively used since the 1990s and represent an ideal tool to construct genetic maps of medium density because of their high polymorphism and genotyping accuracy (Iyengar et al., 2000). However, it was not until 2007 that enough microsatellite markers were developed in turbot (Pardo et al., 2007) to build the first genetic map (Bouza et al., 2007), which comprised of 242 anonymous (nonlinked to genes, type II markers) microsatellite markers grouped and ordered in 26 LGs (Table 6.1) (Fig. 6.2).

Conversely, type I markers are located within genes, usually in their noncoding regions (3' and 5' untranslated regions, UTR). These markers additionally provide functional information and, therefore, are particularly useful for identifying candidate genes associated to productive traits. Furthermore, because they show a higher evolutionary conservation than anonymous markers, type I markers constitute the preferred target for comparative mapping studies. Hence, the turbot map was subsequently enriched with 202 type I markers (Vera et al., 2011, 2013; Navajas-Pérez et al., 2012), which included a set of SNPs, the most popular genetic markers used nowadays. SNPs are much more frequent and evenly distributed in the genome (in turbot: 1 SNP every ~100 bp on average; Vera et al., 2011) and the genotyping cost is much lower than that of microsatellites. The new map contained 438 markers, 180 type I and 258 type II, grouped in 24 LGs, thus representing a reduction from the previous 26 ones (Table 6.1).

Meanwhile, another genetic map was published by Ruan et al. (2010), based on 158 new anonymous microsatellite markers. This new map, along with new candidate genes related to immunity and gonad differentiation identified by Rodríguez-Ramilo et al. (2011) and Viñas et al. (2012) were integrated with previous data (Bouza et al., 2007, 2012) in the last turbot consensus map produced by Hermida et al. (2013). The map finally converged into the 22 chromosomes of the karyotype after confirming the merging of two LG pairs (LG8 + LG18 and LG21 + LG24) into single ones as reported by fluorescent in situ hybridization (FISH) using BAC probes (Taboada et al., 2014a). Next generation sequencing (NGS) technologies have provided new tools to identify and genotype thousands of SNPs, thus facilitating the construction of high-density genetic maps. Wang et al. (2015) reported a high-density map in turbot using 6,647 SNPs obtained through restriction site associated DNA (RAD) sequencing linked and ordered into the expected 22 LGs. The reported map showed a high genomic coverage (99.3%) and an average marker interval of 0.4 cM, constituting a powerful platform for QTL identification and physical mapping.

Genetic maps constitute essential organizational tools for genomic research, but one of the most important applications is mapping monogenic traits or QTL associated with productive traits. Turbot maps have been used to identify QTL and to look for candidate genes associated with sex, growth, and resistançe to diseases, as described in the several sections of this book. Also, these genetic maps provided a suitable support for investigating genome organization (location of centromeres; Martínez et al., 2008), and for evolutionary and applied studies through comparative mapping (Bouza et al., 2012). In spite of the mapping progress achieved in turbot, it is important to go further for even more dense genetic maps to enable a finer mining around the markers associated with productive traits to identify responsible genes for a more efficient selection. It is expected that the ongoing EU projects AQUATRACE and FISHBOOST enhance the current genetic maps to integrate all genomic resources in the near future.

Whole genome sequencing

The whole genome sequence of a species is the most ambitious and likely useful tool to understand the genetic basis of productive traits to be applied in breeding programs. Although genome sequencing today does not represent a big problem due to its lowering costs, the bioinformatic work for the assemblage of millions of small pieces and for exploiting the huge amount of information contained in a single genome means a great effort only affordable by large research consortiums. The recently assembled turbot genome is one of the highest quality among aquaculture species (219× sequence coverage; 95% assemblage over estimated size) and it is among the most compact fish genomes (568 Mb) (Figueras et al., 2016). The turbot genome was integrated with previous transcriptomic resources to identify turbot genes (both protein and RNA coding), as well as with mapping resources (Table 6.1) to establish the correspondence between the physical (genome sequence) and the genetic maps. The transcriptome was used to identify orthologous genes with other well annotated fish genomes (>10,000) and paralogous genes (> 1,800) within the turbot genome, which represent raw information for interpreting gene expression profiles and for comparative mapping (Figueras et al., 2016). Orthologous are genes of different species that have derived from the same ancestral gene and usually retain the same function, while paralogous are genes originated by duplication within a single species and can follow functional specialization.

The predictive mapping of orthologous genes enables us to study the evolution of fish genomic organization in the context of the two duplication rounds occurring in vertebrate and fish (Kai et al., 2011). A significant conservation of genomic organization was observed along fish evolution and, within Percomorpha (turbot, stickleback, or tilapia, among others), a nearly 1:1 relationship could be established between chromosomes of different species. For example, the LG1 of turbot is orthologous to the chromosome (CHR)-III of stickleback and to the CHR-17 of medaka based on syntenic conservation of orthologous sequences (Bouza et al., 2012; Figueras et al., 2016). This fact is relevant from an applied perspective since it is possible to "fish" genes from one species with important genomic resources (model fish) for other species of commercial interest. On the other hand, minor genomic expansions related to gene families involved in adaptation to demersal lifestyle of flatfish were also identified in turbot (Figueras et al., 2016).

Finally, the turbot genome was exploited to understand the genetic basis of the three main productive traits (growth, sex determination, and resistance to pathologies) with the aim of identifying candidate genes useful for MAS programs. Previous work had detected specific genomic regions associated with these traits and several QTL and specific genetic markers, explaining a significant proportion of the phenotypic variance of the trait (> 10%) had been reported (Martínez et al., 2009; Ruan et al., 2010;

Rodríguez-Ramilo et al., 2011, 2013, 2014; Sánchez-Molano et al., 2011). Mining for candidate genes ± 1 Mb around the highest associated markers was performed and a large set of strong candidates related to sex differentiation (e.g., β-catenin, aromatase, antimullerian hormone), growth (e.g., insulin growth factor, transforming growth factor, growth hormone) and resistance to pathogens (e.g., complement, toll-like receptor, interleukin) comapped with previously reported QTL. Relevant gene clusters were detected for the three traits (e.g., at LG5 for sex differentiation; LG16 for growth; LG9 for disease resistance), which had been previously detected in other fish and vertebrate species supporting their relevance and trans-specific conservation (Figueras et al., 2016). Some genes like, for example, those involved in the ubiquitin-proteasome system were associated with resistance QTL for different pathogens indicating their generalist component essential to get more robust broodstock. Besides, processes related to growth, immunity and gonadal differentiation overlapped at specific genomic regions suggesting the existence of genetic correlations, either positive or negative, between different traits, an issue of major productive importance.

Genetic architecture and function of productive traits

Growth-related traits

Molecular genomics of growth

Fish growth is under complex polygenic control, environmental influences (e.g., temperature, oxygenation, salinity) and interactions with other physiological processes (e.g., development, metabolism, nutrition, reproduction, stress). Many fish species show indeterminate growth (life long), the muscle representing a large fraction of the body mass and the main edible part of the fish (Mommsen, 2001; Johnston et al., 2011). Different processes are critical for understanding growth variation, such as the regulation of early development, energy metabolism, muscle growth and protein degradation. Myogenesis in teleosts involves the formation (hyperplasia) and enlargement (hypertrophy) of muscle fibers through serial complex events (cell proliferation, differentiation, migration and fusion), the hyperplasia having a major contribution to muscle growth (Johnston et al., 2011; Valente et al., 2013). The advent of genomic technologies has contributed to understand the genetic basis for growth traits in finfish. Genetic maps have facilitated the identification of individual genetic factors affecting growth traits in fish, useful for MAS programs (Gjedrem and Baranski 2009; Yue 2014). Functional genomics, enhanced by NGS technologies, and comparative genomics against vertebrates and model fish have increased our knowledge of the molecular basis and regulatory mechanisms of growth processes in fish, however this is still poorly understood as compared to mammals (Johnston et al., 2011; García de la Serrana et al., 2012). The dissection of the genetic component of this quantitative trait has allowed identifying growth-related candidate genes in livestock and fish, to be used for marker-assisted selection programs (De-Santis and Jerry, 2007). Genes within the somatotropic axis, myogenic regulatory factors, and transforming growth factors have received special attention to identify polymorphisms associated with growth traits, although additional genetic networks (e.g., muscle differentiation, cell cycle, protein degradation, muscle structural proteins, information processing, and signaling, sensory system, different metabolism sub-pathways) have also been considered (De-Santis and Jerry, 2007; Valente et al., 2013; Figueras et al., 2016).

In turbot, an important set of QTL, candidate genes and associated markers related to growth traits have been detected by functional, structural, and comparative genomics. Model fish genomes have been a very useful resource for this task. Interestingly, population genomics has suggested a relationship between growth performance and adaptive variation in wild fisheries of turbot. Baltic populations, with a slower growth rate than North Sea populations, showed significant genetic divergence for markers associated to growth-related QTL (Vilas et al., 2010, 2015). NGS transcriptome projects under nutritional stress are being developed to identify growth-related genes and SNP markers, in order to explore their colocalization with QTL for growth traits, and to carry out association studies for growth phenotypes at family and population levels (Robledo et al., 2016).

QTL screening: growth-related QTL

QTL methodology

A QTL is a genomic region responsible for the variation of a quantitative trait. Several statistical approaches have been developed to detect such regions by establishing associations between genetic markers and the phenotypes of the trait. These approaches range from simple statistical tests to models that include multiple markers and interactions. Single-marker tests (e.g., t-test, ANOVA, and simple linear regression statistics) evaluate the statistical association of phenotypes relative to marker genotypes (or alleles). This enables to identify which markers among those used for genomic screening are associated with the quantitative trait of interest and, thus, to detect the existence of a QTL. Single-marker tests use markers independently of each other, without any reference to their position or order within the genome. Additional information is gained when markers are placed in a genetic map. In this scenario: (1) higher power to detect QTL is acquired; (2) the relationship between QTL for different quantitative traits can be established; and (3) candidate genes may also be identified if the genome is available or through comparative mapping with model fish genomes. The interval mapping approach uses a genetic map as framework for QTL detection. The use of pairs of flanking markers for interval mapping provides extra power for detecting a QTL, estimating its location and knowing its weight on trait variance. Different methods have been proposed for interval mapping including those based on maximum likelihood (e.g., QTLMap software) and least squares (e.g., GridQTL software). Both methodologies have shown similar power for QTL detection (Haley and Knott, 1992). Interval mapping is undoubtedly more powerful than single-marker approaches for QTL detection and statistical approaches for locating multiple QTL (e.g., MapQTL 5.0 software) are more powerful than single QTL approaches. These methods include models where a specific QTL can act independently, be linked to another QTL, or interact epistatically with other QTL.

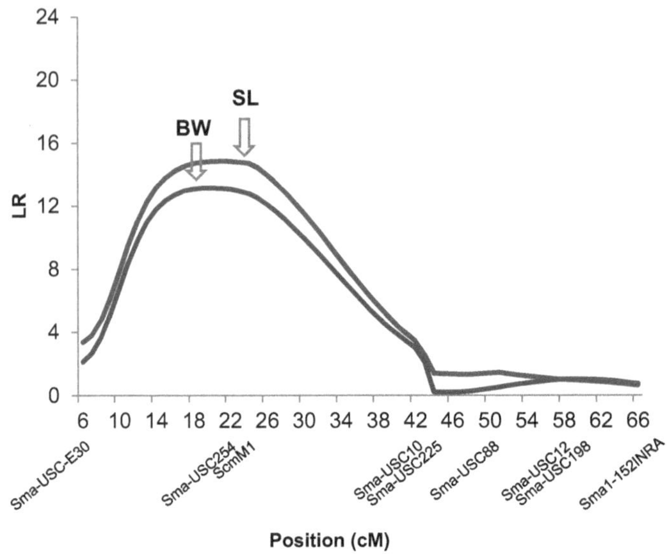

Figure 6.3 Growth-related (*SL*, standard length; *BW*, body weight) QTL at LG5 in turbot (*S. maximus*). In abscises genetic markers and in ordinates the degree of statistical association. Arrows show the points with highest association.

Growth-related QTL

The heritability for growth-related traits is usually from moderate to high. Thus, selection for these traits allows a rather quick rate of genetic gain. Increasing growth rate reduces production costs because it decreases the rearing time at farm facilities, and thus, increases revenues. Consequently, growth-related traits constitute one of the main goals of turbot breeding programs. Classical breeding programs based on best linear unbiased predictor (BLUP) and pedigree information have been applied in the breeding programs of the three main turbot companies with satisfactory results (10–20% growth increase per generation). However, the long generation interval of turbot delays improvement using only BLUP. MAS, using highly informative markers, could accelerate genetic gain by enhancing the prediction of the breeding values and reducing the age for evaluating the candidates for selection (Fig. 6.3).

Different studies have identified growth-related QTL in turbot providing a useful starting point to validate genetic markers for MAS programs. Sánchez-Molano et al. (2011) using ~100 microsatellite markers homogeneously distributed across the genetic map and covering all LGs detected five significant QTL in different LGs for body weight (BW) and six for standard length (SL). Four of these QTL showed association with both traits reflecting a common genetic basis, which would explain the high correlation observed between BW and SL (Fig. 6.3). Additionally, four QTL were also identified for Fulton's condition factor (FK), one of them explaining up to 25.8% of the phenotypic variance. FK is a measure of a fish fatness computed as $100 \times BW \times SL^{-3}$, where BW is measured in grams and SL in centimeters. Other authors using a similar number of markers from a different genetic map identified a SL-QTL explaining 12.4% of the phenotypic variation at a similar genomic position as in the previous study, thus confirming its consistency (Ruan et al., 2010).

An important task after identifying a set of QTL and associated markers for the traits of interest is to validate them, or in other words, to demonstrate that they are not false positives consequence of the type I error associated to statistical inference. In turbot, Sciara et al. (2015) reanalyzed the results from Sánchez-Molano et al. (2011) in a large panel of 19 families using the 39 strongest candidate markers from 11 LGs. They could validate 17 of them because they showed very high statistical association in a single family or a lower association in several families. Furthermore, these authors showed that 20 markers among the 39 evaluated jointly explained 47.5% of the phenotypic variance for BW and 18 markers explained up to 57.7% of the phenotypic variance for SL. The explained phenotypic variance when considering alleles, instead of genotypes, was higher (87.3% for BW, 89.1% for SL, and 11.4% for FK). This useful set of markers is now available to be applied in a pilot MAS program for improving growth and go beyond the BLUP approach performed up to date.

A new high density genetic linkage map for turbot has been recently reported (Wang et al., 2015). These authors used the RAD approach for identifying and genotyping thousands of SNPs. RAD takes advantage of the digestion fidelity of restriction enzymes for concentrating the sequencing reads at a reduced portion of the genome. This map was applied to identify a total of 14 growth-related QTL involving 11 LGs. The explained phenotypic variance for these QTL associated to SL and BW ranged between 14.4% and 57.8%. The results from the above studies suggest that several markers from different LGs should be used to efficiently increase the genetic gain for growth-related traits in turbot breeding programs.

Identification of candidate genes through comparative mapping

Syntenies are homologous genomic regions between two or more species derived from the same single ancestral region and conserved along evolution. These regions are identified through comparative evaluation of the sequences of genetic markers anchored to genetic maps. The broad conservation of homologous genomic regions across fish species has made feasible the transfer of genomic information from the well-studied model fish to farmed species (Bouza et al., 2012), facilitating the identification of candidate genes underlying QTL regions (Li et al., 2011). In turbot, comparative mapping has been useful to identify homologous sequences associated with productive traits using model Acanthopterygii species, such as *Tetraodon nigroviridis*, medaka (*Oryzias latipes*) and stickleback (*Gasterosteus aculateus*), the last one showing the highest syntenic conservation with turbot (Bouza et al., 2012; Figueras et al., 2016). Accordingly, the application of data mining tools enabled to explore the gene content within syntenic blocks surrounding turbot QTL in model fish genomes. Relevant growth-related genes have been located, such as *igf1r* within a FK-QTL or *fgfr2* associated with BW- and SL-QTL, which represent candidate genes for explaining growth variation in this species, as reported in other fish and vertebrates (Sánchez-Molano et al., 2011). Genome mining (Biomart tool; Ensembl) has also provided predictive gene information as, for example, in the BW-QTL at LG16 (Fig. 6.4), where

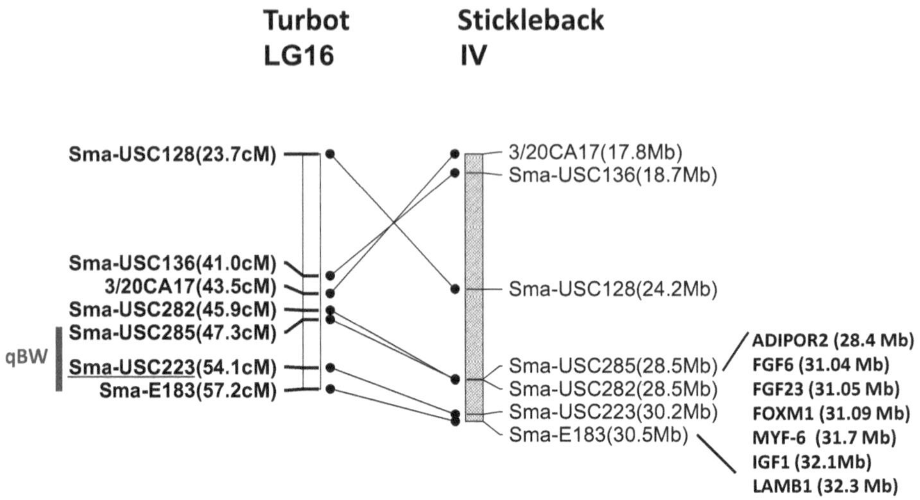

Figure 6.4 Comparative mapping between the turbot LG16 and stickleback chromosome IV to identify candidate genes underlying a BW-QTL interval using data mining tools (Ensembl.org/BioMart). Underlined is the QTL associated marker within the syntenic block between both species (Sánchez-Molano et al., 2011).

several growth-related genes were extracted from the collinear microsyntenic region identified against the stickleback CHR-IV, Tetraodon CHR-19, and medaka CHR-23 (Sánchez-Molano et al., 2011; Bouza et al., 2012). Many of these genes are involved in growth regulation and cell proliferation, and have been associated with growth-related QTL in other fish species (e.g., *lamb1*; *fgf6*; *igf1*; *myf-6*; Johnston et al., 2011; Sánchez-Molano et al., 2011; Tsai et al., 2014). Testing their association with growth across families and populations will be essential to use them as candidates for MAS in aquaculture industry, but also to obtain a functional explanation of growth variation (Taboada et al., 2014b) (Fig. 6.4).

Genomics of the resistance of turbot to the main industrial diseases

Main diseases in turbot farms

Several pathogens, including bacteria, viruses, and parasites, threaten the health status of farmed fish causing important economic losses due to morbidity or mortality episodes. The fast development of turbot aquaculture has also been associated with a concomitant increase in pathologies. Although nowadays there are effective treatments or vaccines against a variety of turbot pathogens, other diseases, especially those produced by endoparasites, do not have a straightforward solution.

Several bacterial pathogens like *Tenacibaculum maritimum* (tenacibaculosis) and two gram negative bacteria, *Vibrio anguillarum* (vibriosis) and *Aeromonas salmonicida* (*subsp. salmonicida*) (furunculosis), currently represent an important threat for the industry. Successful vaccines have been developed for tenacibaculosis and vibriosis, although sometimes the complementary use of antibiotics is necessary (Avendaño-Herrera et al., 2006). In contrast, a successful vaccine against furunculosis is not available for turbot yet, and several epizootic outbreaks of this disease have been reported in turbot farms (Lillehaug et al., 2003).

Parasitic diseases represent the main threat for turbot industry and can cause up to 100% mortality rates. Although ectoparasites are usually easily treated with freshwater or formalin baths, no definitive solutions have been developed for endoparasites. *Philasterides dicentrarchi* (scuticociliatosis) has been reported as the causative agent of several mortality episodes in farmed turbot (100% in many cases; Iglesias et al., 2001), and although encouraging results have been achieved with a specific vaccine (Palenzuela et al., 2009), the high variability among parasite strains has prevented a general protection. In fact, this species has been targeted by the industry to evaluate genomic selection within the FISHBOOST EU project to achieve a more resistant or tolerant broodstock. Finally, the absence of an effective treatment for the parasite *Enteromyxum scophthalmi* (enteromyxosis) represents a big challenge since mortalities usually reach 100% (Robledo et al., 2014; Ronza et al., 2016).

Among viruses, only sporadic cases of nodavirus infection (viral encephalophaty and retinopathy) with high mortality rates have been reported in turbot (Barja, 2004). The susceptibility of flatfish to nodavirus is not negligible as demonstrated in experimental infections (Montes et al., 2010). Occasional infections have also been reported with other viruses like infectious pancreatic necrosis virus (IPNV), which causes variable mortality rates depending on the serotype (Barja, 2004), and with viral haemorrhagic septicaemia virus (VHSV; Ross et al., 1994), which is a rhabdovirus included within the OIE list of reportable diseases. Although the mortality rate in natural infection cases is relatively low, when an outbreak is detected in a fish farm all the stock must be sacrificed as part of a contingency plan.

High-throughput gene expression: microarrays and RNAseq

The study of gene expression profiles is indispensable to understand host-pathogen interaction, critical to devise successful disease prevention strategies and for achieving more robust broodstock. While expression profiles can be accurately addressed by qPCR (quantitative PCR), only a low number of genes can be analyzed, which greatly limits understanding the complex immune response. Microarray technology enabled to analyze the expression profiles of thousands of genes by immobilizing DNA probes on appropriate surfaces, which facilitate the specific hybridization with transcripts of a particular tissue. The fluorescence levels of the labeled targets are then correlated with the amount of transcripts to estimate the expression gene profiles (Fig. 6.5). Microarrays evolved from the whole cDNA probes up to the more sophisticated and accurate oligo-microarrays, where each gene or exon is represented by a short oligoprobe (20–60 bp), also achieving a higher spot density. However, a more accurate and informative methodology emerged with the lowering costs of NGS technologies. The RNA-seq methodology estimates gene expression by correlating the number of reads (~50–150 bp

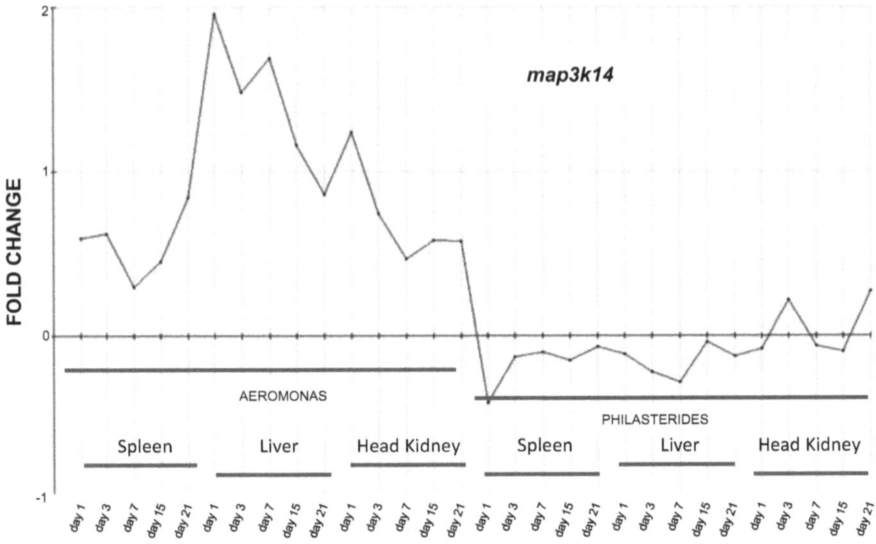

Figure 6.5 Expression profile of the *map3k14* gene from the turbot (*S. maximus*) 8 × 15 K oligomicroarray along a time series (1–21 days) in three immune-related organs in response to intraperitoneal challenges with the bacteria *A. salmonicida* and the parasite *P. dicentrarchi*, responsible for furunculosis and scuticociliatosis, respectively. This gene encodes a mitogen-activated protein, which is critical for the alternative NF-κB activation pathway, thus playing a critical role in the immune response. Notice the very different profiles of this gene in response to both pathogens, being up-regulated mainly in liver just after *A. salmonicida* challenge.

sequences) of each transcript with its expression, so avoiding the problems in the lowest and highest expression range associated with the microarray fluorescence-based technology. Additionally, since RNA-seq is based on sequencing, it provides information on annotation and genetic variants for each gene. Although the cost of RNA-seq is still higher than microarrays, the difference is narrowing, and therefore, RNA-seq will likely replace microarrays, especially for those species with a low level of genomic resources (Fig. 6.5).

Functional genomics of the response to the main turbot diseases

Initial microarray-based studies were performed in turbot to analyze gene expression profiles in response to the bacteria *Aeromonas salmonicida* (AS; Millán et al., 2011) and the parasite *Philasterides dicentrarchi* (PD; Pardo et al., 2012). A custom oligomicroarray, which included 2,176 different gene probes designed from an immune-enriched expressed sequence tag (EST) database was used for this analysis. The aim was to identify genes, functional categories, and gene pathways regulated in response to each pathogen that could aid to understand the specific turbot immune response and also to identify general mechanisms of response against different pathogens. A different approach was carried out later to evaluate the response to the VHS virus using the same microarray in turbot families with different resistance to the disease (Díaz-Rosales et al., 2012). In this work, the interest was focused on identifying candidate genes that may explain the differential response of families against the virus, a highly interesting issue for achieving a more resistant broodstock. Finally, the more powerful RNA-seq technique has been applied to understand the causes underlying the high mortality of turbot to enteromyxosis (Robledo et al., 2014; Ronza et al., 2016).

Furunculosis versus scuticociliatosis

Gene expression profiles in response to AS and PD were evaluated in several immune-related organs (spleen, head kidney, and liver) in a time course study along the infection process. A stronger response to the bacterial than to the parasite disease was detected as reflected by the higher number of differentially expressed (DE) genes 471 (17.3% of the whole microarray) versus 221 genes (8.1%), respectively. Although highly specific DE gene sets were detected at each organ, the functions involved both in up- and down-regulated genes showed notable similarities between organs in both studies, suggesting a coordinated plan of response in the whole organism to rather different pathogens. Response against AS, although decreasing over time, showed an important amount of DE genes until the end of the experiment, while in the case of PD, an intense reactivation of the response was detected at the end according to head kidney and liver profiles suggesting a period of latency along the infection process followed by a reactivation.

Up-regulated genes in response to AS and PD were mainly associated to immunity, indicating an activation of the immune response in infected fishes, while transport and, especially metabolic processes, were by large the main categories among down-regulated genes. This may be related to energy saving associated to the lower activity and food intake of infected fish. Many DE genes and gene families were associated to essential innate and acquired immunity functions including acute-phase response, inflammation, cell recruitment and differentiation, antigen identification and presentation, cytotoxicity, and agglutination and precipitation, among others. Also, relevant DE immune genes were detected including chemokines, chemotaxins, complement, immunoglobulins, major histocompatibility complex, interferon, and lectins, among others. These genes were later located in the turbot genome to ascertain their position in the genetic map with regard to reported QTL (Figueras et al., 2016). This approach has led to the identification of more consistent candidates that could be involved in the differential response of individuals to a specific disease.

Many DE genes identified in response to AS and PD had also been detected in response to other pathogens such as nodavirus in turbot (Park et al., 2009), suggesting a general role on immunity. On the other hand, other DE genes or families had been previously reported as regulated in response to these pathogens in different species, supporting the existence of common routes of response among species. It has been suggested that PD activates antioxidative enzymes to prevent oxidative damage mediated by host phagocytes (Piazzon et al., 2011). Pardo et al. (2012) detected a significant up-regulation of glutathione peroxidase in turbot liver after PD infection, not observed in AS response (Millán et al., 2011), which could reflect higher oxidative activity to counterbalance parasite defenses.

VHSV

Four turbot families with differential survival rates to VHSV infection were analyzed to find candidate genes that may explain the observed between-family differential response (Díaz-Rosales et al., 2012). DE relevant immune genes were identified in head kidney before and after VHSV challenge depending on the susceptibility of the families. Resistant families showed a higher expression of two antimicrobial peptides (Nk-lysin and "antimicrobial peptide precursor") before the infection, suggesting that their preimmune state may be stronger than that of fish from susceptible families. Nk-lysin is mainly produced by cytotoxic T lymphocytes and natural killer cells and stored in cytolytic granules, but its role in viral clearance is not yet fully understood. Nevertheless, recent studies have revealed that higher levels of Nk-lysins are positively related to higher resistance to VHSV (Pereiro et al., unpublished data). The "antimicrobial peptide precursor" was later characterized as a novel turbot hepcidin (hepcidin-2) (Pereiro et al., 2012a). Hepcidins are antimicrobial peptides, which also play a central role in the iron metabolism. Hepcidin-2 was DE after pathogen administration but not after iron overload, suggesting a more specialized role in the immune defense than hepcidin-1.

Considering the poor information available on turbot antiviral sequences in public databases, high-throughput transcriptome sequencing of cDNA from several organs at different times poststimulation with VHSV and substances mimicking viral infections, was recently conducted (Pereiro et al., 2012b). This approach led to the identification of a multitude of immune genes in turbot, which were used for the construction of a more powerful microarray (~40,000 probes). This microarray was used to analyze the transcriptomic profiles after the administration of a highly effective DNA vaccine (encoding the VHSV G glycoprotein) to turbot and to compare the response of vaccinated versus nonvaccinated fish (Pereiro et al., 2014a). The results revealed that the vaccine is able to induce a very similar response to that induced by the virus, and moreover, a broadly different profile was observed between vaccinated and nonvaccinated fish after the infection. A huge amount of DE immune-related genes were found after vaccination, as reflected by the hierarchical analysis of the main immune pathways (toll-like receptor cascade, interferon-related molecules, apoptosis, MHC-I antigen presentation process, coagulation, and complement cascades), as well as gene markers of activation and proliferation of different cell types. For example, two different type I interferons (IFNs), the main responsible molecules orchestrating the antiviral response, were differentially affected by the infection; while one of them was up-regulated after VHSV challenge, the other one was down-regulated. A refined analysis was carried out on these IFNs showing that *ifn1* response was the standard of type I IFN (induction of IFN-stimulated genes and protection against VHSV), whereas *ifn2* was more implicated in immunomodulatory functions mainly with a local effect and without any relevant protective effect against the virus (Pereiro et al., 2014b). Therefore, *ifn1* could be a good antiviral treatment in case of infection and represents an interesting candidate for MAS.

Enteromyxosis

Recently, the powerful RNA-seq methodology was used to evaluate turbot response to enteromyxosis (Robledo et al., 2014; Ronza et al., 2016). Spleen, head kidney and pyloric caeca response was evaluated in light and heavily infected fish classified according to their histological lesion level. Around 20% DE genes were detected in the heavily infected fish while only ~4% in the lightly infected ones. The most interesting DE genes were categorized in: (1) immune and defence response; (2) apoptosis and cell proliferation; (3) iron and erythropoiesis; (4) cytoskeleton and extracellular matrix; and (5) metabolism and digestive function. A putative connection failure was observed between the innate and adaptive immune responses in the heavily infected fish, which could explain the devastating effects of this parasite. Furthermore, the exacerbated local immune response at the intestinal level contrasted with the down-regulation of many immune-related genes in kidney and spleen genes, likely related to the huge lymphohematopoietic depletion in these organs.

QTL screening: disease resistance related QTL

One of the main objectives of aquaculture breeding programs is to achieve more robust broodstock that can face the main diseases currently affecting production and those new ones that may emerge as a result of high density rearing conditions at farms. Disease resistance traits are difficult to manage because phenotypes for disease-related traits cannot be directly measured in the candidates for selection and decisions have to be taken on information from siblings of the candidates. Family-based breeding programs have been implemented in several aquaculture species and in this scenario only inter-family genetic variation is exploited. However, intra-family genetic variation could be used if markers and allelic variants associated to quantitative traits could be identified. On the other hand, disease resistance usually shows low heritability, although notable exceptions have been reported in aquaculture species (Gjedrem, 2005). MAS or genomic selection programs are especially recommended for low heritability traits like these, because response to selection can be notably enhanced using these strategies. QTL mapping for disease resistance traits in aquaculture has been the focus of several studies and the characterization of disease resistance QTL can reveal whether the disease is controlled by one or few loci with large effect or many loci with smaller effect.

In turbot, QTL screening for resistance and survival time to the bacteria *A. salmonicida* (furunculosis), the parasite *P. dicentrarchi* (scuticociliatosis) and the VHS virus has been carried out (Rodríguez-Ramilo et al., 2011, 2013, 2014). Eleven full-sib families of ~100 individuals each were genotyped for ~100 microsatellite markers and analyzed with the same consensus genetic map. Some QTL showed association with resistance or survival to the three pathogens, suggesting genomic regions involved in general immunity, highly relevant for MAS programs. Other genomic regions shared common QTL for furunculosis and scuticociliatosis, and for scuticociliatosis and VHS. Individual QTL were also detected for resistance to furunculosis, scuticociliatosis and VHS explaining up to 16.8%, 22.3%, and 14.0% of the phenotypic variance, respectively. It has been suggested that increasing tolerance to one pathogen could be associated with sensitivity to others, however, no conclusive data has been reported to date on this regard, and thus, pathogen-specific QTL should be considered for MAS programs.

The Japanese flounder (*Paralichthys olivaceus*) is another flatfish where disease resistance QTL have been reported. A major locus for resistance to lymphocystis disease virus explaining up to 50% of the phenotypic variation was detected by genomic screening with 50 microsatellite markers (Fuji et al., 2006). Also, Wang et al. (2014) using 170 microsatellite markers identified a QTL related to resistance to *Vibrio anguillarum* in Japanese flounder explaining more than 60% of the phenotypic variance. The high weight on the phenotypic variance of the traits of QTL reported in Japanese flounder strongly suggest their localization in the turbot genome through comparative mapping to identify potential candidates for selection.

Crossing information from functional and association studies

Functional evaluation of turbot response to the main diseases has provided a set of candidate genes, while association studies have provided a set of markers and QTL for the same pathogens, all of them located in the turbot genome (Figueras et al., 2016). Crossing both information sources has enabled to identify strongly DE genes from functional studies located on resistance QTL for the same pathology. In this way, viral defense and clearance related genes were detected within VHSV-QTL (e.g., genes implicated in T-cell proliferation, differentiation, maturation or activation; genes involved in the blood coagulation cascade; or genes related to iron homeostasis and scavenging); typical antibacterial and bacterial recognition genes were found around AS-QTL (e.g., peptidoglycan recognition, lysozyme and macrophage mannose receptor); and important immune genes were also detected at PD-QTL (e.g., interleukin-, galectin-, perforin- and TLR-related genes). An important task to be done in the near future is to develop markers for these candidates and confirm their association in a large set of families for their application in gene assisted selection (GAS) programs.

Reproduction and sex control: general revision, QTL, functional genomics, and ongoing research

Sex differentiation in fish

Fish represent the most diverse group of vertebrates and show all reproduction systems known in this group. Understanding reproduction is key for aquaculture, and several reproduction-related traits such as growth sex dimorphism, sex ratio, and puberty control are essential to improve production.

The sex of an individual is established by two developmental processes: sex determination (SD) and gonad differentiation (GD). SD establishes the gender of the fish at the beginning of development and depends upon the action of a single gene, several loci or environmental factors or a combination of them. The two extreme SD systems have been traditionally classified as genetic sex determination (GSD) and environmental sex determination (ESD). Among them, a wide range of mechanisms exists with variable influence upon both components. On the other hand, GD is the developmental process responsible for the formation of an ovary or a testis once the gonadal fate has been established in the primordium (Martínez et al., 2014).

GSD has been widely explored in fish to identify the "master" SD gene (SDg), following previous models from birds and mammals where *dmrt1* and *sry* are the genes that trigger the process of sex differentiation, respectively. Up to six different SDg have been discovered in fish (Martínez et al., 2014) and, additionally, several nonhomologous SD regions have been reported through association analysis or QTL screening in different fish species (Martínez et al., 2009). Thus, unlike mammals and birds, fish show a huge variety of GSD mechanisms, as a result of a high evolutionary turnover in such a way that species of the same genus as, for example, different tilapias or pufferfish show different SD mechanisms. Furthermore in most fish species, in addition to genetic factors, environmental cues may play an important role, with temperature playing a central role (Baroiller et al., 2009).

SD and GD were traditionally considered two independent consecutive developmental processes influenced by observations in mammals and birds. However, nowadays sex is seen as a threshold-like character in fish, where the development of a male or a female depends on the balance between several genetic and environmental cues during early development, where SD and GD overlap (Heule et al., 2014). According to this information, SD is considered a quantitative threshold trait and as such, heritability has been estimated and QTL screening performed (Martínez et al., 2014).

Sexual dimorphism for growth is commonly found in cultured fish, and, for example, in turbot and sea bass, two of the main EU aquaculture species, females largely outgrow males. Since turbot shows a GSD system (ZZ/ZW) and higher female growth rate, obtaining all-female populations has been addressed to satisfy the demands of turbot industry (see later). Conversely, European sea bass shows a polygenic SD system with important temperature influence, and in this case, a specific larval rearing temperature protocol has been established to increase the proportion of females. The wide variety of SD systems in fish makes it necessary to understand this process at a species level, in order to control sex ratio in those species with sex dimorphism for productive traits (Martínez et al., 2014).

Genetic architecture of sex determination in turbot

Available data support that turbot SD is based on a ZZ/ZW system of recent evolutionary origin, and with a minor influence of other genetic and environmental factors. This information has been gathered over several years following different approaches. Cytogenetic analysis on mitotic and meiotic chromosomes showed that, in accordance to most fish species, no sex-associated heteromorphism exists in the turbot karyotype (Bouza et al., 1994; Cuñado et al., 2002). Analysis of gynogenetic progenies showed the presence of both males and females discarding a XX/XY system (Martínez et al., 2009). The evaluation of progenies of sex-reversed parents suggested for the first time a ZZ/ZW mechanism, since nearly 100% male progenies were obtained in crosses between normal (ZZ) and reversed-males (ZZ). Slight deviations from this model in some families indicated that minor factors and some temperature influence may also be

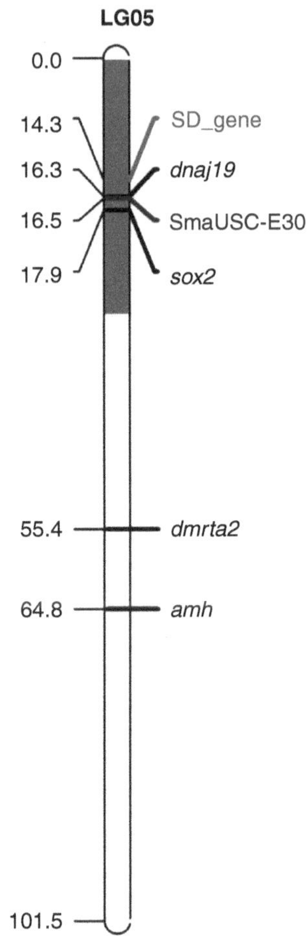

Figure 6.6 Representation of turbot LG5 containing the major SD region. The estimated position of the SD gene and the highest associated sex marker (Sma-USC30) are shown. Genetic distance in centimorgans on the left.

involved (Haffray et al., 2009). As described in other fish, SD behaves like a complex trait in turbot (Martínez et al., 2014). Important progress has been achieved by QTL screening (Martínez et al., 2009; Hermida et al., 2013) and a major QTL was detected at LG5 explaining more than 85% of the phenotypic variance of the trait (Fig. 6.6), but also three minor QTL were identified in other LGs. The previously outlined high-density genetic linkage map by Wang et al. (2015) was also used to detect SD-QTL. A main SD region was also identified, although the lack of correspondence between both maps does not allow us to know if the genomic region is the same in both studies.

A microsatellite marker at the SD region (SmaUSC-E30) allowed the correct sexing of 98.4% of the offspring in most families (Fig. 6.6). This marker was used to develop a molecular tool for precocious sex identification in turbot, applied to obtain all-female populations by industry (patent number: 2354343; Martínez et al., 2009). Recently, this marker was revealed to be part of the *fxr1* gene, related to female gametogenesis in pig (Yang et al., 2012). The position of the SDg was estimated very close to this marker (2.6 cM or 1.2 Mb) and six candidate genes in that region, including two strong candidates (*sox2* and *dnajc19*), were mapped and investigated for their association with sex. None of them showed association with sex at the population level and were discarded as SDg (Taboada et al., 2014b), which highlights the difficulty to identify genes underlying QTL variation, even for those which explain most variance of the trait. A complementary approach to address this issue was reported by Viñas et al. (2012). In this study, several highly relevant candidates were mapped, and although some of them were located within minor SD-QTL, for example, the brain aromatase, *cyp19a1b*, and two genes of the sox family related to development and specifically to SD, *sox9* and *sox17*, none were identified at LG5. This study also enabled to discard genes previously proposed as SDg candidates, such as *amh* and *dmrta2*, which mapped at LG5 but far from the main SD-QTL. The availability of the turbot genome has allowed mining around the SD-QTL and new candidate genes were detected for further evaluation (Figueras et al., 2016) (Fig. 6.6).

Two ongoing projects are providing additional information on this issue. In the EU FISHBOOST project, 30 families have been sexed and genotyped for ~25,000 SNPs providing a dense genomic screening to identify sex-associated markers at species level to be mined in the turbot genome. A second approach is focusing on target sequencing by comparing the 1.2 Mb region including the SDg in ZZ and WW individuals.

Gonad development and differentiation in turbot

As outlined previously, sex is now considered a quantitative threshold character, where the balance among different genetic and environmental factors at the beginning of gonad differentiation determines the development of a testis or an ovary (Martínez et al., 2014). The existence of several SD-QTL in turbot is a reflection of this complex condition and at least four genes or genomic elements have a detectable effect on the male/female balance. To fully understand sex determination, it is necessary to comprehend the sex differentiation process, especially at the initial stages of gonad differentiation. Understanding sex determination and gonad differentiation should aid the turbot industry to control sex proportions and sex maturation (Fig. 6.7).

The study of sex differentiation throughout turbot development began recently, mainly through the study of gene expression by real-time PCR and microarrays, although a preliminary identification of DE genes between sexes was previously carried out using the cDNA-AFLP technique in gonad, liver and brain (Taboada et al., 2012). The implementation of NGS strategies has played an essential role for obtaining genomic resources in model and nonmodel animals, such as turbot, where Ribas et al. (2013) sequenced brain and gonad transcripts at different development stages. A total of 1,410 sequences related to reproduction and key genes involved in sex differentiation and maturation were identified for the first time, including androgen receptor (*ar*), antimüllerian hormone (*amh*), gonadal aromatase (*cyp19a1a*), follicle stimulating hormone receptor (*fshr*) and several *sox*-related genes, among others (Fig. 6.7).

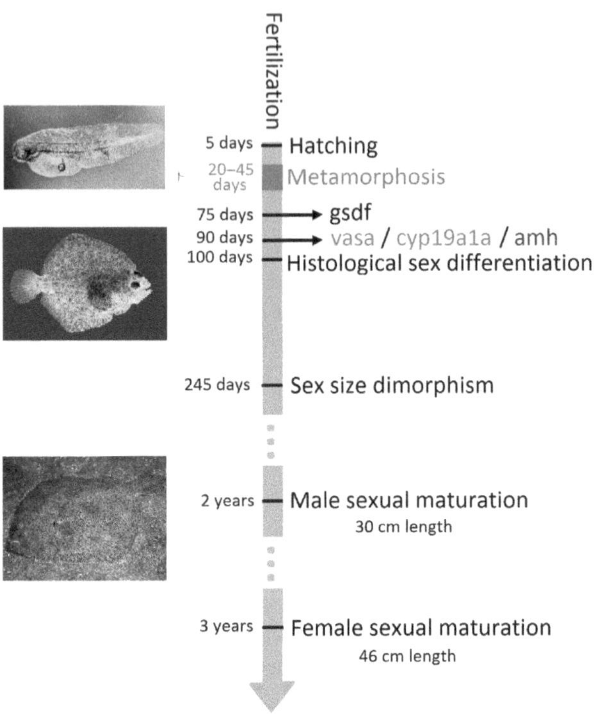

Figure 6.7 Main genetic, histological, and physiological events along sexual differentiation in turbot (*S. maximus*)

These authors reported the most representative (50–90% coverage) reproduction-related pathways in the brain-hypophysis-gonadal axis including oocyte meiosis and progesterone-mediated oocyte maturation. These genomic resources were the basis for the design of a reproduction and immune enriched oligomicroarray for high-throughput gene expression evaluation (Ribas et al., 2016), but also for the evaluation of gene expression on a large set of candidates along gonad development by qPCR (Robledo et al., 2015).

The first genetic signs of gonad differentiation in turbot were detected at 75 dpf related to the sex-independent expression of the gonadal somatic cell derived factor (*gsdf1*), a gene related to the proliferation of primordial germ cells in rainbow trout (Sawatari et al., 2007) (Fig. 6.7). Primordial germ cells seem to play a critical role in sex determination in some fish species (Siegfried and Nüsslein-Volhard, 2008), where the difference in the amount of germ cells between sexes is among the first detectable sign of gonad differentiation. The higher expression of *vasa*, a specific germ cell marker, in females at 90 dpf in turbot suggests a higher amount of primordial cells in females at the initial stages of development. Additionally, the DE of *cyp19a1a* in females, responsible for the synthesis of estrogens, and *amh* in males, which inhibits the formation of the Müllerian ducts in higher vertebrates, constitutes, with *vasa*, the first genetic cues of sex differentiation in turbot (Robledo et al., 2015).

After 90 dpf, which can be considered the onset of gonad differentiation, many genes become differentially expressed between sexes. Gene expression profiles revealed that in turbot females sex differentiation starts sooner than in males (~5 vs ~10 cm length). Further, global transcriptome analysis revealed that while male gonads remain rather similar to undifferentiated gonads, female gonad differentiation occurs earlier and produces deep changes involving the regulation of a large numbers of genes. However, when juvenile testis and ovaries were compared, more DE up-regulated genes were detected in males, as previously documented in other fish (Tao et al., 2013). Interestingly, different gonad development stages were observed for individuals of the same sex and comparable age and size, suggesting a genetic component underlying the rhythm of sex differentiation. If so, it may be possible to delay sex maturation in turbot, as in the salmon industry, to avoid the undesired effects of maturation (stopping growth, disease resistance, or flesh quality).

Concluding remarks and perspectives

The turbot is one of the marine aquaculture species with highest genomic resources and counts upon a well-organized and research-active industrial sector. In spite of its relatively small production, a significant investment in upgrading breeding programs in this species has been carried out. The concentration of the industrial sector within a restricted geographic area, and especially, its high commercial value have determined the interest and commitment for this species. Many genomic resources and tools have been developed and integrated in turbot both from structural (genetic maps, genomic libraries, whole genome) and functional (transcriptome databases, microarrays) perspectives. These resources have been anchored to other fish and vertebrates by comparative genomics, and although further work must be addressed, the information gathered represents a very rich body of resources for genomic applications. Results from ongoing EU projects (FISHBOOST and AQUATRACE), where these resources are being applied, should provide more operative tools, especially regarding the genetic map density, and also the identification of genomic regions explaining resistance to pathologies, domestication and local adaptation. This information will be useful for MAS projects to improve turbot production and for fishery management. Still, a huge body of information remains to be analyzed especially to confirm the many relevant candidate genes identified so far at particular genomic regions for growth, resistance, and sex determination, the main traits of interest for industry. Crossing the information from association and functional studies should provide new data to disclose new candidates that may be not evident for some traits. The most prominent results should be expected on SD, where the master SDg may soon be identified considering the narrow genomic area where it is located, and on growth-related traits, where an important effort has been done for validating associated QTL markers some of them linked to strong candidates. Emerging genomic areas should focus the investigation for new resources and tools and their application. An effort should also be done to characterize and validate small and long noncoding RNA genes taking into account their role in gene regulation and development, and a preliminary project has already been carried out complemented by the in silico analysis performed on turbot genome. Epigenetic factors, which are the basis to understand nonstructural heritable phenotypic changes are important from a practical perspective, and, for example, in European sea bass they have aided in the development of a tool for sex control to increase the proportion of females. Optimization of nutrition and diet composition through genome wide association analysis and functional genomics will enhance growth performance and flesh quality and attain a more sustainable aquaculture using alternative food products. Finally, the important resources accumulated in several flatfish species, should be integrated to facilitate the transference of information between species to confirm common QTL and functional signatures to identify new candidate genomic regions and gene pathways of productive interest for further verification and experimental evaluation.

References

Avendaño-Herrera, R., Toranzo, A.E., Magariños, B., 2006. Tenacibaculosis infection in marine fish caused by *Tenacibaculum maritimum*: a review. Dis. Aquat. Org. 71, 255–266.

Barja, J.L., 2004. Report about fish viral diseases. In: Alvarez-Pellitero, P., Barja, J.L., Basurco, B., Berthe, F., Toranzo, A.E. (Eds.), Mediterranean Aquaculture Diagnostic Laboratories. Zaragoza, CIHEAM, pp. 91–102.

Baroiller, J.F., D'Cotta, H., Saillant, E., 2009. Environmental effects on fish sex determination and differentiation. Sex. Dev. 3, 118–135.

Bouza, C., Sánchez, L., Martínez, P., 1994. Karyotypic characterization of turbot (*Scophthalmus maximus*) with conventional fluorochrome and restriction endonuclease-banding techniques. Mar. Biol. 120, 609–613.

Bouza, C., Hermida, M., Pardo, B.G., Fernández, C., Castro, J., Fortes, G., Sánchez, L., Presa, P., Pérez, M., Sanjuán, A., Comesaña, S., Álvarez, J.A., Calaza, M., Cal, R., Piferrer, F., Martínez, P., 2007. A microsatellite genetic map in the turbot (*Scophthalmus maximus*). Genetics 177, 2457–2467.

Bouza, C., Hermida, M., Pardo, B.G., Vera, M., Fernández, C., de la Herrán, R., Navajas-Pérez, R., Álvarez-Dios, J., Gómez-Tato, A., Martínez, P., 2012. An expressed sequence tag (EST)-enriched genetic map of turbot (*Scophthalmus maximus*): a useful framework for comparative genomics across model and farmed teleosts. BMC Genet. 13, 54.

Bouza, C., Vandamme, S., Hermida, M., Cabaleiro, S., Volkaert, F., Martínez, P., 2014. Turbot. AquaTrace Species Leaflet. Available from: https://aquatrace.eu/leaflets/turbot.

Castro, J., Bouza, C., Presa, P., Pino-Querido, A., Riaza, A., Ferreiro, I., Sánchez, L., Martínez, P., 2004. Potential sources of error in parentage assessment of turbot (*Scophthalmus maximus*) using microsatellite loci. Aquaculture 242, 119–135.

Cuñado, N., Terrones, J., Sánchez, L., Martínez, P., Santos, J.L., 2002. Sex-dependent synaptic behavior in triploid turbot, *Scophthalmus maximus* (Pisces: Scophathalmidae). Heredity 89, 460–464.

Derrien, T., Johnson, R., Bussotti, G., Tanzer, A., Djebali, S., Tilgner, H., Guernec, G., Martin, D., Merkel, A., Knowles, D.G., Lagarde, J., Veeravalli, L., Ruan, X., Ruan, Y., Lassman, T., Carninci, P., Brown, J.B., Lipovich, L., Gonzalez, J.M., Thomas, M., Davis, C.A., Shiekhattar, R., Gingeras, T.R., Hubbard, T.J., Notredame, C., Harrow, J., Guigó, R., 2012. The GENCODE v7 catalog of human long non coding RNAs: analysis of their gene structure, evolution, and expression. Genome Res. 22, 1775–1789.

De-Santis, C., Jerry, D.R., 2007. Candidate growth genes in finfish—where should we be looking? Aquaculture 272, 22–38.

Díaz-Rosales, P., Romero, A., Balseiro, P., Dios, S., Novoa, B., Figueras, A., 2012. Microarray-based identification of differentially expressed genes in families of turbot (*Scophthalmus maximus*) after infection with viral haemorrhagic septicaemia virus (VHS). Mar. Biotechnol. 14, 515–529.

Figueras, A., Corvelo, A., Robledo, D., Hermida, M., Pereiro, P., Gómez, J., Carreté, L., Bello, X., Marcet-Houben, M., Forn-Cuní, G., Abal-Fabeiro, J.L., Pardo, B.G., Taboada, X., Fernández, C., Rubiolo, J.A., Álvarez-Dios, J.A., Gómez-Tato, A., Viñas, A., Maside, X., Gabaldón, T., Novoa, B., Bouza, C., Alioto, T., Martínez, P., 2016. Whole genome sequencing of turbot (*Scophthalmus maximus*; Pleuronectiformes): a fish adapted to demersal life. DNA Res. [e-pub ahead of print].

Fuji, K., Kobayashi, K., Hasegawa, O., Coimbra, M., Sakamoto, T., Okamoto, N., 2006. Identification of a single major genetic locus controlling the resistance to lymphocystis disease in Japanese flounder (*Paralichthys olivaceus*). Aquaculture 254, 203–210.

García de la Serrana, D., Estévez, A., Andree, K., Johnston, I.A., 2012. Fast skeletal muscle transcriptome of the gilthead sea bream (*Sparus aurata*) determined by next generation sequencing. BMC Genomics 13, 181.

Gjedrem, T., 2005. Selection and Breeding Programs in Aquaculture. Springer, Dordrecht, The Netherlands.

Gjedrem, T., Baranski, M., 2009. Selective Breeding in Aquaculture: An Introduction. Springer, Heidelberg, Germany.

Gjerde, B., Roer, J.E., Lein, I., Stoss, J., Refstie, T., 1997. Heritability for body weight 9 in farmed turbot. Aquacult. Int. 5, 175–178.

Haffray, P., Lebegue, E., Jeu, S., Guennoc, M., Guiguen, Y., Baroiller, J.F., Fostier, A., 2009. Genetic determination and temperature effects on turbot *Scophthalmus maximus* sex differentiation: an investigation using steroid sex-inverted males and females. Aquaculture 294, 30–36.

Haley, C.S., Knott, S.A., 1992. A simple regression method for mapping quantitative trait loci in line crosses using flanking markers. Heredity 69, 315–324.

Hermida, M., Bouza, C., Fernández, C., Sciara, A.A., Rodríguez-Ramilo, S.T., Fernández, J., Martínez, P., 2013. Compilation of mapping resources in turbot (*Scophthalmus maximus*): A new integrated consensus genetic map. Aquaculture 414-415, 19–25.

Heule, C., Salzburger, W., Böhne, A., 2014. Genetics of sexual development: an evolutionary playground for fish. Genetics 196, 579–591.

Iglesias, R., Paramá, A., Álvarez, M.F., Leiro, J., Fernández, J., Sanmartín, M.L., 2001. *Philasterides dicentrarchi* (Ciliophora, Scuticociliatida) as the causative agent of scuticociliatosis in farmed turbot *Scophthalmus maximus* in Galicia (NW Spain). Dis. Aquat. Org. 46, 47–55.

Iyengar, A., Piyapattanakorn, S., Heipel, D.A., Stone, D.M., Howell, B.R., Child, A.R., MacLean, N., 2000. A suite of highly polymorphic microsatellite markers in turbot (*Scophthalmus maximus* L.) with potential for use across several flatfish species. Mol. Ecol. 9, 368–371.

Johnston, I.A., Bower, N.I., Macqueen, D.J., 2011. Growth and the regulation of myotomal muscle mass in teleost fish. J. Exp. Biol. 214, 1617–1628.

Kai, W., Kikuchi, K., Tohari, S., Chew, A.K., Tai, A., Fujiwara, A., Hosoya, S., Suetake, H., Naruse, K., Brenner, S., Suzuki, Y., Venkatesh, B., 2011. Integration of the genetic map and genome assembly of fugu facilitates insights into distinct features of genome evolution in Teleosts and mammals. Genome Biol. Evol. 3, 424–442.

Li, J., Boroevich, K.A., Koop, B.F., Davidson, W.S., 2011. Comparative genomics identifies candidate genes for infectious salmon anemia (ISA) resistance in Atlantic salmon (*Salmo salar*). Mar. Biotechnol. 13, 232–241.

Lillehaug, A., Lunestad, B.T., Grave, K., 2003. Epidemiology of bacterial diseases in Norwegian aquaculture—a description based on antibiotic prescription data for the ten-year period 1991 to 2000. Dis. Aquat. Org. 53, 115–125.

Martínez, P., Hermida, M., Pardo, B.G., Fernández, C., Castro, J., Cal, R.M., Alvarez-Dios, J.A., Gómez-Tato, A., Bouza, C., 2008. Centromere-linkage in the turbot (*Scophthalmus maximus*) through half-tetrad analysis in diploid meiogynogenetics. Aquaculture 280, 81–88.

Martínez, P., Bouza, C., Hermida, M., Fernández, J., Toro, M.A., Vera, M., Pardo, B., Millán, A., Fernández, C., Vilas, R., Viñas, A., Sánchez, L., Felip, A., Piferrer, F., Ferreiro, I., Cabaleiro, S., 2009. Identification of the major sex-determining region of turbot (*Scophthalmus maximus*). Genetics 183, 1443–1452.

Martínez, P., Viñas, A.M., Sánchez, L., Díaz, N., Ribas, L., Piferrer, F., 2014. Genetic architecture of sex determination in fish: applications to sex ratio control in aquaculture. Front. Genet. 5, 340.

Millán, A., Gómez-Tato, A., Pardo, B.G., Fernández, C., Bouza, C., Vera, M., Alvarez-Dios, J.A., Cabaleiro, S., Lamas, J., Lemos, M.L., Martínez, P., 2011. Gene expression profiles of spleen, liver and head kidney in turbot (*Scophthalmus maximus*) along the infection process with *Aeromonas salmonicida* using an immune-enriched oligo-microarray. Mar. Biotechnol. 13, 1099–1114.

Mommsen, T.P., 2001. Paradigms of growth in fish. Comp. Biochem. Physiol. B Biochem. Mol. Biol. 129, 207–219.

Montes, A., Figueras, A., Novoa, B., 2010. Nodavirus encephalopathy in turbot (*Scophthalmus maximus*): inflammation, nitric oxide production and effect of anti-inflammatory compounds. Fish Shellfish Immunol. 28, 281–288.

Morris, K.V., Mattick, J.S., 2014. The rise of regulatory RNA. Nat. Rev. Genet. 15, 423–437.

Navajas-Pérez, R., Robles, F., Molina-Luzón, M.J., de la Herrán, R., Álvarez-Dios, J.A., Pardo, B.G., Vera, M., Bouza, C., Martínez, P., 2012. Exploitation of a turbot (*Scophthalmus maximus* L.) immune-related expressed sequence tag (EST) database for microsatellite screening and validation. Mol. Ecol. Resour. 12, 706–716.

Palenzuela, O., Sitjà-Bobadilla, A., Riaza, A., Silva, R., Arán, J., Álvarez-Pellitero, P., 2009. Antibody responses of turbot (*Psetta maxima*) against different antigen formulations of scuticociliates (Ciliophora). Dis. Aquat. Org. 86, 123–134.

Pardo, B.G., Fernández, C., Hermida, M., Vázquez-López, A., Pérez, M., Presa, P., Calaza, M., Alvarez-Dios, J.A., Comesaña, A.S., Raposo-Guillán, J., Bouza, C., Martínez, P., 2007. Development and characterization of 248 novel microsatellite markers in turbot (*Scophthalmus maximus*). Genome 50, 329–332.

Pardo, B.G., Millán, A., Gómez-Tato, A., Fernández, C., Bouza, C., Alvarez-Dios, A., Cabaleiro, S., Lamas, J., Leiro, J.M., Martínez, P., 2012. Gene expression profiles of spleen, liver and head kidney in turbot (*Scophthalmus maximus*) along the infection process with *Philasterides dicentrarchi* using an immune-enriched oligo-microarray. Mar. Biotechnol. 14, 570–582.

Park, K.C., Osborne, J.A., Montes, A., Dios, S., Nerland, A.H., Novoa, B., Figueras, A., Brown, L.L., Johnson, S.C., 2009. Immunological responses of turbot (*Psetta maxima*) to nodavirus infection or polyriboinosinic polyribocytidylic acid (pIC) stimulation, using expressed sequence tags (ESTs) analysis and cDNA microarrays. Fish Shellfish Immunol. 26, 91–108.

Pereiro, P., Figueras, A., Novoa, B., 2012a. A novel hepcidin-like in turbot (*Scophthalmus maximus* L.) highly expressed after pathogen challenge but not after iron overload. Fish Shellfish Immunol. 32, 879–889.

Pereiro, P., Balseiro, P., Romero, A., Dios, S., Forn-Cuní, G., Fuste, B., Planas, J.V., Beltran, S., Novoa, B., Figueras, A., 2012b. High-throughput sequence analysis of turbot (*Scophthalmus maximus*) transcriptome using 454-pyrosequencing for the discovery of antiviral immune genes. PLoS One 7, e35369.

Pereiro, P., Dios, S., Boltana, S., Coll, J., Estepa, A., Mackenzie, S., Novoa, B., Figueras, A., 2014a. Transcriptome profiles associated to VHS infection or DNA vaccination in turbot (*Scophthalmus maximus*). PLoS One 9, e104509.

Pereiro, P., Costa, M.M., Díaz-Rosales, P., Dios, S., Figueras, A., Novoa, B., 2014b. The first characterization of two type I interferons in turbot (*Scophthalmus maximus*) reveals their differential role, expression pattern and gene induction. Dev. Comp. Immunol. 45, 233–244.

Piazzon, M.C., Wiegertjes, G.F., Leiro, J., Lamas, J., 2011. Turbot resistance to *Philasterides dicentrarchi* is more dependent on humoral than on cellular immune responses. Fish Shellfish Immunol. 30, 1339–1347.

Piferrer, F., Cal, R.M., Alvarez-Blázquez, B., Sánchez, L., Martínez, P., 2000. Induction of triploidy in the turbot (*Scophthalmus maximus*). I. Ploidy determination and the effects of cold shocks. Aquaculture 188, 79–90.

Piferrer, F., Cal, R.M., Gómez, C., Alvarez-Blázquez, B., Castro, J., Martínez, P., 2004. Induction of gynogenesis in the turbot (*Scophthalmus maximus*): Effects of UV irradiation on sperm motility, the Hertwig effect and viability during the first 6 months of age. Aquaculture 238, 403–419.

Ribas, L., Pardo, B.G., Fernández, C., Alvarez-Dios, J.A., Gómez-Tato, A., Quiroga, M.I., Planas, J., Sitjà-Bobadilla, A., Martínez, P., Piferrer, F., 2013. A combined strategy involving Sanger and 454 pyrosequencing increases genomic resources to aid in the management of reproduction, disease control and genetic selection in the turbot (*Scophthalmus maximus*). BMC Genomics 14, 180.

Ribas, L., Robledo, D., Gómez-Tato, A., Viñas, A., Martínez, P., Piferrer, F., 2016. Comprehensive transcriptomic analysis of the process of gonadal sex differentiation in the turbot (*Scophthalmus maximus*). Mol. Cell. Endocrinol. 422, 132–149.

Robledo, D., Ronza, P., Harrison, O.W., Losada, A.P., Bermúdez, R., Pardo, B.G., Redondo, M.J., Sitjà-Bobadilla, A., Quiroga, M.I., Martínez, P., 2014. RNA-seq analysis reveals significant transcriptome changes in turbot (*Scophthalmus maximus*) suffering severe enteromyxosis. BMC Genomics 15, 1149.

Robledo, D., Ribas, L., Cal, R., Sánchez, L., Piferrer, F., Martínez, P., Viñas, A., 2015. Gene expression analysis at the onset of sex differentiation in turbot (*Scophthalmus maximus*) at different rearing temperatures. BMC Genom. 16, 973.

Robledo, D., Fernández, C., Hermida, M., Sciara, A., Álvarez-Dios, J.A., Cabaleiro, S., Martínez, P., Bouza, C., 2016. Integrative transcriptome, genome and quantitative trait loci resources identify single nucleotide polimorphisms in candidate genes for growth traits in turbot. Int. J. Mol. Sci. 17, 243.

Rodríguez Villanueva, J.L., Fernández Souto, B., 2009. Scophthalmus maximus. In: Crespi, V., New, M., (Eds.), Cultured Aquatic Species Fact Sheets. FAO. http://www.fao.org/fishery/culturedspecies/Psetta_maxima/en.

Rodríguez-Ramilo, S.T., Toro, M.A., Bouza, C., Hermida, M., Pardo, B.G., Cabaleiro, S., Martínez, P., Fernández, J., 2011. QTL detection for Aeromonas salmonicida resistance related traits in turbot (*Scophthalmus maximus*). BMC Genomics 12, 541.

Rodríguez-Ramilo, S.T., Fernández, J., Toro, M.A., Bouza, C., Hermida, M., Fernández, C., Pardo, B.G., Cabaleiro, S., Martínez, P., 2013. Uncovering QTL for resistance and survival time to *Philasterides dicentrarchi* in turbot (*Scophthalmus maximus*). Anim. Genet. 44, 149–157.

Rodríguez-Ramilo, S.T., De La Herrán, R., Ruiz-Rejón, C., Hermida, M., Fernández, C., Pereiro, P., Figueras, A., Bouza, C., Toro, M.A., Martínez, P., Fernández, J., 2014. Identification of quantitative trait loci associated with resistance to viral haemorrhagic septicaemia (VHS) in turbot (*Scophthalmus maximus*): a comparison between bacterium, parasite and virus diseases. Mar. Biotechnol. 16, 265–276.

Ronza, P., Robledo, D., Bermúdez, R., Losada, A.P., Pardo, B.G., Sitjà-Bobadilla, A., Quiroga, M.I., Martínez, P., 2016. RNA-seq analysis of early enteromyxosis in turbot (*Scophthalmus maximus*): new insights into parasite invasion and immune evasion strategies. Int. J. Parasitol. [e-pub ahead of print].

Ross, K., McCarthy, U., Huntly, P.J., Wood, B.P., Stuart, D., Rough, E.I., Smail, D.A., Bruno, D.W., 1994. An outbreak of viral haemorrhagic septicaemia (VHS) in turbot (*Scophthalmus maximus*) in Scotland. Bull. Eur. Assoc. Fish. Pathol. 14, 213–214.

Ruan, X., Wang, W., Kong, J., Yu, F., Huang, X., 2010. Genetic linkage mapping of turbot (*Scophthalmus maximus* L.) using microsatellite markers and its application in QTL analysis. Aquaculture 308, 89–100.

Sánchez-Molano, E., Cerna, A., Toro, M.A., Bouza, C., Hermida, M., Pardo, B.G., Cabaleiro, S., Fernández, J., Martínez, P., 2011. Detection of growth-related QTL in turbot (*Scophthalmus maximus*). BMC Genomics 12, 473.

Sawatari, E., Shikina, S., Takeuchi, T., Yoshizaki, G., 2007. A novel transforming growth factor-β superfamily member expressed in gonadal somatic cells enhances primordial germ cell and spermatogonial proliferation in rainbow trout (*Oncorhynchus mykiss*). Dev. Biol. 301, 266–275.

Sciara, A.A., Rodríguez-Ramilo, S.T., Hermida, M., Gómez-Tato, A., Fernández, J., Bouza, C., Martínez, P., 2015. Validation of growth-related QTL for marker assisted selection in turbot (*Scophthalmus maximus*). ISGA XII-International Symposium on Genetics in Aquaculture, Santiago de Compostela, Spain, June 21–27, pp. 254.

Siegfried, K.R., Nüsslein-Volhard, C., 2008. Germ line control of female sex determination in zebrafish. Dev. Biol. 324, 277–287.

Taboada, X., Robledo, D., del Palacio, L., Rodeiro, A., Felip, A., Martínez, P., Viñas, A., 2012. Comparative expression analysis in mature gonads, liver and brain of turbot (*Scophthalmus maximus*) by cDNA-AFLPS. Gene 492, 250–261.

Taboada, X., Pansonato-Alves, J.C., Foresti, F., Martínez, P., Viñas, A., Pardo, B.G., Bouza, C., 2014a. Consolidation of the genetic and cytogenetic maps of turbot (*Scophthalmus maximus*) using FISH with BAC clones. Chromosoma 123, 281–291.

Taboada, X., Hermida, M., Pardo, B.G., Vera, M., Piferrer, F., Viñas, A., Bouza, C., Martínez, P., 2014b. Fine mapping and evolution of the major sex determining region in turbot (*Scophthalmus maximus*). Genes Genomes Genet. 4, 1871–1880.

Tao, W., Yuan, J., Zhou, L., Sun, L., Sun, Y., Yang, S., Li, M., Zeng, S., Huang, B., Wang, D., 2013. Characterization of gonadal transcriptomes from Nile tilapia (*Oreochromis niloticus*) reveals differentially expressed genes. PLoS One 8, e63604.

Tsai, H.Y., Hamilton, A., Guy, D.R., Houston, R.D., 2014. Single nucleotide polymorphisms in the insulin-like growth factor 1 (IGF1) gene are associated with growth-related traits in farmed Atlantic salmon. Anim. Genet. 45, 709–715.

Valente, L.M.P., Moutou, K.A., Conceiçao, L.E.C., Engrola, S., Fernandes, J.M.O., Johnston, I.A., 2013. What determines growth potential and juvenile quality of farmed fish species? Rev. Aquacult. 5, S168–S193.

Vera, M., Alvarez-Dios, J.A., Millán, A., Pardo, B.G., Bouza, C., Hermida, M., Fernández, C., de la Herrán, R., Molina-Luzón, M.J., Martínez, P., 2011. Validation of single nucleotide polymorphism (SNP) markers from an immune expressed sequence tag (EST) turbot, *Scophthalmus maximus*, database. Aquaculture 313, 31–41.

Vera, M., Alvarez-Dios, J.A., Fernandez, C., Bouza, C., Vilas, R., Martínez, P., 2013. Development and validation of single nucleotide polymorphisms (SNPs) markers from two transcriptomic 454-run of turbot (*Scophthalmus maximus*) using high-throughput genotyping. Int. J. Mol. Sci. 14, 5694–5711.

Vilas, R., Bouza, C., Vera, M., Millán, A., Martínez, P., 2010. Variation in anonymous and EST-microsatellites suggests adaptive population divergence in turbot. Mar. Ecol. Prog. Ser. 420, 231–239.

Vilas, R., Vandamme, S.G., Vera, M., Bouza, C., Maes, G.E., Volkcaert, F.A.M., Martínez, P., 2015. A genome scan for candidate genes involved in the adaptation of turbot (*Scophthalmus maximus*). Mar. Genomics 23, 77–86.

Viñas, A., Taboada, X., Vale, L., Robledo, D., Hermida, M., Vera, M., Martínez, P., 2012. Mapping of DNA sex-specific markers and genes related to sex differentiation in turbot (*Scophthalmus maximus*). Mar. Biotechnol. 14, 655–663.

Wang, L., Fan, C., Liu, Y., Zhang, Y., Liu, S., Sun, D., Deng, H., Xu, Y., Tian, Y., Liao, X., Xie, M., Li, W., Chen, S., 2014. A genome scan for quantitative trait loci associated with *Vibrio anguillarum* infection resistance in Japanese flounder (*Paralichthys olivaceus*) by bulked segregant analysis. Mar. Biotechnol. 16, 513–521.

Wang, W., Hu, Y., Ma, Y., Xu, L., Guan, J., Kong, J., 2015. High-density genetic linkage mapping in turbot (*Scophthalmus maximus* L.) based on SNP markers and major sex- and growth-related regions detection. PLoS One 10, e0120410.

Yang, C.X., Wright, E.C., Ross, J.W., 2012. Expression of RNA-binding proteins DND1 and FXR1 in the porcine ovary, and during oocyte maturation and early embryo development. Mol. Reprod. Dev. 79, 541–552.

Yue, G.H., 2014. Recent advances of genome mapping and marker-assisted selection in aquaculture. Fish. Fish. 15, 376–396.

Zebrafish offer aquaculture research their services

7

Peter Aleström, Hanne C. Winther-Larsen***

*Department of Basic Sciences and Aquatic Medicine, Norwegian University of Life Sciences, Oslo, Norway; **Department of Pharmaceutical Biosciences, School of Pharmacy, University of Oslo, Oslo, Norway

Introduction

A few selected species from different levels in the phylogenetic tree have been qualified as laboratory research favourites because of their relative ease of maintenance and study and have dominated basic biological research for decades. Classical model organisms range from bacteria and simple eukaryotic single cell species via gradually more sophisticated multicellular organisms from invertebrates toward vertebrates and at the end mammals for mimicking humans as much as possible. The intestinal Gram negative bacteria *Escherichia coli*, bakers yeast (*Saccharomyces cerevisiae*), the fruit fly (*Drosophila melanogaster*), a toad (*Xenopus laevis*), and the rodents mouse (*Mus musculus*) and rat (*Rattus norwegicus*) are all well-known to biologists. Several fish species have also participated as model organisms in the study of animal physiology and in generating the big picture of functional biology. Among research, favourite fish species are rainbow trout (*Oncorhynchus mykiss*), goldfish (*Carassius auratus*), three-spined stickleback (*Gasterosteus aculeatus*), guppy (*Poecilia reticulata*), platyfish (*Xiphophorus sp.*), pufferfish (*Fugu rubripes, Tetraodon nigroviridis*), and the today more and more dominating small lab fish medaka (*Oryzias latipes*) and zebrafish (*Danio rerio*). Medaka that is less widely used then zebrafish, but shares many attributes and have in addition a wider temperature and osmotic tolerance and the availability of inbred lines (Furutani-Seiki and Wittbrodt, 2004; Spivakov et al., 2014). Although several fish species including medaka are excellent model organisms for different reasons, this article will focus on zebrafish, which largely dominates as a laboratory animal, with a wide range of applications.

Zebrafish are small (4–5 cm) tropical freshwater fish in the family of *Cyprinidae* and of order *Cypriniformes* that live wild in the inland water systems of North-East India and Pakistan (Arunachalam et al., 2013; McClure et al., 2006). They are long established as a popular aquarium fish. The environmental conditions in the ecosystems of zebrafish inhabit vary between fastflowing versus stagnant, and from clear to turbid water, with temperatures ranging from 6.2°C to 41.7°C (Cortemeglia and Beitinger, 2005; Engeszer et al., 2007; Spence et al., 2008). The predation pressure in the wild has led to the development of a shoaling behaviour. It is interesting to note that this behaviour can also be observed among fish that are separated by glass and not in physical contact, and thus reducing the potential of stress and aggression among

Genomics in Aquaculture. http://dx.doi.org/10.1016/B978-0-12-801418-9.00007-X

fish kept in small tanks (Engeszer et al., 2004). In the wild, zebrafish have an estimated life span of 1–2 years (Reed and Jennings, 2011) as compared to around 5 years in captivity. While conditions in the wild vary considerably, in a laboratory environment attempts are made to provide stable and optimal husbandry conditions. To maintain healthy fish in a laboratory setting, detailed instructions have been formulated providing the fish with what is aimed to be at a constant optimal growth environment (Harper and Lawrence, 2012; Reed and Jennings, 2011; Westerfield, 2007).

Zebrafish were first selected as vertebrate model for embryo development because of their attributes of being small, robust, cheap, and easy to breed in the laboratory all year round and for having transparent embryos, which develop to hatch within 48–72 h postfertilization (hpf) at 28°C. This transparency contributes to zebrafish being an excellent tool for research on embryogenesis and developmental biology, as well as greatly facilitating various imaging techniques. Zebrafish produce large clutches of eggs (100–400), have external fertilization, and rapid embryogenesis. George Streisinger at the University of Oregon played an important role as the founder of modern zebrafish research during the late 1970s (Streisinger et al., 1981). He paved the way for the deep understanding we now have of its biology and its development into a leading vertebrate *in vivo* model organism. One important reason for its rapidly growing popularity was that zebrafish researchers were already well organized from the 1980s as a scientific community in which exchange of information and research tools, including mutant fish lines etc., was nurtured. The University of Oregon has played an important international node with The Zebrafish Book by Monte Westerfield (Westerfield, 2007) and the zebrafish model database ZFIN (see zfin.org and Howe et al., 2013). The Zebrafish International Resource Center (see ZIRC; zebrafish.org) is closely linked to ZFIN and provides a wide range of zebrafish lines, probes, and health services.

With 70% of the genes shared with humans and its establishment as robust small laboratory fish with an increasingly growing research toolbox, the zebrafish has climbed to being the second most widely used laboratory animal model in biomedicine, surpassed only by the mouse. Since the 1980s, an increasing amount of data about zebrafish genetics and the biology of a rising number of processes including embryo development, tissue regeneration, neuroscience, pathology, etc. have accumulated (most of it available with reference in the ZFIN database). Despite this deep knowledge about the zebrafish as a model fish for studies of biomedicine, the information has not been yet systematically translated to farm animals, including farmed fish. With an increasing need for more and better food production, this is a largely unexplored resource. While fisheries reached the global roof for sustainable production already in the late 1980s, aquaculture is in constant growth (FAO, 2014). Several review articles describe zebrafish as model for biomedicine (Santoriello and Zon, 2012; Seth et al., 2013; White et al., 2013) and its potential for aquaculture research and aquatic medicine have also been discussed (Aleström et al., 2006; Ribas, 2013; Ulloa et al., 2014). During the Ninth European Zebrafish Meeting in Oslo, Norway 2015, a full session was for the first time devoted to the use of zebrafish as a model for aquaculture research (see zebrafish2015.org). This article will present the latest information on structural and functional perspectives of zebrafish genomics, transcriptomics, and epigenomics. Further, an overview of the research toolbox available and aspects on breeding, reproduction control, nutrition, and health will be provided.

The Genome: diversity, structure and organization

While many mammalian genomes carry around 22,000 genes dispersed in 3 billion base pairs (bps), more variation is seen in fish genomes with salmonids at 30,000/3, zebrafish 26,000/1.5, medaka 20,000/0.7, and pufferfish with 18,500 genes in 0.4 billion bp. Genome duplications, distribution of transposable elements, loss of intergenic sequences, and epigenetic control of gene expression all contribute to a substantial diversity, albeit with a certain important core of conserved mechanisms and functions. With this diversity in mind, we will describe in some detail how the zebrafish genome is organized.

The Genome and the corresponding transcriptome

Comparative perspectives

Teleost fish account for 50% of all vertebrate species and the phylogeny of teleost fish species and other vertebrates demonstrates large evolutionary distances between them (McCluskey and Postlethwait, 2015; Near et al., 2012); this already suggests many functional differences between the target species and model organisms like zebrafish used in various comparative studies. However, despite these large evolutionary distances, biomedical research has generated an unexpectedly long list of successfully established human disease models in the zebrafish (Santoriello and Zon, 2012; Seth et al., 2013; White et al., 2013), which argues that the level of functional conservation must be substantial. This argues that there should be many as yet unexplored applications for the small laboratory adapted model fish in the field for aquaculture related research.

Around 300 million years ago, a teleost ancestor underwent an additional whole genome duplication (WGD) as compared to other vertebrates, named WGD3, or the teleost genome duplication (TGD). Although the TGD caused a duplication of all the genes, a substantial number have reverted to single copy gene status and the present teleost genomes carry around one third of their genes in duplicated versions, or ohnologs (Meyer and Schartl, 1999; Postlethwait et al., 2004; Wolfe, 2000). In addition to the TGD, some teleost lineages including the salmonids went through a fourth genome duplication (Berthelot et al., 2014), which again makes comparative genomics with model species challenging, but at the same time an important facilitator.

Synteny describes the comparative study of blocks of neighbouring genes in the same relative positions in the genome over evolutionary time. Syntenic regions can be localized within WGD genomes or between genomes of more or less distantly related species. The synteny database (Catchen et al., 2009; and see http://syntenydb.uoregon.edu/synteny_db/doc/) was created to aid the study of evolutionary history of gene families across multiple lineages. The Synteny DB compares source genomes within itself and/or with an outgroup genome for detection of regions of conserved synteny (paralogons). Comparative genomics is a powerful tool to transfer information from model species to related non-model species. Important for the use of model organisms is the ability to predict orthologues genes between the model and target species. This prediction is often obstructed by the presence of the frequent incidence of the paralogous and ohnologous genes generated by gene duplication events. An additional important consideration is the quality of annotation of genes. *Ensembl Compara* is an Internet resource which provides cross-species analyses of annotated protein coding and noncoding RNA (ncRNA) genes (Vilella et al., 2009; and see http://www.ensembl.org/info/genome/compara/index.html). For comparative genomics between man and zebrafish it has been calculated that around 70% of genes are shared, or orthologues (Howe et al., 2013).

One level of comparative genomics has focus on enhancer elements, with the aim of uncovering their roles in the control of gene expression and the interrelationships between specific enhancer elements and the corresponding genes. The Japanese pufferfish (*Fugu rubripes*) was the first fully sequenced fish genome (Aparicio et al., 2002) and shown to have a very compact genome of 365 Mbp. Unlike most vertebrate genomes, it has lost a large proportion of intergenic repeat elements found in other vertebrates and is left with 90% unique sequences. Based on the assumption that essential cis regulatory elements are conserved (Brenner et al., 1993), zebrafish has been made transgenic with pufferfish intronic or intergenic sequence reporter gene constructs for characterization of cis regulatory elements *in vivo* (Muller et al., 2002).

DNA and genome sequencing

The Sanger sequencing method (Sanger et al., 1977) revolutionized molecular biology and its inventor was awarded a Nobel Prize in chemistry in 1980. The Zebrafish Genome Project (https://www.sanger.ac.uk/resources/zebrafish/genomeproject.html) was launched at the Sanger Institute in Cambridge, UK, in 2001. This was around the time point when at the time controversial and very expensive Human Genome Project eventually delivered a full (rough) genome draft sequence. After nine consecutive versions, Zv1–Zv9, an assembled whole genome sequence was finally published (Howe et al., 2013). During the last decade "next generation sequencing (NGS)" methods have been developed that only take a fraction of the time and cost to generate long assembled stretches of DNA sequence information. NGS has thereby revolutionized biology to allow a wide variety of species, including commercially important fish, and individuals within a species genomes to be deciphered (87 eukaryotic species are now listed at http://www.ensembl.org/info/about/species.html). Since 2014, the Zebrafish Genome Project joined the Genome Reference Consortium (GRC) for continuous maintenance and the GRC released a 10th reference genome assembly named GRCz10 (http://www.ncbi.nlm.nih.gov/assembly/GCA_000002035.3/). In GRCz10, chromosome 4 has gained about 15 Mb in length after 94 of 112 previously unplaced contigs were placed in the genome puzzle and the total length of the zebrafish genome is at present 1,464,443,456 bp long. The 25 pairs of zebrafish chromosomes are quite homogeneously sized, with lengths varying between 37 and 77 million bp. The protein encoding gene count predicts 26,206 genes encoding around 50,000 proteins composed of 225,000 exons, and when compared to the human genome it discloses approximately 70% similarity in terms of number of gene orthologues (Howe et al., 2013). In addition there are 293 loss of function pseudogenes and the noncoding RNA (ncRNA) genes, in total 6,008, of which 3,172 are small ncRNA and 2,741 are long ncRNA genes. The zebrafish mitochondrial genome is 16,596 bp and encodes 13 protein genes, 22 transfer RNAs and 2 ribosomal RNAs (Broughton et al., 2001). A comprehensive and regularly updated overview of resources and tools developed for zebrafish genomics is available at the ZFIN Community Wiki—Genomic Resources for Zebrafish (https://wiki.zfin.org/display/general/Genomic+Resources+for+Zebrafish).

Single nucleotide polymorphism (SNP) is an important class of genetic molecular markers that have profited immensely from the widespread use of NGS. The first SNP map of the zebrafish genome comprised 2,035 SNPs and 178 small insertions/deletions (Stickney et al., 2002). Five years later 645,088 SNPs were mapped (Bradley et al., 2007). Now the numbers are several million SNPs in each of the three strains of zebrafish analyzed and in which approximately half were common to all three and the other half were strain specific (Butler et al., 2015). All these SNPs are accessible through "SNPfisher," a searchable online database (snpfisher.nichd.nih.gov). ENCODE (ENCyclopedia Of DNA Elements) and modENCODE are two big coordinated efforts for annotating functional elements in the human genome and the genomes of the model organisms *D. melanogaster* and the nematode *Caenorhabditis elegans,* respectively. They have resulted in numerous publications and NGS generated datasets on genome-wide transcription profiles, epigenetic landscapes, RNA-binding protein and transcription factor binding patterns (Dunham et al., 2012; Gerstein et al., 2010; Roy et al., 2010). In summary these analyses have demonstrated that the majority of genome sequence, once called "junk DNA," can be ascribed specific functions and comparative studies of a handful of deep annotated genomes can sort out elements functions that have been conserved in evolution (Gerstein et al., 2014). The zebrafish community now follows up with a similar endeavour, DANIO-CODE, to supplement the emerging genome-wide picture and further characterize the zebrafish genome (Tan et al., 2016).

Expressing the genome—the transcriptome defines all transcripts produced

Although many "house-keeping" genes and functions are common between fish, there is one distinct transcriptome for each cell type and developmental stage with a specific spatial and temporal expression pattern in all of them. Also, environmental factors like drugs, pollutants, pathogens etc. can induce changes in a transcriptome and these changes can in principle be mapped to deciphering of modes of action for each effector. In general, the total transcriptome is composed of a vast number of different types of RNA molecules, which can be subdivided into protein coding (mRNAs) and noncoding RNAs (ncRNAs). With 25,642 protein coding genes and 57,369 annotated gene transcripts (mRNAs) in GRCz10, there is an average of 2.2 isoforms transcribed from each gene, with the isoform diversity arising from the use of alternative start, stop, and or splicing choices. Different isoforms from one gene may express different proteins, or proteins with altered composition of functional domains, for example, generated from exon skipping or intron retention events (Aanes et al., 2013; Aanes et al., 2011). In an attempt to increase the annotation quality of zebrafish genomics datasets, five tissues and seven developmental stages were selected for RNA sequencing of the protein coding transcriptome (Collins et al., 2012). The maternal transcriptome stored in the female germline controls the early embryo onset of the zygotic genome (maternal-zygotic transition; MZT) and the concomitant initiation of differentiation into the three germ layers endo-, ecto-, and mesoderm in the gastrulating embryo. In zebrafish there are >10,000 maternal mRNAs of which 70% are dormant in the egg and becomes translationally activated by cytoplasmic polyadenylation between the 1-cell and 16-cell stage (Aanes et al., 2011). The old paradigm for MZT occurring during the midblastula transition at the 10th symmetric cell division around 3hpf is now being modified with more sensitive methods used and reveals that the activation of the zygotic genome rather takes place more gradually from around 2hpf and the 64- to 128-cell state (Heyn et al., 2014). In the maternal transcriptome, many genes are expressed as multiple isoforms which further add to its complexity (Aanes et al., 2013).

To serve as a model for other important but less well-characterized aquaculture relevant species, such as turbot, catfish, and salmonids, a good annotation and clear understanding of the genetic and epigenetic control of the biology is essential. Experimental functional genomics is needed to link genotypes and phenotypes. The zebrafish mutation project at the Welcome Trust Sanger Institute aims at establishing a resource consisting of a complete library of zebrafish mutants in which each cloned line of fish has one gene knockout mutation, together making up the zebrafish *phenome* (Kettleborough et al., 2013) and see http://www.sanger.ac.uk/sanger/Zebrafish_Zmpbrowse). In conclusion, extensive annotation of a genome is needed in order to fully make use of its encoded information for depicting its translation into functional terms. Hence deep sequencing of model organisms plays an important role to bridge the resolution of genomes to applications via comparative genomics.

Epigenetics—three levels of non-Mendelian heritage

The intrinsic picture of epigenetic mechanisms for control of gene expression is continuously getting more and more complex. From originally being represented by DNA methylation patterns, the histone code introduced a second, in itself complex, level of chromatin structure dynamics, allowing or hindering the RNA polymerase and transcription factors to bind and transcribe the genes. Now the short and long noncoding RNAs (ncRNA; lncRNA) with a not yet finally defined repertoire add more complexity to the delicate control of chromatin structure and organization through epigenetic landscapes (Fig. 7.1). Taking into account that the epigenetic control mechanisms are sensitive to external factors like environmental pollutants, drugs, radiation etc., it has been referred to as the non-Mendelian heritage, since an increasing number of cases have been reported in which epigenetic changes have been transferred to daughter cells and inherited to offspring via the germline (Skinner, 2016; Szyf, 2015; Varona et al., 2015).

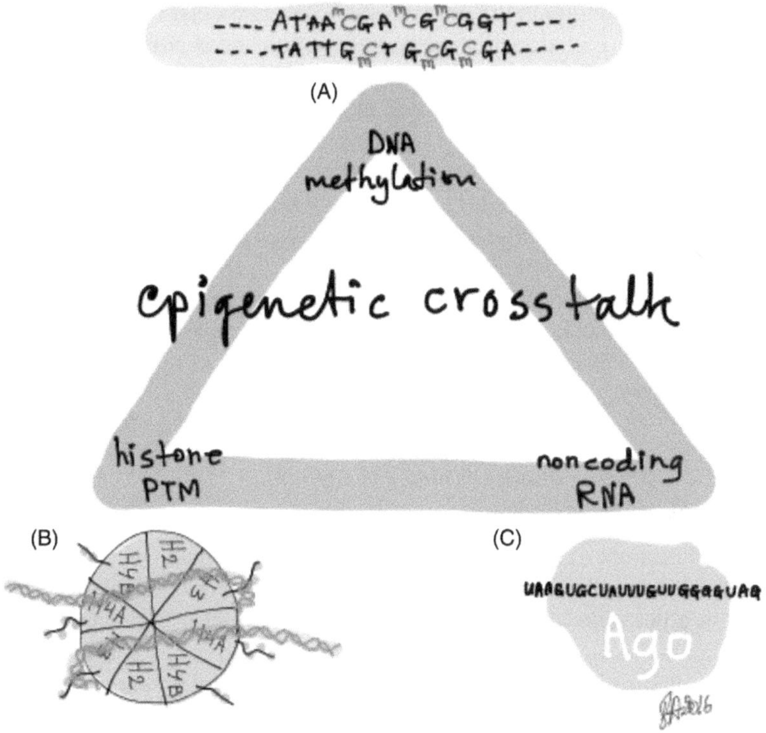

Figure 7.1 Epigenetic cross-talk. Chromatin structure determines what genes can be expressed using a more or less dynamic set of chemical modification variants of DNA and histone proteins of the nucleosome cores. Histone proteins expose their N-terminal tails as targets for enzymatic posttranslational modification (PTM). Noncoding RNAs are involved at several levels of gene expression control, including chromatin 3-D structures. The specific composition of the three levels of epigenetic marks is referred to as the epigenetic landscape and the interaction between them can be described as epigenetic cross-talk. (A) DNA with 5-methylcytocine (5mC) marked in gray. (B) Nucleosome, wrapped around with two turns of DNA double helix, with histone N-terminal chemically modified tails sticking out. (C) Argonaute protein (Ago) with a 22-mer microRNA sequence attached.

Level 1: DNA methylation and hydroxymethylation

Chemical modification through methylation of deoxycytidine in DNA (5mC), predominantly found in CG dinucleotide contexts, is a well-characterized epigenetic mark that has been correlated to numerous functions. Examples of DNA methylation associated processes are (1) inactivation of X chromosomes, (2) transcriptional repression of tissue-specific genes through promoter methylation, (3) gene body methylation linked increases in gene expression, and (4) enhancer transcription factor binding (Jones, 2012; Smith and Meissner, 2013). The first intermediate in TET enzyme-mediated demethylation of DNA is 5-hydroxymethyldeoxycytidine (5hmC) which, in addition to being a metabolic intermediate, is presumed to also serve as an independent epigenetic mark (Pfeifer et al., 2013). Two studies suggest that during the epigenome reprograming after fertilization, passive demethylation and *de novo* methylation occur simultaneously and that the maternal methylome pattern is reprogramed to that of the paternal genome (Jiang et al., 2013; Potok et al., 2013). Quantification of the global 5mC profiles during zebrafish embryo development revealed levels of around 2% at the 1 to 2-cell stage that increased through zygotic remethylation to above 8% after 8hpf (gastrula stage). This is similar level to that found in adult tissues, with the highest 5mC levels seen in brain and the lowest in testis. Levels of 5hmC were undetectable in the embryo until after 12hpf and climbed to 0.23% in larval stages (96hpf). In adult zebrafish tissues, the highest levels were found in the brain (0.49%) and the lowest in testis (0.01%), with muscle, liver, and intestine ranging between 0.13% and 0.06%, a similar distribution and levels of hmC in different tissues to that found in mammals (Kamstra et al., 2015a,b).

Level 2: histone protein modifications are dynamic

The language of covalent histone modifications (Strahl and Allis, 2000) describes a multitude of alteration of chromatin states, which can modulate DNA-histone interactions to promote or repress gene transcription. It is the N-terminal tails of the histone proteins, which protrude from the nucleosomes of chromatin fibres and are exposed for the enzymatically catalyzed chemical post-translational modifications (PTMs) that harbour the dynamic histone code. By aid of mammalian antibodies raised against the specific epitopes of a histone sequence with its chemical modifier attached, the method of zebrafish chromatin immunoprecipitation (ChIP) was established (Collas and Dahl, 2008; Lindeman et al., 2009). Among histone PTMs examined (methylation, acetylation, ADP-ribosylation, ubiquitination, citrullination and phosphorylation), trimethylation of histone H3 lysine 4 (H3K4me3) marks the promoter of most genes that are, or about to become, expressed. Acetylation (H3K9- and H3K27Ac) marks are associated with actively expressed genes whereas H3K9me3 and H3K27me3 are associated with repression of genes and found within a facultative heterochromatin environment. ChIP combined with NGS allows genome-wide analysis of protein epitopes crosslinked to DNA and subsequent correlations to gene expression data, which again can be correlated to DNA methylation patterns that are all related to exact coordinates at the nucleotide level in the genome. Such superimposed genomic maps are often referred to as epigenetic landscapes and aim at deciphering the mechanisms behind the epigenetic control of the resulting phenotypes (Barski et al., 2007; Lindeman et al., 2011; Mikkelsen et al., 2007; Vastenhouw and Schier, 2012).

Level 3: Noncoding RNAs dominate the transcriptomes

The consortium behind ENCODE has taught us that >80% of the genome is functional and biochemically active and to a larger extent transcribed, mainly into nonprotein coding RNAs (Dunham et al., 2012). For a portion of the noncoding transcriptomes, ncRNAs can be ascribed epigenetics related functions as reviewed (Holoch and Moazed, 2015; Sadakierska-Chudy and Filip, 2015).

In addition to the 57,369 mRNAs expressed in zebrafish, approximately 6,000 ncRNAs are transcribed (GRCz10). The noncoding part of the transcriptome consists of short (18–40 nt) and long ncRNAs (>200 nt; lncRNA), many with well-characterized functions and again many with little or no known functions. Ribosomal RNA (rRNA) and transfer RNA (tRNA) have central functions in translation of proteins. Small nuclear RNA moieties of the spliceosome (snRNAs) participate in mRNA splicing and small nucleolar RNAs (snoRNAs) guide chemical modifications of rRNA, tRNA, and snRNA. Again other ncRNAS have functions specified as part of ribo-protein complexes of telomerase enzyme, docking of ribosomes to endoplasmatic reticulum etc. Epigenetic long noncoding RNAs (lncRNAs) can interact with chromatin and play an epigenetic role by regulating the three-dimensional genome architecture (Engreitz et al., 2013).

One feature that short ncRNAs have in common is that they are embedded in a member of the Argonaute (Ago) protein family (Ender and Meister, 2010), where they act as guides to specific target molecules (Ketting, 2011). Among the well-known short ncRNAs are the microRNAs (miRs) which act as transacting posttranscriptional regulators of mRNA translation and stability by guiding the binding of the RNA-induced silencing complex (RISC) to the 3'UTR of mRNAs (Bizuayehu and Babiak, 2014). In zebrafish today there are around 400 miR genes and over 600 mature miRs reported (Desvignes et al., 2014; and see miRBase v20). The piRNAs represent a predominantly germ-cell-specific class of small RNAs that interact with a specific class of Argonaut proteins, known as Piwi proteins, and both Piwi proteins and piRNAs are essential for germ cell development and function (Ketting, 2011). Ziwi, the zebrafish homologue of the Drosophila Piwi, colocalizes with Vasa at the embryonic genital ridge and gonad-specific expression in the adults (Tan et al., 2002). The lncRNA resembles mRNAs apart from the lack of open reading frames (ORFs) for protein coding capacity (Holoch and Moazed, 2015; Pauli et al., 2012; Ulitsky et al., 2011). A common feature for regulatory RNAs is that they can base-pair with complimentary sequences and they represent an alternative to DNA-binding proteins in sequence specific targeting of determinants for transcription control and epigenetic regulation of gene expression.

Levels 1, 2, and 3 cross-talk in epigenetic landscapes

The picture of epigenetic landscapes gradually uncovers the interplay between superimposed levels of epigenetic marks and suggests a complex mechanistic understanding of how effectors act on chromatin structure, its control of gene expression and eventual transfer to daughter cells and offspring (Fig. 7.1; reviewed by (Holoch and Moazed, 2015; Sadakierska-Chudy and Filip, 2015)).

The toolbox: a set of advanced methods

For genetic, genomic, and functional studies, the zebrafish toolbox offers a wide range of possibilities. However, for space limiting reasons only a few highlighted methods from the toolbox are presented. Within the zebrafish community many resources are open source and often links and references can be found at zfin.org.

Genome and epigenome editing

The next generation sequencing today in principle allows one to monitor genomics, transcriptomics, and epigenomics of basically any species. The usefulness of using such powerful tools is however a function of how well annotated the genome is and what functional validation studies can be executed at reasonable time and cost. During the last decade, genome editing has developed from forward to reverse genetics. The methods for targeted genome editing have recently become available for many *in vitro* and *in vivo* models, and are feasible for ordinarily equipped molecular biology laboratories. Via zinc-finger nucleases and TALENs, the CASPR/Cas9 system has revolutionized genome biology during the last few years (Hwang et al., 2013a; Hwang et al., 2013b). This allows fine-tuned research on optimization of as well protein-structure functions as of regulatory DNA elements, "*in situ* of an *in vivo* model." The Zebrafish Mutation Project (Kettleborough et al., 2013; and see http://www.sanger.ac.uk/resources/zebrafish/zmp/) aims at generating a library of zebrafish lines, each with one of the 26,000 genes carrying a knock-out mutation, and will as such make up a valuable resource for the study of phenomics in a vertebrate species. Another powerful perspective opens with the CASPR/Cas9 technology moving towards editing of the epigenome. Genetic engineering of Cas9 aiming at transforming the endonuclease into a DNA methylase/demethylase or histone acetylase/deacetylase has recently been reported (Hilton et al., 2015).

High-throughput imaging—screening

There has been a tremendous development in microscopy techniques combined with automated monitoring of phenotypic changes during embryo and larval development. Their transparency and small size make zebrafish ideal for advanced image analysis. Automatic high-throughput bright field and fluorescence microscopy using capillaries or 96- to 384-well plate formats allow registration and documentation of phenotypic changes taking place at tissue and sub-cellular levels during days of embryo and larval development. Both wild-type and transgenic reporter gene equipped zebrafish lines allow a wide array of biological and toxic effect phenomics with applications in drug development and environmental pollution research. Selective plane illumination microscopy (SPIM), also referred to as light sheet microscopy systems allow real time visualizing embryo development at individual cell division and migration level, combined with multiple fluorophore detection (Keller et al., 2008; Schmid et al., 2013; Weber and Huisken, 2011). Examination of neural circuit functions can be carried out with serial block face scanning electron microscopy that can also be combined with calcium-imaging techniques (Friedrich et al., 2013). Nevertheless, advanced imaging techniques generate large amounts of data and much effort is needed to develop systems to significantly reduce the memory needed for storage of the images acquired by real time microscopy (Schmid et al., 2013). High-throughput video-based monitoring of behaviour patterns, heartbeat activity etc. of individual zebrafish larvae/fish in multi-well plates is another expanding analytical platform. Parameters studied include movements with rest and wake dynamics (locomotor activity response—LMR) and light stimuli provoked specific motor behavioural pattern (photomotor response—PMR) (Kluver et al., 2015). Altogether, the small model fish allow the design of studies at a new level where the genome, and eventually the epigenome, can be edited basically without limits. The resulting phenotypic effects can be monitored at high resolution and applied for further studies of functional genomics and the corresponding epigenomics, and eventually *transformed into target species of economical interest.*

Functional and applied aspects of systems biology

Despite obvious dissimilarities, there are many important similarities to focus on when conducting comparative studies using established model species like zebrafish with other commercially important aquaculture organisms. Since the zebrafish is accepted as a model for human medicine, it is reasonable to argue that the small fish model must be an acceptable model for other fish species. There can be many important areas of basic biology that can be turned into translational research in aquaculture species by using the zebrafish. Areas for using the potentials of the small laboratory fish models include applied research on breeding strategies, reproduction control, health, effects on environment and nutrition. The small size makes high-throughput screening for biological effects feasible in testing new drug candidates, food ingredients, and sensitivity to environmental factors like climate changes. The following sections will elaborate on this options and Fig. 7.2 discloses a SWOT analysis of pros and cons for the applicability of the zebrafish model in aquaculture research.

Breeding model and reproduction control

What can we learn from history?

A major reason for the success of Norwegian farmed salmon is that farm fish, like other farm animals, should be introduced into systematic animal breeding programs. This understanding was pushed into application by Harald Skjervold, a professor in animal genetics and breeding at the Agricultural University of Norway in the late 1960s and early 1970s (NOU 1977: 39 Fiskeoppdrett; NOU 1988: 3 Nasjonal handlingsplan for havbruksforskning, in Norwegian). In 1975 the first generation of farmed Atlantic salmon, originating from the successive crosses of wild salmons caught in over 40 Norwegian rivers, and selected for good growth performance. Today, 11 generations later, the growth performance has doubled and other features such as food conversion and disease resistance have been substantially improved. Although classical breeding has resulted in a great success, it is a fact that only relatively few traits, for which effects of interest can be monitored, has been driving the genetic improvements. However, using zebrafish with their short generation cycles, the important principles from 40 years history of salmon selection breeding could be mimicked in 2.5 years and, with today's toolbox, whole genome sequencing of a relevantly group of offspring from each generation can be conducted. Such a study would give insight in how the genome in detail undergoes selection patterns and as such also model for other farm animal breeding programs. The ambitious aim to understand genetic factors across

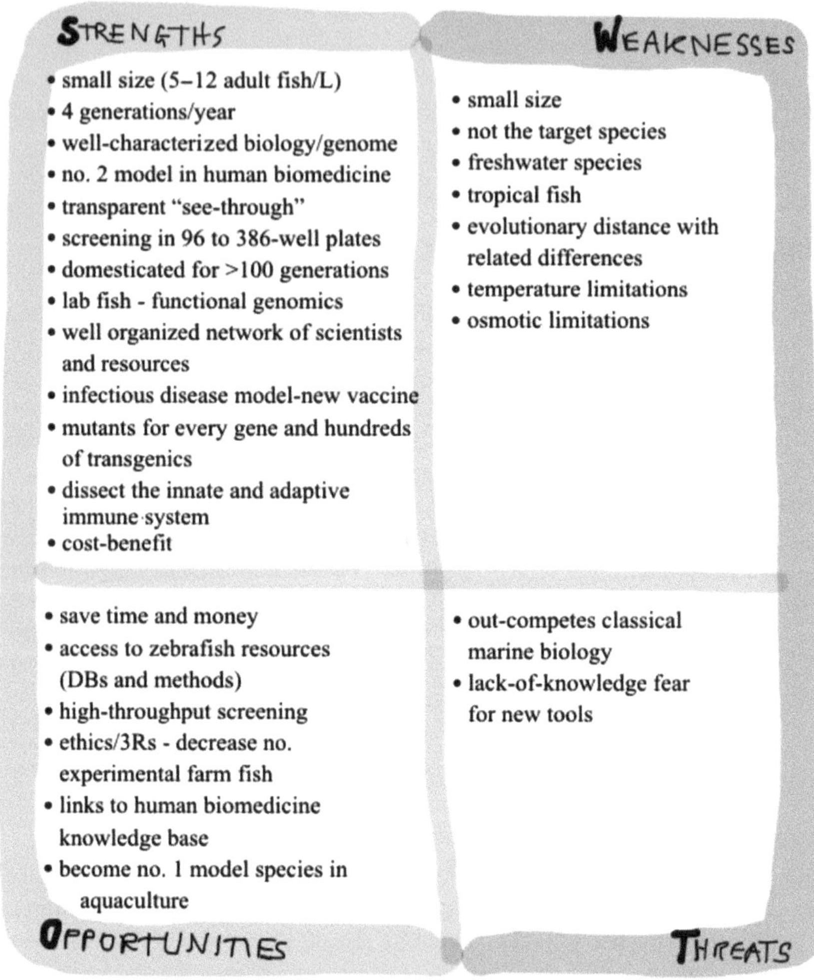

Figure 7.2 Strengths, weaknesses, opportunities and threats (SWOT) analysis for the zebrafish model in aquaculture related research. Benefits, pros and cons for the zebrafish model with respect to its value(s) relative to aquaculture research is summarized.

many generations might further be expanded and complemented with epigenetic NGS studies investigating the epigenetic profiles between generations in a selection breeding program.

Reproduction control

Teleost fish sex determination and sex differentiation is quite diverse and complex, and controlled by genetic, physiological, and environmental factors (inferring epigenetic control mechanisms) and will not be covered in this article (Devlin and Nagahama, 2002; Piferrer, 2013). The diversity can be illustrated with a study reporting differences in zebrafish with genetic determinants in chromosome 4 in wild strains which have been lost in the domesticated lab strains (Wilson et al., 2014). Reproduction control is however an important trait for fish farming; it has always been an economical obstacle for salmon culture that male fish frequently undergo early sexual maturation, during which both quality and volume of the meat are diminished because of the premature gonad development. In addition, during the last decades there has been increased focus on how wild salmon populations are affected by the relative large number of farmed fish that escape from captivity, and how to reduce their side effects. Whereas selection breeding has not been able to solve the issue of reproduction control, one solution has been to generate triploid mono sex, mainly all female, stocks of fish. This is an approved method for trout in Europe and North America but has not been accepted for Atlantic salmon because of the elevated levels of malformations, poorer survival, and poor production results in sea cages (O'Flynn et al., 1997). Therefore, focus has been turned to attempts at manipulation of the early development. One target has been gonadotropin releasing hormone (Gnrh) the main vertebrate sex hormone produced in neurons of the hypothalamus that controls the pituitary and its resulting synthesis and release of the gonadotropic hormones Gth1 and Gth2 (also known as FSH and LH); these in turn control gonad development and sex steroid hormone biosynthesis. Another mechanism in focus for manipulation of fertility involves the development of primordial germ cells (PGCs) giving rise to the germ line. Two proteins, the transcription factors Nanos and Tdrd7, are both central for germ line development and are protected from microRNA-430 targeted mRNA degradation by the germ cell-specific RNA binding protein Dnd1 (Kedde et al., 2007). Targeted molecular strategies to achieve sterility have been to: (1) block of the Gnrh-mediated activation of sex hormone production by induced autoimmunity (Andersen et al., 1992) or transgenic driven Gnrh mRNA knockdown (Aleström et al., 1992; Uzbekova et al., 2000); (2) block of *dnd1* by morpholino gene knock-down in Atlantic cod (Skugor et al., 2014) and CRISPR/Cas9 genome editing *dnd1* gene knock-out in salmon for hindering germline development (Wargelius et al., 2016). Although there are promising indications, there is still no molecular fertility control strategy that has been approved for fish farming. It appears that more basic and systematic experimental research is needed and again the small fish models, such as zebrafish with short generation cycles can help speed up finding robust and sustainable solutions, which do not have health or environmental risk potentials.

Fish health

Zebrafish also proves itself as an excellent tool for studies of farm fish health. Investigations of skeletal deformities in teleost fish and antiviral immune responses of the interferon system (IFN) are reviewed previously (Langevin et al., 2013) and will not be discussed here.

In infection biology, adult and zebrafish larva as well as zebrafish cell lines are applied to study infection mechanisms and to develop vaccines against viral and bacterial pathogens of importance for aquaculture. The field of fish parasite infection biology, however, is not as well developed. A possibility to isolate and study zebrafish wild life parasites as a model for infection mechanisms, treatment, and vaccination strategies were suggested by Gratacap and Wheeler (Gratacap and Wheeler, 2014). For example, parasitic fungal and metazoan infections of adult zebrafish could potentiate such studies (Kent et al., 2002; Matthews, 2004; Matthews et al., 2001). In infection studies, zebrafish has typically been exposed with various infection routes to a specific pathogen relevant for aquaculture and has been reviewed previously (Sullivan and Kim, 2008). One main focus has been to determine the host specific immune responses to the pathogen (Sullivan and Kim, 2008). Further, vaccine and/or immune-stimulant experimental trials have been set up where accumulated survival and the subsequent immune responses were assessed (Kavaliauskis et al., 2015). The expression of immune relevant genes in adult zebrafish are normally investigated by transcription profiling of important immune organs such as spleen, kidney, and heart (Kavaliauskis et al., 2015; Meijer et al., 2004; Pressley et al., 2005) or in whole fish larva (Brudal et al., 2014). The pathogens genome sequences are utilized for design and construction of mutations in virulence factors and the resulting effect(s) determined first in zebrafish and then in the natural host (Santander et al., 2012). Established transgenic zebrafish lines can be used to follow the infection by the various imaging techniques. Examples are transgenic zebrafish expressing green or red fluorescent macrophages or neutrophils, which are exposed to fluorescence bacteria (Brudal et al., 2014; Harvie and Huttenlocher, 2015; Rougeot et al., 2014). Alternatively gene knock-downs, established zebrafish mutants or zebrafish mutant screen can be used to explore important determinants in host-pathogens interactions (Meijer and Spaink, 2011). Such cross-talk investigations between hosts and pathogens are only made possible by utilizing the advanced toolbox already established from the zebrafish genomics research.

Bacterial infections and vaccine development

Zebrafish models have been established for a range of bacterial diseases relevant for aquaculture. Of these, *Edwardsiella tarda* is a clear example of the use of the zebrafish model for infections studies applicable in aquaculture (Gao et al., 2014; Pressley et al., 2005). Other examples are *Streptococcus iniae* (Harvie et al., 2013; Harvie and Huttenlocher, 2015; Neely et al., 2002; Saralahti and Ramet, 2015) and the newly established mucosal infection model caused by *Vibrio angularium* (Liu et al., 2014b). *Mycobacterium* infections have been reported in a range of cultured fish as exemplified by hybrid striped bass (Bozzetta et al., 2010), turbot (dos Santos et al., 2002), and Atlantic salmon (Brocklebank et al., 2003; Zerihun et al., 2011). By the use of microarrays, the Mycobacterium-zebrafish model has also addressed the role of the innate and adaptive immune responses of the host against the infection (van der Sar et al., 2009). A particular advantage of the zebrafish model is that the adaptive immune system is not fully developed until 5–6 weeks post fertilization, enabling studies of the innate immune system without the interference of the adaptive immunity. NGS technology using a tag-based transcriptome sequencing method revealed a high degree of transcriptional complexity of zebrafish response to the *Mycobacterium* infection (Hegedus et al., 2009). Both adult and the embryonic stage zebrafish models have been established for infections by *Francisella noatunensis* (Brudal et al., 2014, 2015;

Vojtech et al., 2009). In the embryos, macrophages were identified as the main site for infection. Bacterial replication was demonstrated by the use of GFP or mCherry reporter gene expressing bacteria in transgenic zebrafish embryos expressing either fluorescently labelled macrophages or neutrophils *(Tg(mpeg1:EGFP)gl22 and Tg(mpx:EGFP)i114,* respectively) (Brudal et al., 2014).

Mutants and attenuated strains of *E. tarda* and *V. angularium* have been explored in zebrafish for vaccine development and in comparative studies with aquaculture relevant fish (Gao et al., 2014; Liu et al., 2015; Santander et al., 2012). Both the mechanism of several mutations designed and constructed in the *E. tarda* genome (Okuda et al., 2014; Xiao et al., 2011) and host immune responses have been studied by zebrafish metabolomics and transcriptomics (Liu et al., 2014a; Yang et al., 2012, 2013; Zhao et al., 2014). Recently, vaccine effectiveness against *Flavobacterium psychrophilum* in salmonids was evaluated by neutrophil migration in transgenic zebrafish *(Tg(mpx:GFP)i114)* (Solis et al., 2015). A potential vaccine against *Mycobacterium* infections in fish was developed based on an attenuated *M. marinum* strain (Live attenuated vaccine: LAV) and shown to protect zebrafish against subsequent challenges with pathogenic bacteria (Cui et al., 2010). However, with relevance to aquaculture the documentation requirements for approved utilization of GMOs as vaccines can be substantial and for live attenuated vaccine strains the reversion to virulence is a concern (Bechah et al., 2010).

Furthermore, the importance of mucosal immunity (skin, gills, and intestine) of teleost fish has received increased focus (Salinas, 2015). Interestingly, in a study comparing LAV of *V. anguillarum* bath vaccinated zebrafish and turbot, relative survival rates were higher by mucosal vaccination compared to the injected vaccine formulation in zebrafish. The bath vaccination increased the protection of turbot from subsequent bath challenge with *V. anguillarum* (Zhang et al., 2014). In this study, the gene expression profile performed by RT-qPCR revealed an increased stimulation of the T helper cell 17 (Th17) specific genes in the mucosal tissue in bath vaccination zebrafish compared to vaccination by injection. These results have not yet been recapitulated in turbot. Using transcriptomics, the immune response to an infection of *Aeromonas hydrophila* through the skin was mapped in zebrafish where 388 genes were up- and 150 genes down-regulated (Lu et al., 2015).

Due to the failure in producing vaccines against many bacterial diseases with the concomitant use of antibiotics and problems with antibiotic resistance development in aquaculture, phage therapy as a treatment strategy for aquaculture has been considered (Nakai and Park, 2002; Oliveira et al., 2012). Here, zebrafish has been used to validate various phage therapy strategies (Laanto et al., 2012; Silva et al., 2014). Phage therapy was tested in a comparative study with zebrafish and rainbow trout against columnaris disease caused by *Flavobacterium columnare,* which resulted in an increase from 0% to 100% survival in zebrafish and from 8% to 40% in trout (Laanto et al., 2015). The phage *de novo* genome assembly sequencing revealed that the protective FCL-2 phage to be 47 kb long with 74 ORFs. While 30% of the ORFs did not give any matches in blast search, ORFs with predicted functions resembled those of known phage packing and function and some matched other known *Flavobacterium* phage. Two ORFs were identical to genes found in the *F. columinare* genome.

Viral infection genomics

Viral infections are among the major challenges to sustainable aquaculture and the potential of using zebrafish as a model to elucidate molecular mechanisms in response to viral infection has been reviewed (Novoa et al., 2010). There are long traditions of infection and vaccine models in zebrafish, from work on viral haemorrhagic virus (VHSV) that, together with spring viraemia carp virus (SVCV) cause infections in salmon and trout, and also zebrafish (Novoa et al., 2006; Sanders et al., 2003). A number of other viral disease models are established in zebrafish: Nervous necrosis virus (NNV) (Chen et al., 2015; Lu et al., 2008), infectious haematopoietic necrosis virus (IHNV) (Ludwig et al., 2011), and kidney necrosis virus (ISKNV) which infect aquaculture species worldwide (Xiong et al., 2011). Snakehead fish vesiculovirus (SHVV) infections were established in the zebrafish embryonic fibroblast cell line ZF4 (Wang et al., 2015). For VHSV, early immune-related responses of fins and organs of zebrafish after immersion infection were investigated using proteomics and transcriptomics (Encinas et al., 2010). The use of 2D-gel electrophoresis detected very few immune-related proteins, while RT-qPCR and microarrays proved more sensitive for detection of pathogen induced changes in gene expression. Further RT-qPCR is used for detecting the host responses against various immune stimulants and nanocarrieres for viral vaccination and drug delivery both in adult zebrafish and in zebrafish derived cell lines (Kavaliauskis et al., 2015; Ruyra et al., 2013; Ruyra et al., 2015). By site-directed mutagenesis, a mutant VHSV virus was created with reduced total positive-strand RNA synthesis. This delayed the cytopathic effect in fish cells and reduced virulence of the VHSV mutant was demonstrated in the zebrafish model (Kim et al., 2015).

Husbandry and nutrition

Husbandry guidelines

Good husbandry protocols are essential for ensuring the best standards of welfare for laboratory animals. To define an optimal laboratory environment for a species, its biology must be known and it is common practice to set standards related to the ecosystems of wild populations. However, since the lab strains of zebrafish have >25 years times 4 generations of domestication and adaption to the artificial ecosystems of aquarium tanks, this fact must be considered in the process of ensuring zebrafish animal welfare based on scientific knowledge. A key goal for animal welfare is described by the 3Rs: Replacement, Reduction, and Refinement (Russell, 1995). The 3R principals have laid the ground for the European Parliament to prepare the "Directive 2010/63/EU on the protection of animals used for scientific or educational purposes" to be implemented by the member states. For fish, it regulates parameters like water supply and quality, oxygen, nitrogen compounds, pH, salinity, temperature, lighting, noise, stocking density and environmental complexity, feeding and handling. Although the use of zebrafish in research is increasing and as such Reduction is not met, zebrafish do Replace mice and other mammalian *in vivo* models. One strategy to achieve Reduction is to lower the live stocks by increased use of cryopreservation (Draper and Moens, 2009; Yang et al., 2007). Moreover, during the first 6 dpf, the zebrafish embryo are in many instances an excellent initial screening model. Finally, with good husbandry standards and design of experimental protocols, reproducible results can be achieved with the use of minimal numbers of animals, and may result in a situation where Refinement leads to Reduction. Since environmental factors involved in husbandry can influence the epigenetic landscapes of a genome, fundamental differences in husbandry practices in different laboratories, or countries, may cause situations where results are not reproducible between studies and thus reduce the scientific value of the model. For this reason, it is a goal and an opportunity for an internationally well-organized research community to set common standards for zebrafish husbandry guidelines (Varga et al., 2016).

Nutritional omics

Fish feed has major influence on fish health and quality of the end product. For economical reasons and a decreased availability of fish based feed alternative protein sources are increasingly used in fish diets. Unfortunately non-optimal feed may cause side effects such as reduced growth, reproduction, and immunity as well as inflammation of the gut epithelium in several aquaculture relevant fish species (reviewed by Ulloa et al., 2014). The effects of the gut microbiota has received increased attention also with regards to diet (Al-Hisnawi et al., 2015; Bakke-McKellep et al., 2007; Gómez and Balcázar, 2008). With its availability of biotechnological and high-throughput tools zebrafish serve as a cost-effective model platform to screen and investigate responses to various nutrients like the composition of essential fatty acids, macro and micro nutrients on fat metabolism, growth, reproduction and inflammation (de Vrieze et al., 2014; Powell et al., 2015; Robison et al., 2008; Seiliez et al., 2013). The role of macro and micro nutrients and the ratios between them affects the expression of important developmental gene cascades in various aqua-cultured fish and was previously reviewed (Lall and Lewis-McCrea, 2007).

Nutritional genomics investigate the relationship between the host genome, transcriptome and diets of different fish (Ulloa et al., 2014). The field encompasses both nutrigenomics and nutrigenetics. The holistic approach offers an advancement to the previous practices where host responses measured by various gene expression analysis are selected more at random. Thereby the interpretation of data from one study to another are difficult to compare (Ulloa et al., 2014). With zebrafish as a model for nutritional investigations, it is equally important to improve the reproducibility of these scientific studies and the development of dietary standards for zebrafish are proposed (Lawrence, 2011). Rotifers (*Brachionus plicatilisare*) are commonly used as diet for fish larvae in the aquaculture industry and have been introduced as a rearing feed at the early developmental stages of zebrafish instead of Artemia and Paramecium (Lawrence et al., 2015). Discussion of the use of pro- and prebiotics and their influence of the microbiome has also reached the aquaculture industry and is discussed for teleost fish (Llewellyn et al., 2014). The taxonomic distribution of bacteria has been characterized after exposure to different diets by 16S RNA-sequencing in zebrafish (Rurangwa et al., 2015). As a basis, methods for using gnotobiotic zebrafish are established (Rawls et al., 2004) that facilitates further studies of the effect of the host microbiome also with regards to nutrition (Semova et al., 2012).

Effects of stressors

Zebrafish embryo toxicity test

The attributes of small size and transparent embryo development have made zebrafish increasingly used in screening for, and classification of toxic chemical compounds, gradually replacing mice and rats as test animals. This is reflected in the OECD guideline *Fish embryo acute toxicity test* (OECD Test No. 236, 2013) that describes measures of acute toxicity as determined by defined phenotypic effects of zebrafish early development. Newly fertilized zebrafish eggs are exposed to the test chemical for a period of 96 h. At every 24 h, four observations are recorded as indicators of lethality: (1) coagulation of the embryos, (2) lack of somite formation, (3) lack of detachment of the tail-bud from the yolk sac, and (4) lack of heartbeat; and the LC50 is calculated.

High-throughput screening and transgenic reporter fish

Toxicology and ecotoxicology differs in that the toxicants found in natural ecosystems exist as complex mixtures of compounds, including their own metabolites generated by degradation in microbiota etc. Among groups of pollutant chemicals of health concern are persistent organic pollutants (POPs), oestrogen-like compounds, and plastic additive chemicals like phthalates and bisphenol A (Nourizadeh-Lillabadi et al., 2009). In the complex chemical scenarios of environmental exposures, the different constituent compounds may display antagonistic, synergistic or additive toxic effects on each other. Sorting out such multiple stressor effects is only possible if high-throughput systems for screening are available. Like the concept in the OECD 336 test, complex mixtures and concentrations can be analyzed using zebrafish embryos and larvae exposed in 96- or 384-well plates. Phenotypic changes can be automatically screened for using high resolution bright field or fluorescent microscopy. Further, transgenic reporter fish, "Glofish," carrying different fluorescent protein genes fused to chemical sensitive control elements can turn fish colours on and off in real time in the presence/absence of the compound (Gong et al., 2003). Today six different fluorescence emission patterns are offered (red, orange, yellow, green, blue, and purple) allowing simultaneous, multicolor reporter gene expression. In addition to phenotypes observable by microscopy and reporter gene expression, effects on transcriptomes and epigenetic landscapes can be measured by NGS for genome-wide and quantitative PCR for targeted changes, which can uncover pathways and mechanisms giving rise to the observed phenotypes.

Transgenerational effects

Chemical pollutants, as well as radioactive sources radiation can cause DNA damage and mutations which, if manifested in the germ line, can be transferred to offspring generations. Increasing evidence demonstrate that epigenetic changes, induced by environmental factors, also can be inherited from progenitor to daughter cells. It is suggested, although still debated, that altered epigenetic landscapes and corresponding phenotypes can be transferred via germline to offspring generations (Anway et al., 2005; Anway et al., 2008; Heard and Martienssen, 2014; Skinner et al., 2015).

The question whether scenarios with long term environmental exposures with low dose rate radiation can be manifested as differential epigenetic patterns in zebrafish offspring generations are presently addressed in the laboratory within the Norwegian Centre for Environmental Radioactivity (http://cerad.nmbu.no). Global climate changes with alterations in temperature, pH and UV add more complexity to the natural exposure scenarios where, for example, an increase in UV may lead to photo enhanced toxicity (Barron et al., 2003). In summary, in order to conduct mechanistic studies and distinguish between mutations and epigenetic alterations a well annotated genome and deep knowledge of the molecular biology of the organism is mandatory.

Future directions future perspectives

Genomics in aquaculture is profiting from the technological refinements with the last two decades of "omics" now being combined with next generation sequencing. With access to nucleotide level genetics, animal and fish breeding is now being revolutionized by the introduction of sequence based parameters for selection programs. NGS has allowed a long list of wild and domesticated animals, fish and plants to have their genomes fully sequenced, albeit in many cases to a much lower level of annotation as compared to the lead model organisms. Although the annotation of genomes is still far from finalized, the emerging knowledge of the dynamic, but still relative stable, epigenomic landscapes opens the process of systematic mapping and understanding of their roles in gene expression control. Again laboratory *in vivo* models will be a step ahead and facilitate functional and comparative epigenomics availability for the commercial target species. There are reasons to postulate that zebrafish, together with a handful of other lead model organisms, will play an important role in this development. With the aim to summarize the pros and cons of the zebrafish model and what cost-benefits to aquaculture research it can offer, we have carried out a SWOT analysis with the form filled in from our perspective (Fig. 7.2). Since the answer to what are the pros and cons may vary with applications, the readers are invited to revise the SWOT analysis of Fig. 7.2 to fit their own interests.

Acknowledgments

We thank professor Gareth Griffiths at the Department of Biosciences, University of Oslo, for critically reading of and valuable comments to the manuscript.

References

Aanes, H., Winata, C.L., Lin, C.H., Chen, J.P., Srinivasan, K.G., Lee, S.G., Lim, A.Y., Hajan, H.S., Collas, P., Bourque, G., Gong, Z., Korzh, V., Alestrom, P., Mathavan, S., 2011. Zebrafish mRNA sequencing deciphers novelties in transcriptome dynamics during maternal to zygotic transition. Genome Res. 21, 1328–1338.

Aanes, H., Ostrup, O., Andersen, I.S., Moen, L.F., Mathavan, S., Collas, P., Aleström, P., 2013. Differential transcript isoform usage pre- and post-zygotic genome activation in zebrafish. BMC Genomics 14, 331.

Aleström, P., Kisen, G., Klungland, H., Andersen, O., 1992. Fish gonadotropin-releasing hormone gene and molecular approaches for control of sexual maturation: development of a transgenic fish model. Mol. Mar. Biol. Biotechnol. 1, 376–379.

Aleström, P., Holter, J.L., Nourizadeh-Lillabadi, R., 2006. Zebrafish in functional genomics and aquatic biomedicine. Trends Biotechnol. 24, 15–21.

Al-Hisnawi, A., Ringø, E., Davies, S.J., Waines, P., Bradley, G., Merrifield, D.L., 2015. First report on the autochthonous gut microbiota of brown trout (*Salmo trutta* Linnaeus). Aquacult. Res. 46, 2962–2971.

Andersen, Ø., Larsen, H.J.S., Aleström, P., 1992. Immunization of rainbow trout (*Oncorhynchus mykiss*) against gonadotropin releasing hormone: a possible approach to the control of sexual maturation in fish. Aquaculture 106, 195–200.

Anway, M.D., Cupp, A.S., Uzumcu, M., Skinner, M.K., 2005. Epigenetic transgenerational actions of endocrine disruptors and male fertility. Science 308, 1466–1469.

Anway, M.D., Rekow, S.S., Skinner, M.K., 2008. Transgenerational epigenetic programming of the embryonic testis transcriptome. Genomics 91, 30–40.

Aparicio, S., Chapman, J., Stupka, E., Putnam, N., Chia, J.M., Dehal, P., Christoffels, A., Rash, S., Hoon, S., Smit, A., Gelpke, M.D., Roach, J., Oh, T., Ho, I.Y., Wong, M., Detter, C., Verhoef, F., Predki, P., Tay, A., Lucas, S., Richardson, P., Smith, S.F., Clark, M.S., Edwards, Y.J., Doggett, N., Zharkikh, A., Tavtigian, S.V., Pruss, D., Barnstead, M., Evans, C., Baden, H., Powell, J., Glusman, G., Rowen, L., Hood, L., Tan, Y.H., Elgar, G., Hawkins, T., Venkatesh, B., Rokhsar, D., Brenner, S., 2002. Whole-genome shotgun assembly and analysis of the genome of Fugu rubripes. Science 297, 1301–1310.

Arunachalam, M., Raja, M., Vijayakumar, C., Malaiammal, P., Mayden, R.L., 2013. Natural history of zebrafish (*Danio rerio*) in India. Zebrafish 10, 1–14.

Bakke-McKellep, A.M., Penn, M.H., Salas, P.M., Refstie, S., Sperstad, S., Landsverk, T., Ringø, E., Krogdahl, A., 2007. Effects of dietary soyabean meal, inulin and oxytetracycline on intestinal microbiota and epithelial cell stress, apoptosis and proliferation in the teleost Atlantic salmon (Salmo salar L.). Br. J. Nutr. 97, 699–713.

Barron, M.G., Carls, M.G., Short, J.W., Rice, S.D., 2003. Photoenhanced toxicity of aqueous phase and chemically dispersed weathered Alaska North Slope crude oil to Pacific herring eggs and larvae. Environ. Toxicol. Chem. 22, 650–660.

Barski, A., Cuddapah, S., Cui, K., Roh, T.Y., Schones, D.E., Wang, Z., Wei, G., Chepelev, I., Zhao, K., 2007. High-resolution profiling of histone methylations in the human genome. Cell 129, 823–837.

Bechah, Y., El Karkouri, K., Mediannikov, O., Leroy, Q., Pelletier, N., Robert, C., Medigue, C., Mege, J.L., Raoult, D., 2010. Genomic, proteomic, and transcriptomic analysis of virulent and avirulent *Rickettsia prowazekii* reveals its adaptive mutation capabilities. Genome Res. 20, 655–663.

Berthelot, C., Brunet, F., Chalopin, D., Juanchich, A., Bernard, M., Noel, B., Bento, P., Da Silva, C., Labadie, K., Alberti, A., Aury, J.M., Louis, A., Dehais, P., Bardou, P., Montfort, J., Klopp, C., Cabau, C., Gaspin, C., Thorgaard, G.H., Boussaha, M., Quillet, E., Guyomard, R., Galiana, D., Bobe, J., Volff, J.N., Genet, C., Wincker, P., Jaillon, O., Roest Crollius, H., Guiguen, Y., 2014. The rainbow trout genome provides novel insights into evolution after whole-genome duplication in vertebrates. Nat. Commun. 5, 3657.

Bizuayehu, T.T., Babiak, I., 2014. MicroRNA in teleost fish. Genome Biol. Evol. 6, 1911–1937.

Bozzetta, E., Varello, K., Giorgi, I., Fioravanti, M.L., Pezzolato, M., Zanoni, R.G., Prearo, M., 2010. Mycobacterium marinum infection in a hybrid striped bass farm in Italy. J. Fish Dis. 33, 781–785.

Bradley, K.M., Elmore, J.B., Breyer, J.P., Yaspan, B.L., Jessen, J.R., Knapik, E.W., Smith, J.R., 2007. A major zebrafish polymorphism resource for genetic mapping. Genome Biol. 8, R55.

Brenner, S., Elgar, G., Sandford, R., Macrae, A., Venkatesh, B., Aparicio, S., 1993. Characterization of the pufferfish (Fugu) genome as a compact model vertebrate genome. Nature 366, 265–268.

Brocklebank, J., Raverty, S., Robinson, J., 2003. Mycobacteriosis in Atlantic salmon farmed in British Columbia. Can. Vet. J. 44, 486–489.

Broughton, R.E., Milam, J.E., Roe, B.A., 2001. The complete sequence of the zebrafish (*Danio rerio*) mitochondrial genome and evolutionary patterns in vertebrate mitochondrial DNA. Genome Res. 11, 1958–1967.

Brudal, E., Ulanova, L.S., E, O.L., Rishovd, A.L., Griffiths, G., Winther-Larsen, H.C., 2014. Establishment of three Francisella infections in zebrafish embryos at different temperatures. Infect. Immun. 82, 2180–2194.

Brudal, E., Lampe, E.O., Reubsaet, L., Roos, N., Hegna, I.K., Thrane, I.M., Koppang, E.O., Winther-Larsen, H.C., 2015. Vaccination with outer membrane vesicles from *Francisella noatunensis* reduces development of francisellosis in a zebrafish model. Fish Shellfish Immunol. 42, 50–57.

Butler, M.G., Iben, J.R., Marsden, K.C., Epstein, J.A., Granato, M., Weinstein, B.M., 2015. SNPfisher: tools for probing genetic variation in laboratory-reared zebrafish. Development 142, 1542–1552.

Catchen, J.M., Conery, J.S., Postlethwait, J.H., 2009. Automated identification of conserved synteny after whole-genome duplication. Genome Res. 19, 1497–1505.

Chen, H.Y., Liu, W., Wu, S.Y., Chiou, P.P., Li, Y.H., Chen, Y.C., Lin, G.H., Lu, M.W., Wu, J.L., 2015. RIG-I specifically mediates group II type I IFN activation in nervous necrosis virus infected zebrafish cells. Fish Shellfish Immunol. 43, 427–435.

Collas, P., Dahl, J.A., 2008. Chop it, ChIP it, check it: the current status of chromatin immunoprecipitation. Front. Biosci. 13, 929–943.

Collins, J.E., White, S., Searle, S.M., Stemple, D.L., 2012. Incorporating RNA-seq data into the zebrafish Ensembl genebuild. Genome Res. 22, 2067–2078.

Cortemeglia, C., Beitinger, T.L., 2005. Temperature Tolerances of Wild-Type and Red Transgenic Zebra Danios. T. Am. Fish. Soc. 134, 1431–1437.

Cui, Z., Samuel-Shaker, D., Watral, V., Kent, M.L., 2010. Attenuated *Mycobacterium marinum* protects zebrafish against mycobacteriosis. J. Fish Dis. 33, 371–375.

de Vrieze, E., Moren, M., Metz, J.R., Flik, G., Lie, K.K., 2014. Arachidonic acid enhances turnover of the dermal skeleton: studies on zebrafish scales. PLoS One 9, e89347.

Desvignes, T., Beam, M.J., Batzel, P., Sydes, J., Postlethwait, J.H., 2014. Expanding the annotation of zebrafish microRNAs based on small RNA sequencing. Gene 546, 386–389.

Devlin, R.H., Nagahama, Y., 2002. Sex determination and sex differentiation in fish: an overview of genetic, physiological, and environmental influences. Aquaculture 208, 191–364.

dos Santos, N.M., do Vale, A., Sousa, M.J., Silva, M.T., 2002. Mycobacterial infection in farmed turbot *Scophthalmus maximus*. Dis. Aquat. Organ. 52, 87–91.

Draper, B.W., Moens, C.B., 2009. A high-throughput method for zebrafish sperm cryopreservation and in vitro fertilization. J. Vis. Exp.

Dunham, I., Kundaje, A., Aldred, S.F., Collins, P.J., Davis, C.A., Doyle, F., et al., 2012. An integrated encyclopedia of DNA elements in the human genome. Nature 489, 57–74.

Encinas, P., Rodriguez-Milla, M.A., Novoa, B., Estepa, A., Figueras, A., Coll, J., 2010. Zebrafish fin immune responses during high mortality infections with viral haemorrhagic septicemia rhabdovirus. A proteomic and transcriptomic approach. BMC Genomics 11, 518.
Ender, C., Meister, G., 2010. Argonaute proteins at a glance. J. Cell Sci. 123, 1819–1823.
Engeszer, R.E., Ryan, M.J., Parichy, D.M., 2004. Learned social preference in zebrafish. Curr. Biol. 14, 881–884.
Engeszer, R.E., Patterson, L.B., Rao, A.A., Parichy, D.M., 2007. Zebrafish in the wild: a review of natural history and new notes from the field. Zebrafish 4, 21–40.
Engreitz, J.M., Pandya-Jones, A., McDonel, P., Shishkin, A., Sirokman, K., Surka, C., Kadri, S., Xing, J., Goren, A., Lander, E.S., Plath, K., Guttman, M., 2013. The Xist lncRNA exploits three-dimensional genome architecture to spread across the X chromosome. Science 341, 1237973.
Friedrich, R.W., Genoud, C., Wanner, A.A., 2013. Analyzing the structure and function of neuronal circuits in zebrafish. Front Neural Circuits 7, 71.
Furutani-Seiki, M., Wittbrodt, J., 2004. Medaka and zebrafish, an evolutionary twin study. Mech. Dev. 121, 629–637.
Gao, Y., Wu, H., Wang, Q., Qu, J., Liu, Q., Xiao, J., Zhang, Y., 2014. A live attenuated combination vaccine evokes effective immune-mediated protection against *Edwardsiella tarda* and *Vibrio anguillarum*. Vaccine 32, 5937–5944.
Gerstein, M.B., Lu, Z.J., Van Nostrand, E.L., Cheng, C., Arshinoff, B.I., Liu, T., Yip, K.Y., Robilotto, R., Rechtsteiner, A., Ikegami, K., Alves, P., Chateigner, A., Perry, M., Morris, M., Auerbach, R.K., Feng, X., Leng, J., Vielle, A., Niu, W., Rhrissorrakrai, K., Agarwal, A., Alexander, R.P., Barber, G., Brdlik, C.M., Brennan, J., Brouillet, J.J., Carr, A., Cheung, M.-S., Clawson, H., Contrino, S., Dannenberg, L.O., Dernburg, A.F., Desai, A., Dick, L., Dosé, A.C., Du, J., Egelhofer, T., Ercan, S., Euskirchen, G., Ewing, B., Feingold, E.A., Gassmann, R., Good, P.J., Green, P., Gullier, F., Gutwein, M., Guyer, M.S., Habegger, L., Han, T., Henikoff, J.G., Henz, S.R., Hinrichs, A., Holster, H., Hyman, T., Iniguez, A.L., Janette, J., Jensen, M., Kato, M., Kent, W.J., Kephart, E., Khivansara, V., Khurana, E., Kim, J.K., Kolasinska-Zwierz, P., Lai, E.C., Latorre, I., Leahey, A., Lewis, S., Lloyd, P., Lochovsky, L., Lowdon, R.F., Lubling, Y., Lyne, R., MacCoss, M., Mackowiak, S.D., Mangone, M., McKay, S., Mecenas, D., Merrihew, G., Miller, D.M., Muroyama, A., Murray, J.I., Ooi, S.-L., Pham, H., Phippen, T., Preston, E.A., Rajewsky, N., Rätsch, G., Rosenbaum, H., Rozowsky, J., Rutherford, K., Ruzanov, P., Sarov, M., Sasidharan, R., Sboner, A., Scheid, P., Segal, E., Shin, H., Shou, C., Slack, F.J., Slightam, C., Smith, R., Spencer, W.C., Stinson, E.O., Taing, S., Takasaki, T., Vafeados, D., Voronina, K., Wang, G., Washington, N.L., Whittle, C.M., Wu, B., Yan, K.-K., Zeller, G., Zha, Z., Zhong, M., Zhou, X., Ahringer, J., Strome, S., Gunsalus, K.C., Micklem, G., Liu, X.S., Reinke, V., Kim, S.K., Hillier, L.W., Henikoff, S., Piano, F., Snyder, M., Stein, L., Lieb, J.D., Waterston, R.H., 2010. Integrative analysis of the Caenorhabditis elegans Genome by the modENCODE project. Science 330, 1775–1787.
Gerstein, M.B., Rozowsky, J., Yan, K.K., Wang, D., Cheng, C., Brown, J.B., Davis, C.A., Hillier, L., Sisu, C., Li, J.J., Pei, B., Harmanci, A.O., Duff, M.O., Djebali, S., Alexander, R.P., Alver, B.H., Auerbach, R., Bell, K., Bickel, P.J., Boeck, M.E., Boley, N.P., Booth, B.W., Cherbas, L., Cherbas, P., Di, C., Dobin, A., Drenkow, J., Ewing, B., Fang, G., Fastuca, M., Feingold, E.A., Frankish, A., Gao, G., Good, P.J., Guigo, R., Hammonds, A., Harrow, J., Hoskins, R.A., Howald, C., Hu, L., Huang, H., Hubbard, T.J., Huynh, C., Jha, S., Kasper, D., Kato, M., Kaufman, T.C., Kitchen, R.R., Ladewig, E., Lagarde, J., Lai, E., Leng, J., Lu, Z., MacCoss, M., May, G., McWhirter, R., Merrihew, G., Miller, D.M., Mortazavi, A., Murad, R., Oliver, B., Olson, S., Park, P.J., Pazin, M.J., Perrimon, N., Pervouchine, D.,

Reinke, V., Reymond, A., Robinson, G., Samsonova, A., Saunders, G.I., Schlesinger, F., Sethi, A., Slack, F.J., Spencer, W.C., Stoiber, M.H., Strasbourger, P., Tanzer, A., Thompson, O.A., Wan, K.H., Wang, G., Wang, H., Watkins, K.L., Wen, J., Wen, K., Xue, C., Yang, L., Yip, K., Zaleski, C., Zhang, Y., Zheng, H., Brenner, S.E., Graveley, B.R., Celniker, S.E., Gingeras, T.R., Waterston, R., 2014. Comparative analysis of the transcriptome across distant species. Nature 512, 445–448.

Gómez, G.D.1., Balcázar, J.L., 2008. A review on the interactions between gut microbiota and innate immunity of fish. FEMS Immunol. Med. Microbiol. 52, 145–154.

Gong, Z., Wan, H., Tay, T.L., Wang, H., Chen, M., Yan, T., 2003. Development of transgenic fish for ornamental and bioreactor by strong expression of fluorescent proteins in the skeletal muscle. Biochem. Biophys. Res. Commun. 308, 58–63.

Gratacap, R.L., Wheeler, R.T., 2014. Utilization of zebrafish for intravital study of eukaryotic pathogen-host interactions. Dev. Comp. Immunol. 46, 108–115.

Harper, C., Lawrence, C., 2012. The Laboratory Zebrafish. CRC Press, Boca Raton, Florida.

Harvie, E.A., Huttenlocher, A., 2015. Non-invasive imaging of the innate immune response in a zebrafish larval model of *Streptococcus iniae* infection. J. Vis. Exp.

Harvie, E.A., Green, J.M., Neely, M.N., Huttenlocher, A., 2013. Innate immune response to *Streptococcus iniae* infection in zebrafish larvae. Infect. Immun. 81, 110–121.

Heard, E., Martienssen, R.A., 2014. Transgenerational epigenetic inheritance: myths and mechanisms. Cell 157, 95–109.

Hegedus, Z., Zakrzewska, A., Agoston, V.C., Ordas, A., Racz, P., Mink, M., Spaink, H.P., Meijer, A.H., 2009. Deep sequencing of the zebrafish transcriptome response to mycobacterium infection. Mol. Immunol. 46, 2918–2930.

Heyn, P., Kircher, M., Dahl, A., Kelso, J., Tomancak, P., Kalinka, A.T., Neugebauer, K.M., 2014. The earliest transcribed zygotic genes are short, newly evolved, and different across species. Cell Rep. 6, 285–292.

Hilton, I.B., D'Ippolito, A.M., Vockley, C.M., Thakore, P.I., Crawford, G.E., Reddy, T.E., Gersbach, C.A., 2015. Epigenome editing by a CRISPR-Cas9-based acetyltransferase activates genes from promoters and enhancers. Nat. Biotechnol. 33, 510–517.

Holoch, D., Moazed, D., 2015. RNA-mediated epigenetic regulation of gene expression. Nat. Rev. Genet. 16, 71–84.

Howe, K., Clark, M.D., Torroja, C.F., Torrance, J., Berthelot, C., Muffato, M., Collins, J.E., Humphray, S., McLaren, K., Matthews, L., McLaren, S., Sealy, I., Caccamo, M., Churcher, C., Scott, C., Barrett, J.C., Koch, R., Rauch, G.J., White, S., Chow, W., Kilian, B., Quintais, L.T., Guerra-Assuncao, J.A., Zhou, Y., Gu, Y., Yen, J., Vogel, J.H., Eyre, T., Redmond, S., Banerjee, R., Chi, J., Fu, B., Langley, E., Maguire, S.F., Laird, G.K., Lloyd, D., Kenyon, E., Donaldson, S., Sehra, H., Almeida-King, J., Loveland, J., Trevanion, S., Jones, M., Quail, M., Willey, D., Hunt, A., Burton, J., Sims, S., McLay, K., Plumb, B., Davis, J., Clee, C., Oliver, K., Clark, R., Riddle, C., Elliot, D., Threadgold, G., Harden, G., Ware, D., Begum, S., Mortimore, B., Kerry, G., Heath, P., Phillimore, B., Tracey, A., Corby, N., Dunn, M., Johnson, C., Wood, J., Clark, S., Pelan, S., Griffiths, G., Smith, M., Glithero, R., Howden, P., Barker, N., Lloyd, C., Stevens, C., Harley, J., Holt, K., Panagiotidis, G., Lovell, J., Beasley, H., Henderson, C., Gordon, D., Auger, K., Wright, D., Collins, J., Raisen, C., Dyer, L., Leung, K., Robertson, L., Ambridge, K., Leongamornlert, D., McGuire, S., Gilderthorp, R., Griffiths, C., Manthravadi, D., Nichol, S., Barker, G., Whitehead, S., Kay, M., Brown, J., Murnane, C., Gray, E., Humphries, M., Sycamore, N., Barker, D., Saunders, D., Wallis, J., Babbage, A., Hammond, S., Mashreghi-Mohammadi, M., Barr, L., Martin, S., Wray, P., Ellington, A., Matthews, N., Ellwood, M., Woodmansey, R., Clark, G., Cooper, J., Tromans, A., Grafham, D., Skuce, C., Pandian, R., Andrews, R., Harrison,

E., Kimberley, A., Garnett, J., Fosker, N., Hall, R., Garner, P., Kelly, D., Bird, C., Palmer, S., Gehring, I., Berger, A., Dooley, C.M., Ersan-Urun, Z., Eser, C., Geiger, H., Geisler, M., Karotki, L., Kirn, A., Konantz, J., Konantz, M., Oberlander, M., Rudolph-Geiger, S., Teucke, M., Lanz, C., Raddatz, G., Osoegawa, K., Zhu, B., Rapp, A., Widaa, S., Langford, C., Yang, F., Schuster, S.C., Carter, N.P., Harrow, J., Ning, Z., Herrero, J., Searle, S.M., Enright, A., Geisler, R., Plasterk, R.H., Lee, C., Westerfield, M., de Jong, P.J., Zon, L.I., Postlethwait, J.H., Nusslein-Volhard, C., Hubbard, T.J., Roest Crollius, H., Rogers, J., Stemple, D.L., 2013. The zebrafish reference genome sequence and its relationship to the human genome. Nature 496, 498–503.

Hwang, W.Y., Fu, Y., Reyon, D., Maeder, M.L., Kaini, P., Sander, J.D., Joung, J.K., Peterson, R.T., Yeh, J.R., 2013a. Heritable and precise zebrafish genome editing using a CRISPR-Cas system. PLoS One 8, e68708.

Hwang, W.Y., Fu, Y., Reyon, D., Maeder, M.L., Tsai, S.Q., Sander, J.D., Peterson, R.T., Yeh, J.R., Joung, J.K., 2013b. Efficient genome editing in zebrafish using a CRISPR-Cas system. Nat. Biotechnol. 31, 227–229.

Jiang, L., Zhang, J., Wang, J.J., Wang, L., Zhang, L., Li, G., Yang, X., Ma, X., Sun, X., Cai, J., Zhang, J., Huang, X., Yu, M., Wang, X., Liu, F., Wu, C.I., He, C., Zhang, B., Ci, W., Liu, J., 2013. Sperm, but not oocyte. DNA methylome is inherited by zebrafish early embryos. Cell 153, 773–784.

Jones, P.A., 2012. Functions of DNA methylation: islands, start sites, gene bodies and beyond. Nat. Rev. Genet. 13, 484–492.

Kamstra, J.H., Aleström, P., Kooter, J.M., Legler, J., 2015a. Zebrafish as a model to study the role of DNA methylation in environmental toxicology. Environ. Sci. Pollut. Res. Int. 22, 16262–16276.

Kamstra, J.H., Loken, M., Aleström, P., Legler, J., 2015b. Dynamics of DNA hydroxymethylation in zebrafish. Zebrafish 12, 230–237.

Kavaliauskis, A., Arnemo, M., Kim, S.H., Ulanova, L., Speth, M., Novoa, B., Dios, S., Evensen, O., Griffiths, G.W., Gjoen, T., 2015. Use of poly(I:C) stabilized with Chitosan as a vaccine-adjuvant against viral hemorrhagic septicemia virus infection in zebrafish. Zebrafish 12, 421–431.

Kedde, M., Strasser, M.J., Boldajipour, B., Oude Vrielink, J.A., Slanchev, K., le Sage, C., Nagel, R., Voorhoeve, P.M., van Duijse, J., Orom, U.A., Lund, A.H., Perrakis, A., Raz, E., Agami, R., 2007. RNA-binding protein Dnd1 inhibits microRNA access to target mRNA. Cell 131, 1273–1286.

Keller, P.J., Schmidt, A.D., Wittbrodt, J., Stelzer, E.H., 2008. Reconstruction of zebrafish early embryonic development by scanned light sheet microscopy. Science 322, 1065–1069.

Kent, M.L., Bishop-Stewart, J.K., Matthews, J.L., Spitsbergen, J.M., 2002. Pseudocapillaria tomentosa, a nematode pathogen, and associated neoplasms of zebrafish (*Danio rerio*) kept in research colonies. Comp. Med. 52, 354–358.

Ketting, R.F., 2011. microRNA Biogenesis and Function : an overview. Adv. Exp. Med. Biol. 700, 1–14.

Kettleborough, R.N., Busch-Nentwich, E.M., Harvey, S.A., Dooley, C.M., de Bruijn, E., van Eeden, F., Sealy, I., White, R.J., Herd, C., Nijman, I.J., Fenyes, F., Mehroke, S., Scahill, C., Gibbons, R., Wali, N., Carruthers, S., Hall, A., Yen, J., Cuppen, E., Stemple, D.L., 2013. A systematic genome-wide analysis of zebrafish protein-coding gene function. Nature 496, 494–497.

Kim, S.H., Guo, T.C., Vakharia, V.N., Evensen, Ø., 2015. Specific nucleotides at the 3'-terminal promoter of viral hemorrhagic septicemia virus are important for virulence in vitro and in vivo. Virology 476, 226–232.

Kluver, N., Konig, M., Ortmann, J., Massei, R., Paschke, A., Kuhne, R., Scholz, S., 2015. Fish embryo toxicity test: identification of compounds with weak toxicity and analysis of behavioral effects to improve prediction of acute toxicity for neurotoxic compounds. Environ. Sci. Technol. 49, 7002–7011.

Laanto, E., Bamford, J.K., Laakso, J., Sundberg, L.R., 2012. Phage-driven loss of virulence in a fish pathogenic bacterium. PLoS One 7, e53157.

Laanto, E., Bamford, J.K., Ravantti, J.J., Sundberg, L.R., 2015. The use of phage FCL-2 as an alternative to chemotherapy against columnaris disease in aquaculture. Front. Microbiol. 6, 829.

Lall, S.P., Lewis-McCrea, L.M., 2007. Role of nutrients in skeletal metabolism and pathology in fish—an overview. Aquaculture 267, 3–19.

Langevin, C., Aleksejeva, E., Passoni, G., Palha, N., Levraud, J.P., Boudinot, P., 2013. The antiviral innate immune response in fish: evolution and conservation of the IFN system. J. Mol. Biol. 425, 4904–4920.

Lawrence, C., 2011. Advances in zebrafish husbandry and management. Methods Cell Biol. 104, 429–451.

Lawrence, C., James, A., Mobley, S., 2015. Successful replacement of *Artemia salina nauplii* with marine Rotifers (*Brachionus plicatilis*) in the diet of Preadult zebrafish (*Danio rerio*). Zebrafish 12, 366–371.

Lindeman, L.C., Vogt-Kielland, L.T., Alestrom, P., Collas, P., 2009. Fish'n ChIPs: chromatin immunoprecipitation in the zebrafish embryo. Methods Mol. Biol. 567, 75–86.

Lindeman, L.C., Andersen, I.S., Reiner, A.H., Li, N., Aanes, H., Ostrup, O., Winata, C., Mathavan, S., Muller, F., Alestrom, P., Collas, P., 2011. Prepatterning of developmental gene expression by modified histones before zygotic genome activation. Dev. Cell. 21, 993–1004.

Liu, X., Chang, X., Wu, H., Xiao, J., Gao, Y., Zhang, Y., 2014a. Role of intestinal inflammation in predisposition of *Edwardsiella tarda* infection in zebrafish (*Danio rerio*). Fish Shellfish Immunol. 41, 271–278.

Liu, X., Wu, H., Chang, X., Tang, Y., Liu, Q., Zhang, Y., 2014b. Notable mucosal immune responses induced in the intestine of zebrafish (*Danio rerio*) bath-vaccinated with a live attenuated *Vibrio anguillarum* vaccine. Fish Shellfish Immunol. 40, 99–108.

Liu, X., Wu, H., Liu, Q., Wang, Q., Xiao, J., Chang, X., Zhang, Y., 2015. Profiling immune response in zebrafish intestine, skin, spleen and kidney bath-vaccinated with a live attenuated *Vibrio anguillarum* vaccine. Fish Shellfish Immunol. 45, 342–345.

Llewellyn, M.S., Boutin, S., Hoseinifar, S.H., Derome, N., 2014. Teleost microbiomes: the state of the art in their characterization, manipulation and importance in aquaculture and fisheries. Front. Microbiol. 5, 207.

Lu, M.W., Chao, Y.M., Guo, T.C., Santi, N., Evensen, Ø., Kasani, S.K., Hong, J.R., Wu, J.L., 2008. The interferon response is involved in nervous necrosis virus acute and persistent infection in zebrafish infection model. Mol. Immunol. 45, 1146–1152.

Lu, A.J., Hu, X.C., Wang, Y., Zhu, A.H., Shen, L.L., Tian, J., Feng, Z.Z., Feng, Z.J., 2015. Skin immune response in the zebrafish, *Danio rerio* (Hamilton), to *Aeromonas hydrophila* infection: a transcriptional profiling approach. J. Fish Dis. 38, 137–150.

Ludwig, M., Palha, N., Torhy, C., Briolat, V., Colucci-Guyon, E., Bremont, M., Herbomel, P., Boudinot, P., Levraud, J.P., 2011. Whole-body analysis of a viral infection: vascular endothelium is a primary target of infectious hematopoietic necrosis virus in zebrafish larvae. PLoS Pathog. 7, e1001269.

Matthews, J.L., 2004. Common diseases of laboratory zebrafish. Methods Cell Biol. 77, 617–643.

Matthews, J.L., Brown, A.M., Larison, K., Bishop-Stewart, J.K., Rogers, P., Kent, M.L., 2001. *Pseudoloma neurophilia* n. g., n. sp., a new microsporidium from the central nervous system of the zebrafish (*Danio rerio*). J. Eukaryot. Microbiol. 48, 227–233.

McClure, M., McIntyre, P., McCune, A., 2006. Notes on the natural diet and habitat of eight danionin fishes, including the zebrafish *Danio rerio*. J. Fish Biol. 69, 553–570.

McCluskey, B.M., Postlethwait, J.H., 2015. Phylogeny of zebrafish, a "model species," within Danio, a "model genus". Mol. Biol. Evol. 32, 635–652.

Meijer, A.H., Spaink, H.P., 2011. Host-pathogen interactions made transparent with the zebrafish model. Curr. Drug Targets 12, 1000–1017.

Meijer, A.H., Gabby Krens, S.F., Medina Rodriguez, I.A., He, S., Bitter, W., Ewa Snaar-Jagalska, B., Spaink, H.P., 2004. Expression analysis of the toll-like receptor and TIR domain adaptor families of zebrafish. Mol. Immunol. 40, 773–783.

Meyer, A., Schartl, M., 1999. Gene and genome duplications in vertebrates: the one-to-four (-to-eight in fish) rule and the evolution of novel gene functions. Curr. Opin. Cell Biol. 11, 699–704.

Mikkelsen, T.S., Ku, M., Jaffe, D.B., Issac, B., Lieberman, E., Giannoukos, G., Alvarez, P., Brockman, W., Kim, T.K., Koche, R.P., Lee, W., Mendenhall, E., O'Donovan, A., Presser, A., Russ, C., Xie, X., Meissner, A., Wernig, M., Jaenisch, R., Nusbaum, C., Lander, E.S., Bernstein, B.E., 2007. Genome-wide maps of chromatin state in pluripotent and lineage-committed cells. Nature 448, 553–560.

Muller, F., Blader, P., Strahle, U., 2002. Search for enhancers: teleost models in comparative genomic and transgenic analysis of cis regulatory elements. Bioessays 24, 564–572.

Nakai, T., Park, S.C., 2002. Bacteriophage therapy of infectious diseases in aquaculture. Res. Microbiol. 153, 13–18.

Near, T.J., Eytan, R.I., Dornburg, A., Kuhn, K.L., Moore, J.A., Davis, M.P., Wainwright, P.C., Friedman, M., Smith, W.L., 2012. Resolution of ray-finned fish phylogeny and timing of diversification. Proc. Natl. Acad. Sci. USA 109, 13698–13703.

Neely, M.N., Pfeifer, J.D., Caparon, M., 2002. Streptococcus-zebrafish model of bacterial pathogenesis. Infect. Immun. 70, 3904–3914.

Nourizadeh-Lillabadi, R., Lyche, J.L., Almaas, C., Stavik, B., Moe, S.J., Aleksandersen, M., Berg, V., Jakobsen, K.S., Stenseth, N.C., Skare, J.U., Alestrom, P., Ropstad, E., 2009. Transcriptional regulation in liver and testis associated with developmental and reproductive effects in male zebrafish exposed to natural mixtures of persistent organic pollutants (POP). J. Toxicol. Environ. Health A 72, 112–130.

Novoa, B., Romero, A., Mulero, V., Rodriguez, I., Fernandez, I., Figueras, A., 2006. Zebrafish (Danio rerio) as a model for the study of vaccination against viral haemorrhagic septicemia virus (VHSV). Vaccine 24, 5806–5816.

Novoa, B., Mackenzie, S., Figueras, A., 2010. Inflammation and innate immune response against viral infections in marine fish. Curr. Pharm. Des. 16, 4175–4184.

O'Flynn, F., McGeachy, S., Friars, G., Benfey, T., Bailey, J., 1997. Comparisons of cultured triploid and diploid Atlantic salmon (Salmo salar L.). ICES J. Mar. Sci. 54, 1160–1165.

Okuda, J., Takeuchi, Y., Nakai, T., 2014. Type III secretion system genes of *Edwardsiella tarda* associated with intracellular replication and virulence in zebrafish. Dis. Aquat. Organ. 111, 31–39.

Oliveira, J., Castilho, F., Cunha, A., Pereira, M., 2012. Bacteriophage therapy as a bacterial control strategy in aquaculture. Aquacult. Int. 20, 879–910.

Pauli, A., Valen, E., Lin, M.F., Garber, M., Vastenhouw, N.L., Levin, J.Z., Fan, L., Sandelin, A., Rinn, J.L., Regev, A., Schier, A.F., 2012. Systematic identification of long noncoding RNAs expressed during zebrafish embryogenesis. Genome Res. 22, 577–591.

Pfeifer, G.P., Kadam, S., Jin, S.G., 2013. 5-hydroxymethylcytosine and its potential roles in development and cancer. Epigenetics Chromatin 6, 10.

Piferrer, F., 2013. Epigenetics of sex determination and gonadogenesis. Dev. Dyn. 242, 360–370.

Postlethwait, J., Amores, A., Cresko, W., Singer, A., Yan, Y.L., 2004. Subfunction partitioning, the teleost radiation and the annotation of the human genome. Trends Genet. 20, 481–490.

Potok, M.E., Nix, D.A., Parnell, T.J., Cairns, B.R., 2013. Reprogramming the maternal zebrafish genome after fertilization to match the paternal methylation pattern. Cell 153, 759–772.

Powell, M.L., Pegues, M.A., Szalai, A.J., Ghanta, V.K., D'Abramo, L.R., Watts, S.A., 2015. Effects of the dietary omega3:omega6 fatty acid ratio on body fat and inflammation in zebrafish (*Danio rerio*). Comp. Med. 65, 289–294.

Pressley, M.E., Phelan Iii, P.E., Eckhard Witten, P., Mellon, M.T., Kim, C.H., 2005. Pathogenesis and inflammatory response to *Edwardsiella tarda* infection in the zebrafish. Dev. Comp. Immunol. 29, 501–513.

Rawls, J.F., Samuel, B.S., Gordon, J.I., 2004. Gnotobiotic zebrafish reveal evolutionarily conserved responses to the gut microbiota. Proc. Natl. Acad. Sci. USA 101, 4596–4601.

Reed, B., Jennings, M., 2011. Guidance on the housing and care of Zebrafish. Southwater: Royal Society for the Prevention of Cruelty to Animals.

Ribas, L.P.F., 2013. The zebrafish (Danio rerio) as a model organism, with emphasis on applications for finfish aquaculture research. Rev. Aquacult. 6, 52.

Robison, B.D., Drew, R.E., Murdoch, G.K., Powell, M., Rodnick, K.J., Settles, M., Stone, D., Churchill, E., Hill, R.A., Papasani, M.R., Lewis, S.S., Hardy, R.W., 2008. Sexual dimorphism in hepatic gene expression and the response to dietary carbohydrate manipulation in the zebrafish (Danio rerio). Comp. Biochem. Physiol. Part D Genomics Proteomics 3, 141–154.

Rougeot, J., Zakrzewska, A., Kanwal, Z., Jansen, H.J., Spaink, H.P., Meijer, A.H., 2014. RNA sequencing of FACS-sorted immune cell populations from zebrafish infection models to identify cell specific responses to intracellular pathogens. Methods Mol. Biol. 1197, 261–274.

Roy, S., Ernst, J., Kharchenko, P.V., Kheradpour, P., Negre, N., Eaton, M.L., Landolin, J.M., Bristow, C.A., Ma, L., Lin, M.F., Washietl, S., Arshinoff, B.I., Ay, F., Meyer, P.E., Robine, N., Washington, N.L., Di Stefano, L., Berezikov, E., Brown, C.D., Candeias, R., Carlson, J.W., Carr, A., Jungreis, I., Marbach, D., Sealfon, R., Tolstorukov, M.Y., Will, S., Alekseyenko, A.A., Artieri, C., Booth, B.W., Brooks, A.N., Dai, Q., Davis, C.A., Duff, M.O., Feng, X., Gorchakov, A.A., Gu, T., Henikoff, J.G., Kapranov, P., Li, R., MacAlpine, H.K., Malone, J., Minoda, A., Nordman, J., Okamura, K., Perry, M., Powell, S.K., Riddle, N.C., Sakai, A., Samsonova, A., Sandler, J.E., Schwartz, Y.B., Sher, N., Spokony, R., Sturgill, D., van Baren, M., Wan, K.H., Yang, L., Yu, C., Feingold, E., Good, P., Guyer, M., Lowdon, R., Ahmad, K., Andrews, J., Berger, B., Brenner, S.E., Brent, M.R., Cherbas, L., Elgin, S.C., Gingeras, T.R., Grossman, R., Hoskins, R.A., Kaufman, T.C., Kent, W., Kuroda, M.I., Orr-Weaver, T., Perrimon, N., Pirrotta, V., Posakony, J.W., Ren, B., Russell, S., Cherbas, P., Graveley, B.R., Lewis, S., Micklem, G., Oliver, B., Park, P.J., Celniker, S.E., Henikoff, S., Karpen, G.H., Lai, E.C., MacAlpine, D.M., Stein, L.D., White, K.P., Kellis, M., 2010. Identification of functional elements and regulatory circuits by Drosophila modENCODE. Science 330, 1787–1797.

Rurangwa, E., Sipkema, D., Kals, J., Ter Veld, M., Forlenza, M., Bacanu, G.M., Smidt, H., Palstra, A.P., 2015. Impact of a novel protein meal on the gastrointestinal microbiota and the host transcriptome of larval zebrafish *Danio rerio*. Front. Physiol. 6, 133.

Russell, W.M., 1995. The development of the three Rs concept. Altern. Lab. Anim. 23, 298–304.

Ruyra, A., Cano-Sarabia, M., Mackenzie, S.A., Maspoch, D., Roher, N., 2013. A novel liposome-based nanocarrier loaded with an LPS-dsRNA cocktail for fish innate immune system stimulation. PLoS One 8, e76338.

Ruyra, A., Torrealba, D., Morera, D., Tort, L., MacKenzie, S., Roher, N., 2015. Zebrafish liver (ZFL) cells are able to mount an anti-viral response after stimulation with Poly (I:C). Comp. Biochem. Physiol. B Biochem. Mol. Biol. 182, 55–63.

Sadakierska-Chudy, A., Filip, M., 2015. A comprehensive view of the epigenetic landscape. Part II: Histone post-translational modification, nucleosome level, and chromatin regulation by ncRNAs. Neurotox. Res. 27, 172–197.

Salinas, I., 2015. The mucosal immune system of teleost fish. Biology (Basel) 4, 525–539.

Sanders, G.E., Batts, W.N., Winton, J.R., 2003. Susceptibility of zebrafish (*Danio rerio*) to a model pathogen, spring viremia of carp virus. Comp. Med. 53, 514–521.

Sanger, F., Air, G.M., Barrell, B.G., Brown, N.L., Coulson, A.R., Fiddes, C.A., Hutchison, C.A., Slocombe, P.M., Smith, M., 1977. Nucleotide sequence of bacteriophage phi X174 DNA. Nature 265, 687–695.

Santander, J., Golden, G., Wanda, S.Y., Curtiss, 3rd, R., 2012. Fur-regulated iron uptake system of *Edwardsiella ictaluri* and its influence on pathogenesis and immunogenicity in the catfish host. Infect. Immun. 80, 2689–2703.

Santoriello, C., Zon, L.I., 2012. Hooked! Modeling human disease in zebrafish. J. Clin. Invest. 122, 2337–2343.

Saralahti, A., Ramet, M., 2015. Zebrafish and Streptococcal infections. Scand. J. Immunol. 82, 174–183.

Schmid, B., Shah, G., Scherf, N., Weber, M., Thierbach, K., Campos, C.P., Roeder, I., Aanstad, P., Huisken, J., 2013. High-speed panoramic light-sheet microscopy reveals global endodermal cell dynamics. Nat. Commun. 4, 2207.

Seiliez, I., Medale, F., Aguirre, P., Larquier, M., Lanneretonne, L., Alami-Durante, H., Panserat, S., Skiba-Cassy, S., 2013. Postprandial regulation of growth- and metabolism-related factors in zebrafish. Zebrafish 10, 237–248.

Semova, I., Carten, J.D., Stombaugh, J., Mackey, L.C., Knight, R., Farber, S.A., Rawls, J.F., 2012. Microbiota regulate intestinal absorption and metabolism of fatty acids in the zebrafish. Cell Host Microbe. 12, 277–288.

Seth, A., Stemple, D.L., Barroso, I., 2013. The emerging use of zebrafish to model metabolic disease. Dis. Model Mech. 6, 1080–1088.

Silva, Y.J., Costa, L., Pereira, C., Mateus, C., Cunha, A., Calado, R., Gomes, N.C., Pardo, M.A., Hernandez, I., Almeida, A., 2014. Phage therapy as an approach to prevent *Vibrio anguillarum* infections in fish larvae production. PLoS One 9, e114197.

Skinner, M.K., 2016. Endocrine disruptors in 2015: epigenetic transgenerational inheritance. Nat. Rev. Endocrinol. 12, 68–70.

Skinner, M.K., Guerrero-Bosagna, C., Haque, M.M., 2015. Environmentally induced epigenetic transgenerational inheritance of sperm epimutations promote genetic mutations. Epigenetics 10, 762–771.

Skugor, A., Tveiten, H., Krasnov, A., Andersen, O., 2014. Knockdown of the germ cell factor Dead end induces multiple transcriptional changes in Atlantic cod (*Gadus morhua*) hatchlings. Anim. Reprod. Sci. 144, 129–137.

Smith, Z.D., Meissner, A., 2013. DNA methylation: roles in mammalian development. Nat. Rev. Genet. 14, 204–220.

Solis, C.J., Poblete-Morales, M., Cabral, S., Valdes, J.A., Reyes, A.E., Avendano-Herrera, R., Feijoo, C.G., 2015. Neutrophil migration in the activation of the innate immune response

to different Flavobacterium psychrophilum vaccines in zebrafish (Danio rerio). J. Immunol. Res. 2015, 515187.
Spence, R., Gerlach, G., Lawrence, C., Smith, C., 2008. The behaviour and ecology of the zebrafish. Danio rerio. Biol. Rev. Camb. Philos. Soc. 83, 13–34.
Spivakov, M., Auer, T.O., Peravali, R., Dunham, I., Dolle, D., Fujiyama, A., Toyoda, A., Aizu, T., Minakuchi, Y., Loosli, F., Naruse, K., Birney, E., Wittbrodt, J., 2014. Genomic and phenotypic characterization of a wild medaka population: towards the establishment of an isogenic population genetic resource in fish. G3 (Bethesda) 4, 433–445.
Stickney, H.L., Schmutz, J., Woods, I.G., Holtzer, C.C., Dickson, M.C., Kelly, P.D., Myers, R.M., Talbot, W.S., 2002. Rapid mapping of zebrafish mutations with SNPs and oligonucleotide microarrays. Genome Res. 12, 1929–1934.
Strahl, B.D., Allis, C.D., 2000. The language of covalent histone modifications. Nature 403, 41–45.
Streisinger, G., Walker, C., Dower, N., Knauber, D., Singer, F., 1981. Production of clones of homozygous diploid zebra fish (Brachydanio rerio). Nature 291, 293–296.
Sullivan, C., Kim, C.H., 2008. Zebrafish as a model for infectious disease and immune function. Fish Shellfish Immunol. 25, 341–350.
Szyf, M., 2015. Nongenetic inheritance and transgenerational epigenetics. Trends Mol. Med. 21, 134–144.
Tan, C.H., Lee, T.C., Weeraratne, S.D., Korzh, V., Lim, T.M., Gong, Z., 2002. Ziwi, the zebrafish homologue of the Drosophila piwi: co-localization with vasa at the embryonic genital ridge and gonad-specific expression in the adults. Mech. Dev. 119 (Suppl 1), S221–224.
Tan, H., Onichtchouk, D., Winata, C., 2016. DANIO-CODE: toward an encyclopedia of DNA elements in zebrafish. Zebrafish 13, 54–60.
Ulitsky, I., Shkumatava, A., Jan, C.H., Sive, H., Bartel, D.P., 2011. Conserved function of lincRNAs in vertebrate embryonic development despite rapid sequence evolution. Cell 147, 1537–1550.
Ulloa, P.E., Medrano, J.F., Feijoo, C.G., 2014. Zebrafish as animal model for aquaculture nutrition research. Front. Genet. 5, 313.
Uzbekova, S., Chyb, J., Ferriere, F., Bailhache, T., Prunet, P., Alestrom, P., Breton, B., 2000. Transgenic rainbow trout expressed sGnRH-antisense RNA under the control of sGnRH promoter of Atlantic salmon. J. Mol. Endocrinol. 25, 337–350.
van der Sar, A.M., Spaink, H.P., Zakrzewska, A., Bitter, W., Meijer, A.H., 2009. Specificity of the zebrafish host transcriptome response to acute and chronic mycobacterial infection and the role of innate and adaptive immune components. Mol. Immunol. 46, 2317–2332.
Varga, Z.M., Wilson, C., Alestrøm, P., 2016. European Zebrafish Meeting 2015 husbandry session report. Zebrafish 13, 230–231.
Varona, L., Munilla, S., Mouresan, E.F., Gonzalez-Rodriguez, A., Moreno, C., Altarriba, J., 2015. A Bayesian model for the analysis of transgenerational epigenetic variation. G3 (Bethesda) 5, 477–485.
Vastenhouw, N.L., Schier, A.F., 2012. Bivalent histone modifications in early embryogenesis. Curr. Opin. Cell Biol. 24, 374–386.
Vilella, A.J., Severin, J., Ureta-Vidal, A., Heng, L., Durbin, R., Birney, E., 2009. EnsemblCompara GeneTrees: complete, duplication-aware phylogenetic trees in vertebrates. Genome Res. 19, 327–335.
Vojtech, L.N., Sanders, G.E., Conway, C., Ostland, V., Hansen, J.D., 2009. Host immune response and acute disease in a zebrafish model of Francisella pathogenesis. Infect. Immun. 77, 914–925.

Wang, W., Asim, M., Yi, L., Hegazy, A.M., Hu, X., Zhou, Y., Ai, T., Lin, L., 2015. Abortive infection of snakehead fish vesiculovirus in ZF4 cells was associated with the RLRs pathway activation by viral replicative intermediates. Int. J. Mol. Sci. 16, 6235–6250.

Wargelius, A., Leininger, S., Skaftnesmo, K.O., Kleppe, L., Andersson, E., Taranger, G.L., Schulz, R.W., Edvardsen, R.B., 2016. Dnd knockout ablates germ cells and demonstrates germ cell independent sex differentiation in Atlantic salmon. Sci. Rep. 6, 21284.

Weber, M., Huisken, J., 2011. Light sheet microscopy for real-time developmental biology. Curr. Opin. Genet. Dev. 21, 566–572.

Westerfield, M., 2007. The Zebrafish Book: A Guide for the Laboratory Use of Zebrafish (Danio rerio), fifth ed. University of Oregon Press, Eugene, USA.

White, R., Rose, K., Zon, L., 2013. Zebrafish cancer: the state of the art and the path forward. Nat. Rev. Cancer 13, 624–636.

Wilson, C.A., High, S.K., McCluskey, B.M., Amores, A., Yan, Y.L., Titus, T.A., Anderson, J.L., Batzel, P., Carvan, 3rd, M.J., Schartl, M., Postlethwait, J.H., 2014. Wild sex in zebrafish: loss of the natural sex determinant in domesticated strains. Genetics 198, 1291–1308.

Wolfe, K., 2000. Robustness--it's not where you think it is. Nat. Genet. 25, 3–4.

Xiao, J., Chen, T., Wang, Q., Liu, Q., Wang, X., Lv, Y., Wu, H., Zhang, Y., 2011. Search for live attenuated vaccine candidate against edwardsiellosis by mutating virulence-related genes of fish pathogen *Edwardsiella tarda*. Lett. Appl. Microbiol. 53, 430–437.

Xiong, X.P., Dong, C.F., Xu, X., Weng, S.P., Liu, Z.Y., He, J.G., 2011. Proteomic analysis of zebrafish (*Danio rerio*) infected with infectious spleen and kidney necrosis virus. Dev. Comp. Immunol. 35, 431–440.

Yang, H., Carmichael, C., Varga, Z.M., Tiersch, T.R., 2007. Development of a simplified and standardized protocol with potential for high-throughput for sperm cryopreservation in zebrafish *Danio rerio*. Theriogenology 68, 128–136.

Yang, D., Liu, Q., Yang, M., Wu, H., Wang, Q., Xiao, J., Zhang, Y., 2012. RNA-seq liver transcriptome analysis reveals an activated MHC-I pathway and an inhibited MHC-II pathway at the early stage of vaccine immunization in zebrafish. BMC Genomics 13, 319.

Yang, D., Liu, Q., Ni, C., Li, S., Wu, H., Wang, Q., Xiao, J., Zhang, Y., 2013. Gene expression profiling in live attenuated *Edwardsiella tarda* vaccine immunized and challenged zebrafish: insights into the basic mechanisms of protection seen in immunized fish. Dev. Comp. Immunol. 40, 132–141.

Zerihun, M.A., Nilsen, H., Hodneland, S., Colquhoun, D.J., 2011. *Mycobacterium salmoniphilum* infection in farmed Atlantic salmon, Salmo salar L. J. Fish Dis. 34, 769–781.

Zhang, H., Shen, B., Wu, H., Gao, L., Liu, Q., Wang, Q., Xiao, J., Zhang, Y., 2014. Th17-like immune response in fish mucosal tissues after administration of live attenuated *Vibrio anguillarum* via different vaccination routes. Fish Shellfish Immunol. 37, 229–238.

Zhao, X., Wu, C., Peng, X., Li, H., 2014. Interferon-alpha2b against microbes through promoting biosynthesis of unsaturated fatty acids. J. Proteome Res. 13, 4155–4163.

Current status in other finfish species

8

Description of current genomic resources for the gilthead seabream (*Sparus aurata*) and soles (*Solea senegalensis* and *Solea solea*)

Manuel Manchado, Josep V. Planas**, Xavier Cousin[†], Laureana Rebordinos[‡], Manuel Gonzalo Claros[§,¶]*

*IFAPA Center El Toruño, IFAPA, Consejeria de Agricultura y Pesca, El Puerto de Santa María (Cádiz), Spain; **Department of Physiology and Immunology, School of Biology, University of Barcelona and Institute of Biomedicine (IBUB), Barcelona, Spain; [†]IFREMER, Laboratory of Adaptation and Adaptabilities of Animal and Systems (L3AS), Palavas and INRA, Laboratory of Fish Physiology and Genomics (LPGP), Rennes, France; [‡]Laboratory of Genetics, Faculty of Environmental and Sea Sciences, University of Cadiz, Polígono del Río San Pedro, Puerto Real, Cádiz, Spain; [§]Department of Molecular Biology and Biochemistry, Faculty of Sciences, Campus de Teatinos s/n, University of Málaga, Málaga, Spain; [¶]Plataforma Andaluza de Bioinformática, University of Málaga, Málaga, Spain

Introduction

The gilthead seabream (*Sparus aurata*) (Order Perciformes) and the soles, Senegalese sole (*Solea senegalensis*) and common sole (*Solea solea*) (Order Pleuronectiformes), are important species in Mediterranean aquaculture, although their current status of exploitation is completely different in terms of volume of production, culture technologies, and specific challenges and bottlenecks. Nevertheless, from an evolutionary point of view, these three species belong to the Acanthopterygii superorder, a taxonomic group that exhibits a high degree of conservation at the genome level with a clear differentiated pattern of paralog retention with respect to the Ostariophysi (zebrafish) (Diopere et al., 2014; Garcia de la Serrana et al., 2014; Tsigenopoulos et al., 2014). A better knowledge of their genomic resources, of the way in which their genomes have evolved in relation with closely related species as well as of the molecular pathways behind their functional and biological characteristics will importantly result in the identification of key regulatory mechanisms that can be successfully applied in aquaculture.

The gilthead seabream is the main species in aquaculture by volume production with Greece (41.7% of total production), followed by Turkey (23.2%) and Spain (9.3%), as the main producers (Fig. 8.1) (APROMAR, 2014; FEAP, 2014). The important

Genomics in Aquaculture. http://dx.doi.org/10.1016/B978-0-12-801418-9.00008-1

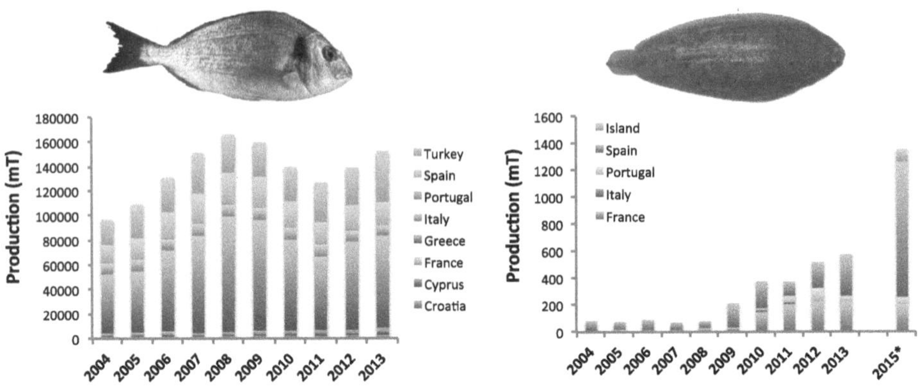

Figure 8.1 Total volume (metric tons) of aquaculture production for gilthead seabream (left) and Senegalese sole (right). Main producer countries are also indicated.

development of seabream aquaculture has been directly associated with the robustness and plasticity of this species that renders it able to adapt to different environments, and to its high resistance to diet changes and microbial outbreaks. These characteristics, in addition to significant technological advances in reproduction using light control, larval rearing conditions, production of high-quality feed, disease prevention, and environmental control, supported the expansion of this industry (Moretti et al., 1999). Recently, genetic breeding programs were initiated in Europe to improve growth, morphology, flesh quality, feed efficiency, and disease resistance traits. Nevertheless, some producer organizations have indicated a drop in the production efficiency of this species as determined by key performance indicators, encouraging the development and optimization of genetics, genomics and epigenetics, optimization of diets and veterinary treatments to sustain the production of this species (EATiP, 2014).

In sole, the aquaculture situation is completely different. Although the first trials for sole aquaculture were performed early in the 1980s (Dinis et al., 1999), the production began much later with the improvement of hatchery and husbandry rearing procedures. The latest advances in feeding and management protocols during larval rearing have resulted in optimal weaning rates and high-quality larvae (Cerda and Manchado, 2013). Moreover, growth-out stages are carried out in dedicated intensive recirculation systems with a high control of environmental and sanitary conditions (Morais et al., 2014). Nevertheless, Senegalese sole aquaculture has to face new challenges. First, reproduction has become one of the major bottlenecks since the males reared in captivity are not able to express proper courtship behavior that is required for successful fertilization. Second, the bottom dwelling and sedentary nature of sole lead to highly hierarchical population structures that result in populations extremely skewed in size. Finally, improvement of growth-out diets that satisfy nutritional and physiological demands while maintaining an optimal welfare status remains a major issue in this industry. Hence, genomic information appears as an essential tool to surpass these bottlenecks and support the growth of this industry.

Transcriptome resources and their main characteristics in gilthead seabream and soles

The transcriptome has been intensively explored in seabream and soles over the last few years as a first approach to improving genomic information in nonmodel species. In the gilthead seabream, one of the first attempts at generating genomic resources for this species was performed within the framework of the EU-funded BRIDGEMAP project (2000–04) (Table 8.1). Through the construction of cDNA libraries from various seabream tissues, a total of 3021 ESTs were generated. Subsequently, the EU-project AQUAFIRST (2004–08) constructed suppression subtractive hybridization cDNA libraries based on samples from seabream exposed to confinement stress or to a parasite (*Enteromyxum leei*) and generated 4851 unique ESTs. On the other hand, the EU-funded Network of Excellence Marine Genomics Europe (2005–09; MGE) made a significant contribution by generating 29,895 ESTs through the 5′ end-Sanger sequencing of 14 normalized cDNA libraries from a broad collection of different tissues from male, female, and hermaphrodite fish (Louro et al., 2010). More recently, the AQUAGENOMICS project, a Spanish-funded project initiative, expanded the MGE-generated genomic resources by reannotating and incorporating MGE gilthead seabream sequences as well as approximately 7000 ESTs obtained from Sanger sequencing of two cDNA libraries from pathogen (bacterial and viral)-challenged cells and tissues into a searchable database (*Aquasea*). Furthermore, sequences from early developmental stages of larval seabream (first 60 days posthatching, dph) were also generated using the Roche 454 technology providing a total of 68,289 assembled contigs, of which 44,977 were annotated (Yufera et al., 2012). Thereafter, additional transcriptomes for skeletal muscle, skeletal tissues, intestine, blood, and head kidney samples from juvenile seabream obtained by 454 sequencing were reported, increasing significantly the number of ESTs available in this species and resulting in a global transcriptome of 113,927 contigs, of which approximately 50% were annotated (Calduch-Giner et al., 2013; Garcia de la Serrana et al., 2012; Vieira et al., 2013).

In soles, the first genomic studies were carried out at a later time coinciding with the development of the aquaculture industry for these species. The PLEUROGENE project produced several normalized libraries from adult tissues (brain, stomach, intestine, liver, ovary, and testis), larval stages (premetamorphosis, metamorphosis), juvenile stages (postmetamorphosis, abnormal fish), and undifferentiated gonads, generating up to 5208 contigs (Cerda et al., 2008). Later, the arrival of NGS technologies triggered an impressive increase in genomic resources for these species. NGS technologies were first applied in *S. solea* using the 454 technology to characterize the transcriptome of larvae from 4 to 33 dph, obtaining 22,252 contigs (when combined with an additional set of muscle specific transcripts), of which 16,731 were annotated (Ferraresso et al., 2013). The AQUAGENET project (www.aquagenet.eu) used both the 454 and Illumina technologies to identify transcriptomes covering processes such as immunity, osmoregulation, reproduction, and larval development in both *S. senegalensis* and *S. solea* (Benzekri et al., 2014). As given in Table 8.1, the number of transcripts in the different tissues used ranged between 55,469 and 117,058

Table 8.1 Transcriptome data in *S. aurata*, *S. senegalensis*, and *S. solea*

Species	Technique	Input sequences	No of transcripts	Tissues	MCL	Annotated	SNPs	SSR
S. aurata	Sanger[a]	3,778	646	Immune cells challenged with bacteria	746	2,545[l]		
S. aurata	Sanger[a]	3,059	504	Immune cells challenged with viruses	785	2,179[l]		
S. aurata	Sanger[b]	29,895	5,268	Adult tissues	676	5,166[l]	575	997
S. aurata	NGS 454[c]	869,077	68,289	Early larval stages	596	44,977		
S. aurata	NGS 454[d]	2,711,149	43,461	White skeletal muscle	454	10,465		
S. aurata	NGS 454[e]	271,613	32,374	Vertebrae	493	10,455		
S. aurata	NGS 454[e]	258,102	28,371	Gill arch	514	10,625		
S. aurata	NGS 454[f]	447,166	7,808	White skeletal muscle	968			
S. aurata	NGS 454[f]	614,899	9,475	Intestine	1,090			
S. aurata	NGS 454[f]	576,796	12,003	Blood	831			
S. aurata	NGS 454[f]	577,112	14,008	Head kidney, early parasite infection	1,059			
S. aurata	NGS 454[f]	729,941	12,474	Head kidney, chronic parasite infection	1,148			
S. aurata	NGS 454[f]	2,765,597	125,263	Several adult tissues	762	63,880		
S. senegalensis	Sanger[g]	10,185	5,208	Larval and adult tissues		2,122		
S. senegalensis	NGS 454[h]	3,774,412		Several adult tissues	336	84,763		
S. senegalensis	NGS 454[i,j]	779,130	117,058	Immune-related organs	259	111,939	153,500	
S. senegalensis	NGS 454[j]	838,997	61,779	Osmoregulatory-related organs		60,768	114,742	10,856
S. senegalensis	NGS 454[j]	934,429	64,637	Gonads (testis and ovary)	374	63,940	115,083	11,246
S. senegalensis	NGS 454[j]	699,765	68,254	Pituitary (male and female)	453	60,663	81,605	15,811

S. senegalensis	NGS 454[j]	348,799	55,469	Hypothalamus (male and female)	269	54,771	82,748	8,944
S. senegalensis	NGS Illumina and 454[h]	$1,503 \times 10^6$	697,125	Several larval stages and adult tissues		147,536		
S. solea	NGS Illumina[h]	$2,101 \times 10^6$	523,637	Several larval stages and adult tissues	89	38,402	381,404	316,388
S. solea	NGS 454[k]	909,466	22,223	Several larval stages	245	16,731		3,612
Whole transcript	**Different orthology**	**Complete different ORFs**	**Transcrip-tome reference**					
S. senegalensis	45,063[h]	22,683	59,514					
S. solea	38,402[h]	18,738	54,005					

The sequencing platform, raw data, number of transcripts assembled, tissues used for library construction, the mean contig length (MCL), number of annotated transcripts and SNP and SSR markers are indicated.

[a] Boltana et al., submitted

[b] Louro et al. (2010).

[c] Yufera et al. (2012).

[d] Garcia de la Serrana et al. (2012).

[e] Vieira et al. (2013).

[f] Calduch-Giner et al. (2013).

[g] Cerdà et al. (2008).

[h] Benzekri et al. (2014).

[i] Ponce et al. (2011).

[j] SoleaDB (www.aquagenet.eu).

[k] Ferraresso et al. (2013).

[l] Annotation results obtained by querying the Swissport database with all unigenes (contigs plus singletons).

in hypothalamus and immune organs, respectively. Moreover, a detailed analysis of the transcriptomes of gonads, pituitary, and hypothalamus from male and female *S. senegalensis* was performed. In the testis, the average number of transcripts (30,655) was higher than in the ovary (17,773), whereas in pituitary the number of transcripts was similar in both sexes (26,271 vs 24,952 in males and females, respectively). Taken together, Illumina sequencing resulted in 697,125 transcripts in *S. senegalensis* and 523,637 transcripts in *S. solea*.

Comparison of transcriptomes based on gene ontology (GO) annotation showed similar category representation in deep-sequenced transcriptomes, with the highest number of annotated transcripts comprised in metabolic (~15%) and cellular (~22%) processes assessed by biological function, in binding (~41–45%) and catalytic activity categories (30–35%) by molecular function and in cell (~36% in soles and up to 59% in seabream) and organelle (22–26%) categories by cellular components (Benzekri et al., 2014; Yufera et al., 2012). Nevertheless, GO term analysis of these transcriptomic sets evidenced enrichment in some other biological categories corresponding to transport, protein modification, response to stress, immune system process and muscle development, differentiation, and contraction according to the origin of the RNAs used for sequencing.

Note that the number of transcripts is in general very high in most assemblies and particularly in soles. Obviously, these assemblies do not accurately represent the actual number of genes in each species since they also include true variants such as gene alleles, alternative spliced transcripts as well as fragmented transcripts, and immature mRNAs among others. Hence, strategies for data mining to filter this redundancy have been developed in soles by selecting (1) the longest transcripts with unique, different orthologous IDs and (2) the putative nonredundant new transcripts. In this way, a total of 18,738 and 22,683 full-length cDNAs in *S. senegalensis* and *S. solea*, respectively, were identified. The number of genes identified is similar to the expected number of protein-coding genes in vertebrates (between 15,000 and 25,000) (Chen et al., 2014; Kettleborough et al., 2013; Prachumwat and Li, 2008). Moreover, a reference transcriptome to be used in RNA-seq studies was established and included these unique transcripts as well other putative-coding sequences as determined by in silico analysis (Benzekri et al., 2014). This resulted in the selection of 59,514 transcripts for *S. senegalensis* (v4.1; 8.5% of initial set of valid transcripts) and 54,005 transcripts for *S. solea* (v1.1; 10.2%). A test representation using certain sets of RNA-seq data showed that mapping of useful reads onto the two reference transcriptomes was positive for 82.3–87.5% of the reads onto transcripts confirming that sole reference transcriptomes adequately represent actual transcriptomes.

All these transcriptomes provide enough genomic resources for the design and application of high-throughput tools for gene expression analysis to be applied in aquaculture. Hence, custom microarrays and openarrays have been reported for seabream and soles in the last few years. The first microarray developed for seabream contained 19,715 unique sequences that were spotted in a total of 39,379 oligonucleotide probes (60-mers) on glass slides based on the Agilent 4 × 44K format (Ferraresso et al., 2008). A second updated version of this microarray was generated after a reassembly of the seabream transcriptome which added 6,412 new transcripts relative to the previous assembly and contained a total of 42,651 probes spotted again on a 4 × 44K format (Pellizzari et al., 2013). Also, a cDNA microarray containing 4876 unique sequences has been developed for seabream (Calduch-Giner et al., 2010). For the Senegalese sole, the first microarray developed contained 5087 unique sequences spotted on glass slides (Cerda et al., 2008). More recently, a much more comprehensive microarray for the Senegalese sole was developed that contained 43,303 probes based on the Agilent 4 × 44K format (Benzekri et al., 2014). In common sole, a microarray containing 14,674 oligonucleotides was also tested (Ferraresso et al., 2013). Moreover, high-throughput tools for quantitative real-time PCR (qPCR) have been developed in soles to study the innate immune system, lipid metabolism, oxidative stress, and osmoregulation (Hachero-Cruzado et al., 2014; Jimenez-Fernandez et al., 2015; Montero et al., 2015). More details about these studies and the application of these custom tools will be explained in the subsequent sections.

Current genetic maps and genome drafts in gilthead seabream and soles

Genetic maps and QTLs

Genetic maps are considered essential tools in those nonmodel species with limited genomic resources available for providing information of existing genes in targeted chromosomal regions and for facilitating the identification of candidate genes associated with traits of interest. Most genetic maps are based on microsatellites (SSRs), single nucleotide polymorphism (SNPs), EST, and bacterial artificial chromosomes (BACs) markers (Fig. 8.2) that taken together represent a powerful tool to validate assemblies, syntenies, and gene location. In particular, high-density SNP maps and physical mapping based on BACs are considered as key steps to validate positioning and anchoring genomic information and linkage data onto physical chromosomes (Ariyadasa and Stein, 2012; Garcia-Cegarra et al., 2013).

Currently, there are genetic maps in gilthead seabream and soles although they were built using different technical approaches. In seabream, genetic maps were built using whole-genome radiation hybrid (RH) panels (Sarropoulou et al., 2007; Senger et al., 2006) and linkage information (Franch et al., 2006; Tsigenopoulos et al., 2014) (Table 8.2). As RH maps are not dependent on loci polymorphism, they allow for the location of EST and neutral polymorphic markers. Therefore, this facilitates the identification of orthologous genes and synteny analyses across model species and the scanning for quantitative traits loci (QTLs) through comparative genomics. The first generation genetic map published by Franch et al. (2006) included 198 SSR markers ordered in 26 linkage groups (LG) with a total map length of 1241.9 cM. The number of markers per group ranged from 2 to 17. The second-generation linkage map increased this number to 321 markers (including genomic SSRs, EST-SSRs, and SNPs) related to 27 LGs and a total length of sex-average map of 1769.7 cM. The average distance between markers was 4.38 and 7.05 cM in males and females, respectively, and with a number of markers in each LG ranging from 2 to 30 (Tsigenopoulos et al., 2014).

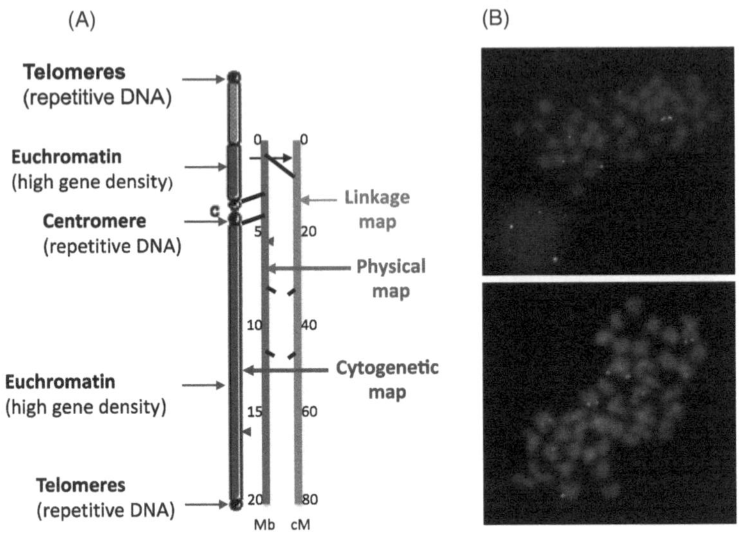

Figure 8.2 Genetic integrated maps. (A) Comparison of genetic, cytogenetic and physical maps; (B, upper) location of two BACs containing lysozyme gene and thyroid hormone receptor beta in Senegalese sole; and (B, lower) location of four BACs containing calreticulin gene, interferon regulatory factor 5, Mx, and toll-like receptor 8 in Senegalese sole.

In addition, two RH maps were released with a total number of 937 molecular markers. Although the first version contained 28 RH groups, only 25 groups were reported in the second draft, one more than the 24 expected chromosomes (Sarropoulou et al., 2007; Senger et al., 2006). Nevertheless, comparative genomics shows that two RH groups (19 and 20) were clearly located onto the same chromosome of *Tetraodon*. Both linkage and RH maps were anchored with a set 26 SSR markers establishing ortholog groups. Interestingly, the RH approach allowed for the location of 510 gene-based markers (82 and 428 in the first and second RG maps, respectively) suitable for comparative genomics and identification of regions with conserved synteny by scanning for QTLs mainly controlling growth, disease resistance, sex determination and reversal, reproduction, as well as environmental tolerance (Sarropoulou et al., 2007).

In the Senegalese sole, two genetic maps based on SSR (Molina-Luzon et al., 2015a) and BAC-FISH (Garcia-Cegarra et al., 2013) were described (Table 8.2). The former was developed using three reference gynogenetic families, two haploid families to position the SSR markers in LGs and a third diploid family to estimate the marker-centromere distances. The final consensus map consisted of 129 SSRs distributed in 27 LG (the whole-genome is structured in 21 pair of chromosomes) with an average density of 4.7 markers per LG that comprised 1004 cM. Otherwise, an integrated BAC map was developed by Garcia-Cegarra et al. (2013) using 10 clones as probes to map the karyotype of the species by FISH and integrating information of 2-color FISH using the 5S rDNA and lysozyme genes, as previously published (Ponce et al., 2011)

Table 8.2 Main characteristics of genetic maps described in gilthead seabream, Senegalese sole, and common sole

	S. aurata	*S. senegalensis*	*S. solea*
RFLPs	11		
SNPs	4		423
SSR	317	129	8
Maps	2 RH 2 Linkage	1 BAC-FISH 1 Linkage	1 Linkage
Chrom/LGs	$2n = 48/27$	$2n = 42/27$	$2n = 42/38$
LG size	1,769.7[a] cM	1,004 cM	1,233.8 cM
Average distance between markers	4.38–7.05 cM	4.7 cM	0–92.1 cM
QTLs in seabream			
Growth		LG1[b], LG9[b,c], LG21[b,d], LG23[b]	
Sex reversal		LG9[b], LG21[b,d], LG23[b]	
Morphology		LG4a[e], LG7[e], LG9[c,e], LG18[e], LGA[e], LGC[e], LGF[e], LGI[e],	
Stress		LG10[e], LGB[e]	
Disease resistance		LG3[f], LG6[f], LG21[f]	

The number of markers, linkage groups (LGs), and distances are indicated. QTLs identified seabream and their LG locations are also shown.
[a] Tsigenopoulos et al. (2014).
[b] Loukovitis et al. (2012).
[c] Loukovitis et al. (2013).
[d] Loukovitis et al. (2011).
[e] Boulton et al. (2011).
[f] Massault et al. (2011).

(Fig. 8.2). Most of the clones localized in 1 chromosome pair whereas only four BACs colocalized in 2 chromosome pairs. NGS sequencing and annotation of these BACs allowed for the localization of 23 genes including the endocrine system-related genes thyroid hormone receptors (*thraa*, *thrab*, *thrb*) and follicle-stimulating hormone, the immune system-related genes g-type lysozyme, mx, hepcidin, and semaphorin 7a. In common sole, only a linkage map that included 423 SNPs derived from ESTs and 8 neutral SSRs has been reported. The total map length is 1233.8 cM organized in 38 LGs with a size varying between 0 and 92.1 cM in each group (Diopere et al., 2014).

Comparative genomics using genetic maps containing EST markers with seabream and common sole (Diopere et al., 2014; Sarropoulou et al., 2007; Tsigenopoulos et al., 2014) with respect to the genomes of model species confirmed a high degree of conserved synteny between Acanthopterygii. Hence, most of the RH groups could be unequivocally mapped onto 16 chromosomes of *Tetraodon nigroviridis* (Sarropoulou et al., 2007). Moreover, macrosynteny relationships between the gilthead seabream and the stickleback's genome showed a clear colocalization for most LGs in the second generation linkage map and a conserved synteny for 132 seabream markers (42.9% of those mapped) (Tsigenopoulos et al., 2014). In the common sole, most of the LGs in the genetic map could be sorted in 21 putative chromosomes with respect to the turbot LGs using the genomes of Acanthopterygii model species such as Tetraodon, medaka, or stickleback as a bridge (Diopere et al., 2014). Interestingly, total length of the syntenic regions was higher (200.1–257.8 Mb) with respect to stickleback, tilapia, and medaka and lower with Tetraodon (89.2 Mb). These data confirm that comparative genomics with highly related species is adequate for marker positioning as will be described later for the characterization of the Senegalese sole genome.

One of the main applications of genetic maps is the identification of QTLs and candidate genomic regions involved in the control and regulation or growth, disease resistance, or sex determination with relevance in aquaculture. Using comparative genomics, two significant QTLs, one for body weight and other for sex reversal, were located in LG21 suggesting an association between both traits in seabream (Loukovitis et al., 2011) (Table 8.2). This association was later confirmed by a genome scan using SSRs covering nine LGs, which identified four QTLs for growth and three pairs for sex determination in the same LGs (LG9, LG21, and LG23) (Loukovitis et al., 2012). Moreover, additional QTLs for growth and morphology were also identified in LG9 and LG21 (Boulton et al., 2011; Loukovitis et al., 2011, 2012, 2013) indicating that these LGs would contain one pleiotropic QTL affecting overall body size.

Further studies that deal with the identification of additional QTLs for stress, morphology, and disease resistance to the pathogen *Photobacterium damselae* subs *piscicida* have been reported (Boulton et al., 2011; Loukovitis et al., 2013; Massault et al., 2011). Interestingly, a targeted approach to find polymorphisms in the promoter and introns of five candidate genes influencing growth such as growth hormone, insulin-like growth factor-1, myostatin, prolactin, and somatolactin identified a significant association for myostatin, a major gene regulating muscle mass (Sanchez-Ramos et al., 2012). All this information, coupled with an estimated medium or high heritability (0.2–0.5) depending on the productive trait, indicate that these QTLs can be successfully applied for the implementation of selective breeding programs (through the combination of marker-assisted and classical selection) in seabream.

Toward the first draft of Senegalese sole genome

A preliminary draft genome for a *S. senegalensis* female has been recently constructed within the AQUAGENET project. Three paired-end Illumina libraries corresponding to two female specimens, containing a total of 2983×10^6 raw reads (300 bp and 3 kbp libraries), and six 454/Roche long paired-end libraries accounting for 8.4×10^6 reads (3, 8, and 20 kb) were used. After sequence cleaning for low quality nucleotides, contamination and other artifacts using the SeqTrimNext software, the total percentage of suitable reads recovered for assembly were 96% for Illumina and 82% for 454/Roche data. Due to the huge number of short reads, some assemblers described in the literature as memory-effective were tested using an Illumina library (Table 8.3). The RAY assembler was selected as the most appropriate according to the number of contigs produced, even though no scaffolds for *k*-mer 31 were obtained due to execution time-out. The other assemblers tested produced a higher number of contigs and scaffolds, providing a quite fragmented genome. To complete the assembly, each library

Table 8.3 Assembling tests using the 828,254,151 useful reads of library AQ1 using two different *k*-mers

Assemblers	*k*-mer 31		*k*-mer 51	
	Contigs	Scaffolds	Contigs	Scaffolds
RAY	263,123	TO	582,421	562,157
Minia	910,625	NA	430,128	NA
SOAPdenovo	21,469,279	1,429,112	10,716,892	2,058,361
Platanus	NA	NA	2,822,377	1,259,857

TO, time-out exception; NA, not available.

Table 8.4 Relevant characteristics of the current *S. senegalensis* draft genome

	Complete set	Positioned on tongue sole chromosomes	Redundant or unpositioned
No scaffolds	34,176	8,538	25.638
No scaffolds > 500 bp	26,368	8,430	17,938
Longest scaffold (nt)	638,263	638,263	460,366
Total length (nt)	600,313,426	466,802,169	133,511,257
No of Ns	14,313,038	9,952,203	4,360,835
Mean length (nt)	14,565	54,673	5,207
N50	85,596	105,233	25,581
N90	13,932	29,646	2,227

Data represent the complete set of scaffolds obtained and those that were positioned and unpositioned on the tongue sole (*C. semilaevis*) chromosomes.

was first separately assembled using RAY with a *k*-mer of 31 and then contigs were extended and combined to obtain a total of 132,710 contigs. The draft was finished by scaffolding using SOAPdenovo and SSPACE, filling gaps with GAP CLOSER, and reconciliation of scaffolds using NUCMER. As a result, the draft genome consisted of 34,176 scaffolds (Table 8.4) with an N50 of 85 kb. Characterization using the CEG-MA mapping protocol to assess the set of core eukaryotic genes (CEG) indicated that the draft genome contained 209 out of the 274 ultraconserved CEG (Parra et al., 2009) with a completeness of 84.3% and an average number of orthologs of 1.31. These figures demonstrate that this current draft genome contains a high proportion of genes in the same contig/scaffold and, hence, that it can be useful for evolutionary, linkage, and association studies.

A further synteny analysis based on the nine longest scaffolds of the Senegalese sole draft genome with respect to the closely related tongue sole (*Cynoglossus semilaevis*) indicated that most genes contained in Senegalese sole scaffolds lie together on tongue sole chromosomes, showing a clear synteny between both species (Fig. 8.3). Moreover, a detailed analysis of apolipoprotein genes, a diversified family with several genes and paralogs, recovered from SoleaDB (Benzekri et al., 2014), confirmed a high conservation of synteny across teleosts (Román-Padilla et al., 2016) that

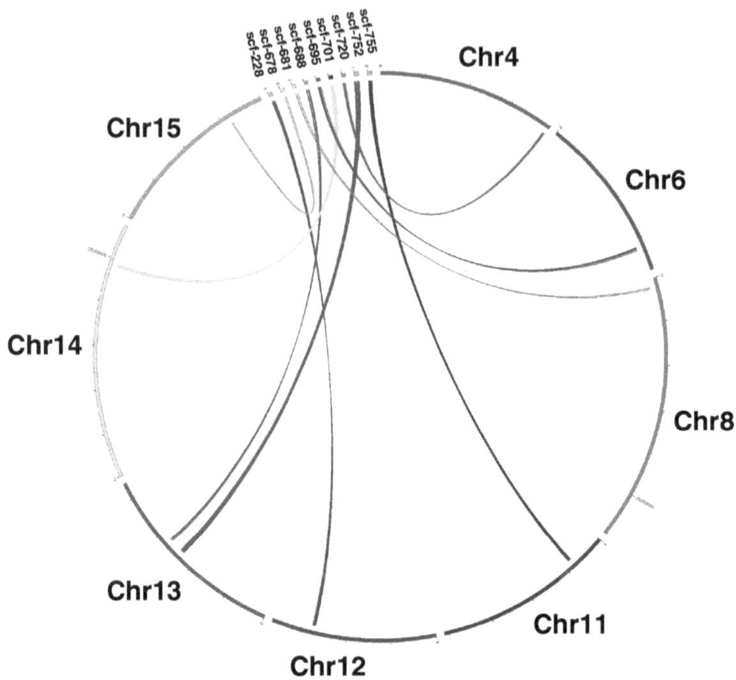

Figure 8.3 Synteny comparison of the largest scaffolds in Senegalese sole (scf) with respect to the chromosomes (Chr) of tongue sole.

was particularly high with *C. semilaevis* and *Gasterosteus aculeatus* (three-spined stickleback) and lower with the most distant *Danio rerio*. These results confirm the conserved synteny observed between Pleuronectiformes and Gasterosteiformes as previously observed in Acanthopterygii (Diopere et al., 2014)

Superscaffolds as putative chromosomes of Senegalese sole

The high degree of synteny mentioned previously was used for the sorting of scaffolds into chromosomes, using as a reference the tongue sole as the closest species, even though this species contains sexual chromosomes that have not been defined on Senegalese sole karyotypes.

In total, 8538 Senegalese sole scaffolds were arranged into superscaffolds according to the tongue sole chromosomes to rebuild 21 chromosomes, the expected karyotype in this species. A total of 25,638 scaffolds were discarded mainly due to overlapping or redundancy and only 7% failed to map onto the tongue sole chromosomes (Table 8.4). These results demonstrate that ~25% of Senegalese scaffolds covered almost completely the tongue sole genome (expected size: 477 Mb; superscaffold size: 466 Mb), significantly increasing the N50 and N90 values (Table 8.4). This superscaffold arrangement was later confirmed by locating 113 SSR markers described in the literature (Molina-Luzon et al., 2012, 2015a) with only 10 SSRs showing a discrepancy in position.

Coverage analysis indicated that tongue sole chromosomes 1–20 and chromosome Z were >100% covered by Senegalese sole scaffolds, which may be explained by the fact that the genome size in the Senegalese sole is greater than that of the tongue sole (Table 8.5). Interestingly, chromosome W was not well covered at all with only 40% of the length covered in spite of the fact that the number of scaffolds mapping in this chromosome is similar to others with a very low N50 (reflecting that most scaffolds are very small; Table 8.5); moreover, these scaffolds are full of repetitive sequences that can map in several places onto this chromosome. These findings seem to be consistent with the known fact that only the homologous chromosome Z is found in our female draft genome due to a homogametic sex determination system (XX) (Molina-Luzon et al., 2015b) or a chromosome rearrangement, although these hypotheses will have to be tested in the future.

Table 8.5 Senegalese sole scaffold distribution on superscaffolds based on the tongue sole chromosomes

Tongue sole chromosome	Length (nt)	Number of scaffolds mapped	Total length of mapped scaffolds	% covered	N50 (nt)
1	34,528,841	657	34,392,510	99.6	98,077
2	20,052,193	355	22,929,498	114.4	112,365
3	16,251,686	249	17,909,809	110.2	136,734
4	20,010,643	361	21,773,075	108.8	112,995
5	19,279,693	391	21,646,339	112.3	100,362
6	18,840,743	338	21,405,415	113.6	114,837
7	13,814,722	258	15,852,235	114.8	105,594
8	30,151,083	577	31,174,684	103.4	103,856
9	19,616,557	411	22,584,492	115.1	93,776
10	21,015,401	357	22,234,735	105.8	111,585
11	20,528,231	421	22,619,669	110.2	97,345
12	18,396,836	342	19,457,520	105.8	122,179
13	21,921,923	427	23,837,735	108.7	128,458
14	28,847,931	622	28,597,901	99.1	100,461
15	20,091,810	405	19,808,757	98.6	104,963
16	18,785,820	337	19,639,786	104.6	113,000
17	16,464,107	312	18,090,037	109.9	101,686
18	15,207,555	278	17,433,808	114.6	112,168
19	17,744,385	311	19,631,992	110.6	102,397
20	15,231,414	213	17,218,034	113.0	125,817
Z	21,910,988	443	21,867,681	40.7	89,912
W	16,446,795	473	6,696,457	40.7	39,578

Databases for genomic resources in gilthead seabream and soles

The significant volume of genomic information generated in the last few years for seabream and soles has made it necessary to build specific databases to properly manage all of this information in order to facilitate data comparison and to build collaborative efforts for the genetic characterization of these species.

In seabream, a transcriptomic database named seabreamdb (http://www.nutrigroup-iats.org/seabreamdb) that hosts transcriptome information (17,809 transcripts) and a microarray database named Fish and Chips that includes 354 microarray datasets from 7 fish species (http://fishandchips.genouest.org/index.php) have been described, although sequences and expression data are not publicly accessible (Calduch-Giner et al., 2013). In contrast, a public database named SoleaDB (http://www.juntadeandalucia.es/agriculturaypesca/ifapa/soleadb_ifapa/) that hosts all transcriptomic information has been released. This database currently houses annotated sequences of 454 and Illumina projects developed in both *S. senegalensis* and *S. solea* and provides putative transcriptomes by tissues as well as global transcriptomes to be used as reference for functional studies. SoleaDB is based on a user-friendly interface that can be browsed anonymously to facilitate researchers to access transcriptome information (transcript description, nucleotide sequences, predicted protein sequence, KEGG pathways, GO terms, InterPro codes, SSR markers, zebrafish orthologs, etc.). The content is structured in buttons and tabs that permit data browsing and mining and provides general information about data analysis, projects, etc. Users can easily access all assemblies and a specific identifier for each sequence and assembly is provided. The most current global transcriptome in *S. senegalensis* (currently v4.1) is selected by default. Nevertheless, alternative transcriptomes (e.g., tissue-specific) can be selected by the users. In addition, blast and keyword-based search engines have been added to easily retrieve information that can be downloaded and visualized in different formats.

Functional and applied aspects using genomic approaches

The availability of genomic resources and the development of species-specific high-throughput expression tools (see earlier section for description) have opened the "genomic era" and the characterization of transcriptomic responses for target biological processes with impact in aquaculture. In this section, the main results obtained using these genomic approaches to decipher the cellular regulatory pathways associated with larval development and different stimuli such as diet components, pathogens, or stress are reviewed.

Transcriptomic studies in larval stages and metamorphosis

The larval stage is a critical period in fish aquaculture. During this period, larvae undergo profound transformations that are particularly evident in soles due to the eye

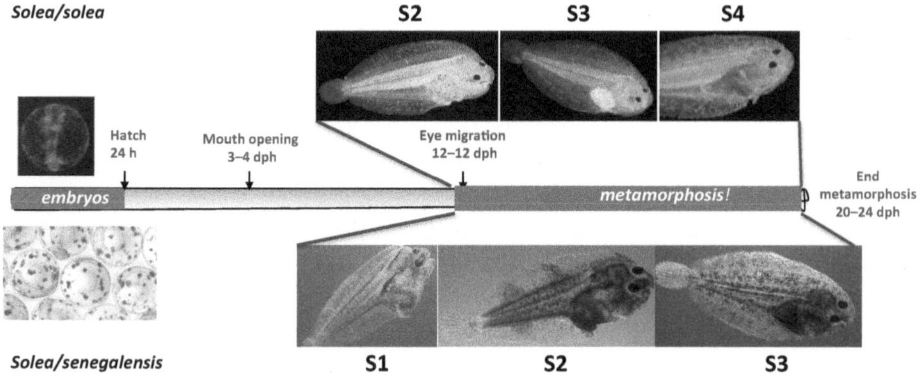

Figure 8.4 Larval development and metamorphosis timing in Senegalese sole and common sole.
Source: S. solea picture courtesy of Patrick Quazuguel.

migration and the transformation from pelagic to benthic fish (Fig. 8.4). In seabream, metamorphosis occurs approximately toward 45 dph whereas in sole it occurs earlier at 12 dph. This transition from larvae to adult involves important physiological, morphological, behavioral, and metabolic changes leading to acquisition of a functional digestive tract, competent immunity responses, mature neuro-muscular skeletal system, skin and pigmentation development, visual performance, and sexual maturation (Cerda and Manchado, 2013; Manchado et al., 2008b; Patruno et al., 1998). In addition, this period is characterized by an impressive plasticity of the individuals that are able to imprint new properties as a function of the environmental factors (Pittman et al., 2013). Nutritional components as well as physical factors such as temperature, oxygen, or pH can modulate the epigenetic markers modulating the transcriptional regulatory pathways and transforming the quality potential of animals.

In seabream, the characterization of the transcriptome from hatching until weaning (60 dph) revealed an overrepresentation of transcripts related to developmental processes such as the formation of the neural system, sensory organs, morphogenesis, tissue remodeling, digestion, muscular system, circulatory system, and epithelia development, identifying the major regulatory pathways in these stages (Yufera et al., 2012). More precise studies revealed that the transition from lecithotrophic to exotrophic larval stages (at mouth opening) implied a modulation of expression of genes related to digestion and energy production. In this way, pancreatic proteolytic enzymes (elastase, chymotrypsinogen) and amylase increased significantly to efficiently process and digest the live preys. This activation of digestive function was concomitant with a reduction of gluconeogenic enzymes and lipid mobilization genes that were highly expressed in the yolk syncytial layer (apolipoproteins B, E, and 14 kDa), reflecting a transition in terms of energy generation from lipid mobilization from yolk to external feeding (Ferraresso et al., 2008; Sarropoulou et al., 2005). Moreover, these expression patterns have been associated with changes in structural and muscle-related genes (cytokeratins I, tropomyosin, myosin light chain 2 and 3, myosin heavy chain) required for larval growth as well as those genes involved in eye pigmentation and vision (green-sensitive, retinal cone arrestin-3) that are necessary to perceive environmental stimuli and capture live preys.

In soles, transcriptome data indicate the existence of similar processes during larval development as in seabream. In *S. solea*, young larvae until 6 dph coordinately activated the expression of several genes related to visual perception and eye morphogenesis (opsins and retinoic acid-related metabolism genes) that have been associated with the onset of external feeding (Ferraresso et al., 2013). Nevertheless, most of these genes switch off later at the beginning of metamorphosis and this change is probably associated with the shift to a dwelling mode of life and nocturnal habits. Until the onset of metamorphosis (around 12–13 dph), the expression of several genes related to the development and functionality of the gastrointestinal system as well as folic acid and retinol metabolism are significantly activated preparing the larvae to feed actively and accumulate enough energy reserves to be later used during metamorphosis (Ferraresso et al., 2013; Hachero-Cruzado et al., 2014; Manchado et al., 2008a).

Metamorphosis in sole is a key developmental stage that, in most flatfish, coincides with a surge of thyroid hormones (THs) (Manchado et al., 2008b, 2009). During this process, severe changes occur in transcriptional pathways involved in the regulation of endocrine function, chromatophore differentiation, eye migration and cell proliferation, apoptosis and asymmetric development and antioxidative status. In this way, the pituitary-thyroid axis appears upregulated with an increase of thyroglobulin expression in the thyroid gland until larva reaches metamorphosis climax (Campinho et al., 2015; Manchado et al., 2008b) as well as TH receptors to transduce the TH signals and to initiate transcriptional responses (Ferraresso et al., 2013; Manchado et al., 2009). During this stage, several genes involved in TH biosynthesis (*tshb*, *tpo*, *tgb*, or *tshr*) have been demonstrated to be regulated by their ligand suggesting the existence of a negative feedback loop, although not at the hypothalamic level (Campinho et al., 2015; Iziga et al., 2010; Ponce et al., 2010). In addition to this endocrine pathway, several other genes involved in growth (IGFs and IGFBP) (Ferraresso et al., 2013), metabolic pathways such as glycolysis and calcium homeostasis (Infante et al., 2011b; Manchado et al., 2007), structural components of skin and gill epithelium such as keratins (Infante et al., 2007, 2011a), sexual hormones (Guzman et al., 2009), and protein biosynthesis including elongation factor 1a (Infante et al., 2008), and chaperonin hsp90AA (Manchado et al., 2008c) change their expression levels during metamorphosis and most of them are thyroid-dependent. Nevertheless, further transcriptome studies are required for a better comprehension of the molecular changes that occur during the drastic transformation from a pelagic to a benthic fish.

Skin pigmentation represents a major issue during flatfish metamorphosis, especially in aquaculture. Several anomalies in skin malpigmentation have been described in sole related to hormonal treatments or dietary components such as lipid levels (Cerda and Manchado, 2013). High levels of arachidonic acid during pre- and prometamorphosis stages induced significant pigmentation disorders (Boglino et al., 2014). The observed pseudo-albino phenotype was characterized by a disrupted expression of genes involved in melanophore differentiation and melanin synthesis (asip, pax3, cKit, mitf, tyr, trp1, and slc24a5) (Darias et al., 2013). The ratios cKit/pax3, tyr/mitf, asip/mc1r, and trp1/tyr have been suggested as markers to evaluate pigmentation regulatory pathways.

Transcriptomic studies in nutrition

Transcriptomic analysis has also become very popular in nutrition studies (referred to as nutrigenomics). In sole, a RNA-seq-based study using validation with middensity qPCR arrays demonstrated that early pelagic larvae were able to establish compensatory mechanisms to correctly manage lipids and store energy to successfully accomplish metamorphosis as a function of dietary triacylglycerol (TAG) content. A coordinated activation of genes related with lipid transport (i.e., apolipoproteins) was observed in larvae fed with high levels of TAG that avoids the detrimental effects of lipid droplet accumulation in the gut. On the contrary, larvae were able to activate metabolic pathways related to fatty acid and TAG biosynthesis and protein digestion under limiting TAG content to compensate for energy expenditure and to sustain growth rates for the upcoming metamorphic process (Hachero-Cruzado et al., 2014). Moreover, distinct expression profiles can be observed in the muscle and liver of sole juveniles depending on the lipid content. Interestingly, the expression levels of two myogenic factors (*myog* and *mrf4*), myosin light chain 2 (*mylc2*), and insulin growth factor I receptor (*igf1r*) appear to be highly correlated with growth and nutrient utilization indices (Campos et al., 2010)

Another major issue in nutrition that benefits from genomic information is the optimization of aquafeed formulations. Several studies have used this strategy to evaluate the cellular responses after replacement of fish-based meals or oils by alternative ingredients available in terrestrial plant sources. Previous studies in seabream and sole demonstrated that an almost complete replacement of marine-derived protein sources is feasible if accompanied with a minimal amino acid supplementation (reviewed in Morais et al., 2014). Nevertheless, there are existing controversies about the adequate blends of vegetable oil that are suitable to substitute fish oils in diets. Although some studies revealed the absence of any disruption in the performance and feed utilization of sole juveniles after substituting up to 100% of fish oil by a blend of rapeseed, soybean, and linseed vegetable oils (Benitez-Dorta et al., 2013; Borges et al., 2014). Transcriptomic responses in the gut of animals fed with a 100% replacement of fish oil by soybean oil (rich in n-6 fatty acids) revealed an overexpression of genes related to the complement pathway, to the recognition of pathogen associated molecular patterns, defense responses against bacteria and viruses, antigen differentiation, cytokines and their receptors, suggesting the development of microbiota dysbiosis and inflammatory processes (Montero et al., 2015). In the same sense, no changes in growth performance and transcriptome profiling in the intestine were determined by microarray analysis in seabream juveniles fed two diets with up to a 66% replacement of fish oil by a blend of vegetable oils (Calduch-Giner et al., 2012; Perez-Sanchez et al., 2013). Nevertheless, these animals displayed a higher sensibility to the parasite *E. leei* with differences in the time course and intensity of responses to the parasite. The expression profiles obtained were proposed as diagnostic markers of disease progression although without a prognostic value able to predict the susceptibility to this pathogen (Calduch-Giner et al., 2012). Moreover, these experimental groups showed different responses to crowding stress with higher plasma cortisol and glucose levels in those specimens fed plant-based diets over those fed control (fish oil–based) diets. Analysis of expression profiles in animals fed fish oil–based diets indicated the activation of genes related to cell-repair machinery that assists mitochondria and endoplasmic reticulum in protein-folding and misfolding, whereas compensatory mechanisms for glucocorticoid receptors, antioxidant enzymes, enzyme subunits of the mitochondrial respiratory chain, and enzymes involved in tissue fatty acid uptake and β-oxidation were detected in vegetable oil–fed fish (Perez-Sanchez et al., 2013).

A further characterization of cardiac and skeletal muscle transcriptomes in seabream subjected to feed restriction revealed that heart, and red skeletal muscle to a lesser extent, were highly responsive to nutritional changes whereas glycolytic white skeletal muscle was resistant to these changes (Calduch-Giner et al., 2014). The main nutritionally regulated genes in heart in response to feed restriction were related to signal transduction and transcriptional regulation, reflecting the metabolic plasticity of these tissues. In contrast, changes in the white muscle were more related to proteolysis and protein ubiquitination. Moreover, a microarray meta-analysis revealed that the three types of muscles shared differentially expressed transcripts related to mitochondrial function, similar to stress-mediated responses (Calduch-Giner et al., 2010; Perez-Sanchez et al., 2013), suggesting the mitochondria as target to obtain reliable markers of growth performance and welfare in this fish species to cope with environmental and nutritional stress stimuli.

Transcriptomic studies in health and welfare

Development of prophylactic strategies and treatments against parasites and bacterial and viral pathogens requires a profound knowledge of the responses of immune and nonimmune tissues to infection. Transcriptomic analyses have recently begun to be applied in seabream to describe immune responses to parasites and bacterial pathogens. For example, the effects of infection to the myxosporean enteric parasite *E. leei*, that is the cause of important economic losses for the aquaculture industry, have been investigated by microarray analysis in the intestine and head kidney (Davey et al., 2011). One of the most important responses to *E. leei* infection was the marked downregulation of immune-related genes (e.g., complement system, lectins, proteases/antiproteases, acute phase response proteins, etc.), a response attributed to the parasite's mechanism of immune evasion. Furthermore, *E. leei* infection also caused an upregulation of apoptosis, that appears to be the hallmark of parasitic infections and that could indicate excessive immune responses in the host in the face of parasitic infections.

As an example of transcriptomic analyses of bacterial infections in seabream, the transcriptional response to infection with live *P. damseale*, a relevant pathogen for the gilthead seabream, has been recently investigated by performing microarray analysis on head kidney samples from control and infected fish (Pellizzari et al., 2013). This study reported increased expression not only of immune genes involved in pathogen recognition and defense response but also of genes involved in negative regulation of immunity, including antiinflammatory cytokines, such as IL-10, and genes responsible for the activation of the polyamine pathway which results in the activation of arginine metabolism toward biosynthesis of L-ornithine and away from nitric oxide production. These immune suppressive responses have been interpreted as protective of tissue damage in the host or, alternatively, as the result of the pathogen action to evade the immune defenses of the host.

Interestingly, the reproductive hormone estradiol has been shown to affect the transcriptome of gilthead seabream macrophages in vitro. By performing suppression substractive hybridization, Liarte et al. (2011) reported on the identification of a number of genes whose expression was upregulated by estradiol, including cytokines, chemokines, antimicrobial peptides, and ion homeostasis proteins. These results suggest that estradiol induces a potent inflammatory response and oxidative stress, providing evidence for the modulation of the immune system by hormones. In contrast to estradiol, cortisol induces a strong transcriptional repression of immune-related genes in the seabream liver, including complement factors, antigen presenting molecules, cytokines and chemokines and, at the same time, also induces carbohydrate catabolism as part of a typical stress response (Teles et al., 2013). Therefore, transcriptomic analyses in seabream tissues have provided interesting clues regarding the modulation of innate immune responses by reproductive and stress hormones in this species.

Microarray analysis has also been applied to describe the effects of stress by confinement on the liver transcriptome in seabream. The results obtained indicate that stress causes a fast metabolic response to provide energy, a strong activation of tissue repair and remodeling processes, the restablishment of redox balance with increased scavenging of reactive oxygen species (ROS), and a general decline of ROS production (Calduch-Giner et al., 2010). These responses highlight the complexity of the stress response in seabream.

One aspect in fish health that has not received sufficient attention is the potential consequence of the activation of the innate immune system and the development of an inflammatory response in tissues outside of the immune system. In particular, the skeletal muscle is one of the most important tissues not only because it represents the edible part of fish but also because it is the most abundant tissue in the animal and plays a fundamental role in locomotion and whole body metabolism. In an approach to model this kind of relationship, the effects of lipopolysaccharide (LPS) administration on the white and red skeletal muscle transcriptomes have been evaluated in the gilthead seabream (Kaitetzidou et al., 2012). Particularly from a metabolic point of view, the transcriptomic effects of LPS administration caused opposite effects in the two types of muscle. That is, LPS increased the expression of genes involved in protein catabolism in white skeletal muscle, similar to the well-known induction of muscle atrophy by inflammation in mammals, and increased the expression of genes involved in protein synthesis in red skeletal muscle. Similarly, LPS administration resulted in an increase in glycolysis in white, but not red, skeletal muscle at 24 h postadministration and these changes were reversed at 72 h. However, a common and robust response in white and red skeletal muscles was the marked downregulation of immune-related genes at 24 and 72 h post-LPS administration, including genes involved in cytokine response, bactericidal activity, and antigen processing and presentation. Therefore, these results strongly suggest that proinflammatory stimuli may have important metabolic consequences in skeletal muscle and provide insight into the potential alterations in muscle quality under disease states in fish.

Future directions

Genomic research is progressing extremely fast and although genomic resources in seabream and soles are in different stages of development, they will progress in the same direction to bring their advantages to the optimization of aquaculture procedures.

High-throughput RNA-seq analyses for transcriptome studies will become one of the most powerful tools to reveal the underlying molecular mechanisms driving the responses to dietary components, vaccination, or reproduction control. Although this technique is currently scarcely applied in sole and seabream, it will become an essential tool in functional studies that will complement the results obtained from other expression tools as microarrays and openarrays. Nevertheless, to successfully develop this technique and allow for comparison across laboratories, devoted transcriptomic databases and curated reference transcriptomes need to be established. This information already exists in soles but information is still somewhat scattered in seabream. The identification of representative transcripts for each species that include spliced forms, tissue-specific transcripts, paralogous genes, and species-specific genes are required for robust RNA-seq analyses. Moreover, this information should be completed with a standardized and updated annotation using GO, KEGG, and other gene identifiers (such as those for species in ENSEMBL) that can easily manage this information, for statistical analysis and to make it compatible with other specific software focused on functional analysis. Although the first steps have been performed in sole, more efforts should be made to bring these essential tools to aquaculture in order to identify regulatory pathways that control reproduction, sex determination and reversal in hermaphroditic species, larval plasticity and metamorphosis, the interaction between diet ingredients, and performance of key indicators and welfare during the aquaculture production.

Although improvement of transcriptomes is the first logical step, the generation of reference genomes is also required to fully understand the biological relevance of an organism and to identify markers for genetic breeding and comparative genomics. The drastic reduction of sequencing costs in second generation sequencing platforms (mainly Illumina) and the fast development of cost-effective third generation sequencing platforms (such as PacBio) with longer reads make it feasible for new projects to obtain complete genomes in these species. Although a preliminary genome draft for the Senegalese sole is available, genome sequence information is quite limited in seabream. This information will play a key role for comparative genomic studies as well as to define better functional and selection markers.

Moreover, high-throughput genotyping platforms devoted to reveal whole-genome variation will acquire strength in the coming years for genetic breeding in aquaculture. The rapid generation of sequencing data and reference genomes will facilitate the identification of SNP markers throughout the entire genome. In this way, the optimization of genotyping-by-sequencing (GBS) technologies for whole-genome profiling will make it possible for SNP discovery and genotyping to be simultaneous and to facilitate genome-wide association studies, QTLs identification, and genomic selection. Although this method does not require a priori knowledge of the species' genome, it can clearly benefit from the abundant information contained in the genomes by reducing its complexity in a practical manner. Moreover, this abundant information on SNPs would facilitate the further development of other genotyping platforms based on microarray or open array format for a routine application of genomic selection in the industry in a cost-effective way.

Nevertheless a major constraint for these studies is bioinformatics. Most of these techniques require expertise and skilled people as well as devoted facilities to be able to process and analyze large amounts of sequence data in an efficient way. It must be noted that as laboratory sequencing protocols become easier and costs decrease, the time and effort required for bioinformatics increase correspondingly. Efforts should be invested in systematizing sequence analyses concerning basic processes (data cleaning, mapping, and statistical analyses) as well as dedicated workflows (identification of markers, association studies, phenotypic and genotypic correlations, etc.) based on supercomputing platforms. Consequently, customized and curated databases and specific tools/workflows would be essential to foster GBS approaches and RNA-seq studies in order to convert these genomic resources in practicable strategies for the aquaculture industry.

A special mention deserves the impact of epigenetics in aquaculture. This additional level of gene expression regulation has emerged over the last few years. Several epigenetic mechanisms exist including chromatin remodeling and genome methylation. In both cases, the resulting modifications in the genome can last throughout the entire life of the organism and underlie long-lasting effects through transmission to the following generations. Such mechanisms could explain how early events (e.g., during embryonic and larval stages) can persistently affect the performance of fish later in life. To date, very few genome-wide methylation profiles have been determined in aquaculture fish while a target-gene approach has been used to reveal environmental and/or sexual imprinting in several species (Chen et al., 2014; Navarro-Martin et al., 2011). In seabream and soles, as in other aquaculture fish, the characterization of methylation regulation may be of great importance in the understanding of how nutritional components and physical factors such as temperature or oxygen can modulate fish performance (e.g., growth, sex differentiation).

Concluding remarks

Overall, significant contributions have been made so far toward the transcriptome of the seabream and sole, and the characterization of the sole genome is ongoing. In both species, high-throughput expression tools such as microarray or openarray have been developed to successfully investigate regulatory pathways associated with diverse physiological, nutritional, and developmental situations of interest in aquaculture.

Studies using these technologies have demonstrated the usefulness of these techniques to generate robust information for the optimization of management and rearing procedures. Nevertheless, major efforts should be made to obtain the complete genomes of these species that will facilitate downstream applications including high-throughput genotyping (GBS) and expression (RNA-seq) techniques. Moreover, the development of species-specific bioinformatic platforms that provide curated transcriptome and genome information will play a key role in the expansion of these technologies. Although a public genomic database is now available for sole species, more effort should be made to define and publicize reference and annotated transcriptomes to be used in RNA-seq analysis and other genomic-based applications.

Acknowledgments

This study has been funded by project AQUAGENET (SOE2/P1/E287) program INTERREG IVB SUDOE and European Regional Development Fund (FEDER/ERDF) and INIA and EU through FEDER 2014-2020 "Programa Operativo de Crecimiento Inteligente" project RTA2013-00023-C02, and MICINN (AGL 2011- 25596 and AGL2014-51860C2-1P).

References

APROMAR, 2014. La Acuicultura en España 2014. ed Asociación Empresarial de Productores de Cultivos Marinos de España (APROMAR) and Asociación Española de Productores de Acuicultura Continental (ESCUA). Available from: www.apromar.es; www.esacua.com

Ariyadasa, R., Stein, N., 2012. Advances in BAC-based physical mapping and map integration strategies in plants. J. Biomed. Biotechnol. 2012, 184854.

Benitez-Dorta, V., Caballero, M.J., Izquierdo, M., Manchado, M., Infante, C., Zamorano, M.J., Montero, D., 2013. Total substitution of fish oil by vegetable oils in Senegalese sole (*Solea senegalensis*) diets: effects on fish performance, biochemical composition, and expression of some glucocorticoid receptor-related genes. Fish Physiol. Biochem. 39, 335–349.

Benzekri, H., Armesto, P., Cousin, X., Rovira, M., Crespo, D., Merlo, M.A., Mazurais, D., Bautista, R., Guerrero-Fernandez, D., Fernandez-Pozo, N., Ponce, M., Infante, C., Zambonino, J.L., Nidelet, S., Gut, M., Rebordinos, L., Planas, J.V., Begout, M.L., Claros, M.G., Manchado, M., 2014. De novo assembly, characterization and functional annotation of Senegalese sole (*Solea senegalensis*) and common sole (*Solea solea*) transcriptomes: integration in a database and design of a microarray. BMC Genomics 15, 952.

Boglino, A., Wishkerman, A., Darias, M.J., de la Iglesia, P., Andree, K.B., Gisbert, E., Estevez, A., 2014. Senegalese sole (*Solea senegalensis*) metamorphic larvae are more sensitive to pseudo-albinism induced by high dietary arachidonic acid levels than post-metamorphic larvae. Aquaculture 433, 276–287.

Borges, P., Valente, L.M., Veron, V., Dias, K., Panserat, S., Medale, F., 2014. High dietary lipid level is associated with persistent hyperglycaemia and downregulation of muscle Akt-mTOR pathway in Senegalese sole (*Solea senegalensis*). PLoS One 9, e102196.

Boulton, K., Massault, C., Houston, R.S., de Koning, D.J., Haley, C.S., Bovenhuis, H., Batargias, C., Canario, A.V.M., Kotoulas, G., Tsigenopoulos, C.S., 2011. QTL affecting morphometric traits and stress response in the gilthead seabream (*Sparus aurata*). Aquaculture 319, 58–66.

Calduch-Giner, J.A., Davey, G., Saera-Vila, A., Houeix, B., Talbot, A., Prunet, P., Cairns, M.T., Perez-Sanchez, J., 2010. Use of microarray technology to assess the time course of liver stress response after confinement exposure in gilthead sea bream (*Sparus aurata* L.). BMC Genomics 11, 193.

Calduch-Giner, J.A., Bermejo-Nogales, A., Benedito-Palos, L., Estensoro, I., Ballester-Lozano, G., Sitja-Bobadilla, A., Perez-Sanchez, J., 2013. Deep sequencing for de novo construction of a marine fish (*Sparus aurata*) transcriptome database with a large coverage of protein-coding transcripts. BMC Genomics 14, 178.

Calduch-Giner, J.A., Echasseriau, Y., Crespo, D., Baron, D., Planas, J.V., Prunet, P., Perez-Sanchez, J., 2014. Transcriptional assessment by microarray analysis and large-scale meta-analysis of the metabolic capacity of cardiac and skeletal muscle tissues to cope with reduced nutrient availability in Gilthead Sea Bream (*Sparus aurata* L.). Mar. Biotechnol. (NY) 16, 423–435.

Calduch-Giner, J.A., Sitja-Bobadilla, A., Davey, G.C., Cairns, M.T., Kaushik, S., Perez-Sanchez, J., 2012. Dietary vegetable oils do not alter the intestine transcriptome of gilthead sea bream (*Sparus aurata*), but modulate the transcriptomic response to infection with *Enteromyxum leei*. BMC Genomics 13, 470.

Campinho, M.A., Silva, N., Roman-Padilla, J., Ponce, M., Manchado, M., Power, D.M., 2015. Flatfish metamorphosis: a hypothalamic independent process? Mol. Cell Endocrinol. 404, 16–25.

Campos, C., Valente, L.M., Borges, P., Bizuayehu, T., Fernandes, J.M., 2010. Dietary lipid levels have a remarkable impact on the expression of growth-related genes in Senegalese sole (*Solea senegalensis* Kaup). J. Exp. Biol. 213, 200–209.

Cerda, J., Manchado, M., 2013. Advances in genomics for flatfish aquaculture. Genes Nutr. 8, 5–17.

Cerda, J., Mercade, J., Lozano, J.J., Manchado, M., Tingaud-Sequeira, A., Astola, A., Infante, C., Halm, S., Vinas, J., Castellana, B., Asensio, E., Canavate, P., Martinez-Rodriguez, G., Piferrer, F., Planas, J.V., Prat, F., Yufera, M., Durany, O., Subirada, F., Rosell, E., Maes, T., 2008. Genomic resources for a commercial flatfish, the Senegalese sole (*Solea senegalensis*): EST sequencing, oligo microarray design, and development of the Soleamold bioinformatic platform. BMC Genomics 9, 508.

Chen, S., Zhang, G., Shao, C., Huang, Q., Liu, G., Zhang, P., Song, W., An, N., Chalopin, D., Volff, J.N., Hong, Y., Li, Q., Sha, Z., Zhou, H., Xie, M., Yu, Q., Liu, Y., Xiang, H., Wang, N., Wu, K., Yang, C., Zhou, Q., Liao, X., Yang, L., Hu, Q., Zhang, J., Meng, L., Jin, L., Tian, Y., Lian, J., Yang, J., Miao, G., Liu, S., Liang, Z., Yan, F., Li, Y., Sun, B., Zhang, H., Zhang, J., Zhu, Y., Du, M., Zhao, Y., Schartl, M., Tang, Q., Wang, J., 2014. Whole-genome sequence of a flatfish provides insights into ZW sex chromosome evolution and adaptation to a benthic lifestyle. Nat. Genet. 46, 253–260.

Darias, M.J., Andree, K.B., Boglino, A., Fernandez, I., Estevez, A., Gisbert, E., 2013. Coordinated regulation of chromatophore differentiation and melanogenesis during the ontogeny of skin pigmentation of *Solea senegalensis* (Kaup, 1858). PLoS One 8, e63005.

Davey, G.C., Calduch-Giner, J.A., Houeix, B., Talbot, A., Sitja-Bobadilla, A., Prunet, P., Perez-Sanchez, J., Cairns, M.T., 2011. Molecular profiling of the gilthead sea bream (*Sparus aurata* L.) response to chronic exposure to the myxosporean parasite *Enteromyxum leei*. Mol. Immunol. 48, 2102–2112.

Dinis, M.T., Ribeiro, L., Soares, F., Sarasquete, C., 1999. A review on the cultivation potential of *Solea senegalensis* in Spain and in Portugal. Aquaculture 176, 27–38.

Diopere, E., Maes, G.E., Komen, H., Volckaert, F.A., Groenen, M.A., 2014. A genetic linkage map of sole (*Solea solea*): a tool for evolutionary and comparative analyses of exploited (flat)fishes. PLoS One 9, e115040.

EATiP, 2014. Performance of the sea bass and sea bream sector in the Mediterranean, Minutes of a Workshop held within Aquaculture Europe 2014 16th October 2014. European Aquaculture Society and the European Aquaculture Technology & Innovation Platform, San Sebastián, Spain.

FEAP, 2014. European Aquaculture Production Report 2004–2013. In: Dehasque, M. (Ed). Federation of European Aquaculture Producers.

Ferraresso, S., Bonaldo, A., Parma, L., Cinotti, S., Massi, P., Bargelloni, L., Gatta, P.P., 2013. Exploring the larval transcriptome of the common sole (*Solea solea* L.). BMC Genomics 14, 315.

Ferraresso, S., Vitulo, N., Mininni, A.N., Romualdi, C., Cardazzo, B., Negrisolo, E., Reinhardt, R., Canario, A.V., Patarnello, T., Bargelloni, L., 2008. Development and validation of a gene expression oligo microarray for the gilthead sea bream (*Sparus aurata*). BMC Genomics 9, 580.

Franch, R., Louro, B., Tsalavouta, M., Chatziplis, D., Tsigenopoulos, C.S., Sarropoulou, E., Antonello, J., Magoulas, A., Mylonas, C.C., Babbucci, M., Patarnello, T., Power, D.M., Kotoulas, G., Bargelloni, L., 2006. A genetic linkage map of the hermaphrodite teleost fish *Sparus aurata* L. Genetics 174, 851–861.

Garcia de la Serrana, D., Estevez, A., Andree, K., Johnston, I.A., 2012. Fast skeletal muscle transcriptome of the gilthead sea bream (*Sparus aurata*) determined by next generation sequencing. BMC Genomics 13, 181.

Garcia de la Serrana, D., Mareco, E.A., Johnston, I.A., 2014. Systematic variation in the pattern of gene paralog retention between the teleost superorders Ostariophysi and Acanthopterygii. Genome Biol. Evol. 6, 981–987.

Garcia-Cegarra, A., Merlo, M.A., Ponce, M., Portela-Bens, S., Cross, I., Manchado, M., Rebordinos, L., 2013. A preliminary genetic map in *Solea senegalensis* (Pleuronectiformes, Soleidae) using BAC-FISH and next-generation sequencing. Cytogenet. Genome Res. 141, 227–240.

Guzman, J.M., Bayarri, M.J., Ramos, J., Zohar, Y., Sarasquete, C., Mañanos, E.L., 2009. Follicle stimulating hormone (FSH) and luteinizing hormone (LH) gene expression during larval development in Senegalese sole (*Solea senegalensis*). Comp. Biochem. Physiol. A 154, 37–43.

Hachero-Cruzado, I., Rodriguez-Rua, A., Roman-Padilla, J., Ponce, M., Fernandez-Diaz, C., Manchado, M., 2014. Characterization of the genomic responses in early Senegalese sole larvae fed diets with different dietary triacylglycerol and total lipids levels. Comp. Biochem. Physiol. D 12, 61–73.

Infante, C., Asensio, E., Canavate, J.P., Manchado, M., 2008. Molecular characterization and expression analysis of five different elongation factor 1 alpha genes in the flatfish Senegalese sole (*Solea senegalensis* Kaup): differential gene expression and thyroid hormones dependence during metamorphosis. BMC Mol. Biol. 9, 19.

Infante, C., Manchado, M., Asensio, E., Canavate, J.P., 2007. Molecular characterization, gene expression and dependence on thyroid hormones of two type I keratin genes (sseKer1 and sseKer2) in the flatfish Senegalese sole (*Solea senegalensis* Kaup). BMC Dev. Biol. 7, 118.

Infante, C., Ponce, M., Asensio, E., Zerolo, R., Manchado, M., 2011a. Molecular characterization of a novel type II keratin gene (*sseKer3*) in the Senegalese sole (*Solea senegalensis*): differential expression of keratin genes by salinity. Comp. Biochem. Physiol. B 160, 15–23.

Infante, C., Ponce, M., Manchado, M., 2011b. Duplication of calsequestrin genes in teleosts: molecular characterization in the Senegalese sole (*Solea senegalensis*). Comp. Biochem. Physiol. B 158, 304–314.

Iziga, R., Ponce, M., Infante, C., Rebordinos, L., Cañavate, J.P., Manchado, M., 2010. Molecular characterization and gene expression of thyrotropin-releasing hormone in Senegalese sole (*Solea senegalensis*). Comp. Biochem. Physiol. B 157, 167–174.

Jimenez-Fernandez, E., Ponce, M., Rodriguez-Rua, A., Zuasti, E., Manchado, M., Fernandez-Diaz, C., 2015. Effect of dietary vitamin C level during early larval stages in Senegalese sole (*Solea senegalensis*). Aquaculture 443, 65–76.

Kaitetzidou, E., Crespo, D., Vraskou, Y., Antonopoulou, E., Planas, J.V., 2012. Transcriptomic response of skeletal muscle to lipopolysaccharide in the gilthead seabream (*Sparus aurata*). Mar. Biotechnol. (NY) 14, 605–619.

Kettleborough, R.N., Busch-Nentwich, E.M., Harvey, S.A., Dooley, C.M., de Bruijn, E., van Eeden, F., Sealy, I., White, R.J., Herd, C., Nijman, I.J., Fenyes, F., Mehroke, S., Scahill, C., Gibbons, R., Wali, N., Carruthers, S., Hall, A., Yen, J., Cuppen, E., Stemple, D.L., 2013. A systematic genome-wide analysis of zebrafish protein-coding gene function. Nature 496, 494–497.

Liarte, S., Cabas, I., Chaves-Pozo, E., Arizcun, M., Meseguer, J., Mulero, V., Garcia-Ayala, A., 2011. Natural and synthetic estrogens modulate the inflammatory response in the gilthead seabream (*Sparus aurata* L.) through the activation of endothelial cells. Mol. Immunol. 48, 1917–1925.

Loukovitis, D., Batargias, C., Sarropoulou, E., Apostolidis, A.P., Kotoulas, G., Magoulas, A., Tsigenopoulos, C.S., Chatziplis, D., 2013. Quantitative trait loci affecting morphology traits in gilthead seabream (*Sparus aurata* L.). Anim. Genet. 44, 480–483.

Loukovitis, D., Sarropoulou, E., Batargias, C., Apostolidis, A.P., Kotoulas, G., Tsigenopoulos, C.S., Chatziplis, D., 2012. Quantitative trait loci for body growth and sex determination in the hermaphrodite teleost fish *Sparus aurata* L. Anim. Genet. 43, 753–759.

Loukovitis, D., Sarropoulou, E., Tsigenopoulos, C.S., Batargias, C., Magoulas, A., Apostolidis, A.P., Chatziplis, D., Kotoulas, G., 2011. Quantitative trait loci involved in sex determination and body growth in the gilthead sea bream (*Sparus aurata* L.) through targeted genome scan. PLoS One 6, e16599.

Louro, B., Passos, A.L., Souche, E.L., Tsigenopoulos, C., Beck, A., Lagnel, J., Bonhomme, F., Cancela, L., Cerda, J., Clark, M.S., Lubzens, E., Magoulas, A., Planas, J.V., Volckaert, F.A., Reinhardt, R., Canario, A.V., 2010. Gilthead sea bream (*Sparus auratus*) and European sea bass (*Dicentrarchus labrax*) expressed sequence tags: characterization, tissue-specific expression and gene markers. Mar. Genomics 3, 179–191.

Manchado, M., Infante, C., Asensio, E., Canavate, J.P., 2007. Differential gene expression and dependence on thyroid hormones of two glyceraldehyde-3-phosphate dehydrogenases in the flatfish Senegalese sole (*Solea senegalensis* Kaup). Gene 400, 1–8.

Manchado, M., Infante, C., Asensio, E., Crespo, A., Zuasti, E., Canavate, J.P., 2008a. Molecular characterization and gene expression of six trypsinogens in the flatfish Senegalese sole (*Solea senegalensis* Kaup) during larval development and in tissues. Comp. Biochem. Physiol. B 149, 334–344.

Manchado, M., Infante, C., Asensio, E., Planas, J.V., Canavate, J.P., 2008b. Thyroid hormones down-regulate thyrotropin beta subunit and thyroglobulin during metamorphosis in the flatfish Senegalese sole (*Solea senegalensis* Kaup). Gen. Comp. Endocrinol. 155, 447–455.

Manchado, M., Infante, C., Rebordinos, L., Canavate, J.P., 2009. Molecular characterization, gene expression and transcriptional regulation of thyroid hormone receptors in Senegalese sole. Gen. Comp. Endocrinol. 160, 139–147.

Manchado, M., Salas-Leiton, E., Infante, C., Ponce, M., Asensio, E., Crespo, A., Zuasti, E., Canavate, J.P., 2008c. Molecular characterization, gene expression and transcriptional regulation of cytosolic HSP90 genes in the flatfish Senegalese sole (*Solea senegalensis* Kaup). Gene 416, 77–84.

Massault, C., Franch, R., Haley, C., de Koning, D.J., Bovenhuis, H., Pellizzari, C., Patarnello, T., Bargelloni, L., 2011. Quantitative trait loci for resistance to fish pasteurellosis in gilthead sea bream (*Sparus aurata*). Anim. Genet. 42, 191–203.

Molina-Luzon, M.J., Hermida, M., Navajas-Perez, R., Robles, F., Navas, J.I., Ruiz-Rejon, C., Bouza, C., Martinez, P., de la Herran, R., 2015a. First haploid genetic map based on microsatellite markers in Senegalese sole (*Solea senegalensis*, Kaup 1858). Mar. Biotechnol. (NY) 17, 8–22.

Molina-Luzon, M.J., Lopez, J.R., Navajas-Perez, R., Robles, F., Ruiz-Rejon, C., De La Herran, R., 2012. Validation and comparison of microsatellite markers derived from Senegalese sole (*Solea senegalensis*, Kaup) genomic and expressed sequence tags libraries. Mol. Ecol. Resour. 12, 956–966.

Molina-Luzon, M.J., Lopez, J.R., Robles, F., Navajas-Perez, R., Ruiz-Rejon, C., De la Herran, R., Navas, J.I., 2015b. Chromosomal manipulation in Senegalese sole (*Solea senegalensis* Kaup, 1858): induction of triploidy and gynogenesis. J. Appl. Genet. 56, 77–84.

Montero, D., Benitez-Dorta, V., Caballero, M.J., Ponce, M., Torrecillas, S., Izquierdo, M., Zamorano, M.J., Manchado, M., 2015. Dietary vegetable oils: effects on the expression of immune-related genes in Senegalese sole (*Solea senegalensis*) intestine. Fish Shellfish Immunol. 44, 100–108.

Morais, S., Aragão, C., Cabrita, E., Conceição, L.E.C., Constenla, M., Costas, B., Dias, J., Duncan, N., Engrola, S., Estevez, A., Gisbert, E., Mañanos, E., Valente, L.M.P., Yufera, M., Dinis, M.T., 2014. New developments and biological insights into the farming of *Solea senegalensis r*einforcing its aquaculture potential. Rev. Aquacult. 6, 1–37.

Moretti, A., Pedini Fernandez-Criado, M., Cittolin, G., Guidastri, R., 1999. Manual on hatchery production of seabass and gilthead seabream, vol. 1. Rome, FAO. 194 p. Full book can be downloaded at: http://www.fao.org/3/a-x3980e.pdf

Navarro-Martin, L., Vinas, J., Ribas, L., Diaz, N., Gutierrez, A., Di Croce, L., Piferrer, F., 2011. DNA methylation of the gonadal aromatase (cyp19a) promoter is involved in temperature-dependent sex ratio shifts in the European sea bass. PLoS Genet. 7, e1002447.

Parra, G., Bradnam, K., Ning, Z., Keane, T., Korf, I., 2009. Assessing the gene space in draft genomes. Nucleic Acids Res. 37, 289–297.

Patruno, M., Radaelli, G., Mascarello, F., Candia Carnevali, M.D., 1998. Muscle growth in response to changing demands of functions in the teleost *Sparus aurata* (L.) during development from hatching to juvenile. Anat. Embryol. (Berl) 198, 487–504.

Pellizzari, C., Krasnov, A., Afanasyev, S., Vitulo, N., Franch, R., Pegolo, S., Patarnello, T., Bargelloni, L., 2013. High mortality of juvenile gilthead sea bream (*Sparus aurata*) from photobacteriosis is associated with alternative macrophage activation and anti-inflammatory response: results of gene expression profiling of early responses in the head kidney. Fish Shellfish Immunol. 34, 1269–1278.

Perez-Sanchez, J., Borrel, M., Bermejo-Nogales, A., Benedito-Palos, L., Saera-Vila, A., Calduch-Giner, J.A., Kaushik, S., 2013. Dietary oils mediate cortisol kinetics and the hepatic mRNA expression profile of stress-responsive genes in gilthead sea bream (*Sparus aurata*) exposed to crowding stress. Implications on energy homeostasis and stress susceptibility. Comp. Biochem. Physiol. D 8, 123–130.

Pittman, K., Yufera, M., Pavlidis, M., Geffen, A.J., Koven, W., Ribeiro, L., Zambonino-Infante, J.L., Tandler, A., 2013. Fantastically plastic: fish larvae equipped for a new world. Rev. Aquacult. 5, S224–S267.

Ponce, M., Infante, C., Manchado, M., 2010. Molecular characterization and gene expression of thyrotropin receptor (TSHR) and a truncated TSHR-like in *Senegalese sole*. Gen. Comp. Endocrinol. 168, 431–439.

Ponce, M., Salas-Leiton, E., Garcia-Cegarra, A., Boglino, A., Coste, O., Infante, C., Gisbert, E., Rebordinos, L., Manchado, M., 2011. Genomic characterization, phylogeny and gene regulation of g-type lysozyme in sole (*Solea senegalensis*). Fish Shellfish Immunol. 31, 925–937.

Prachumwat, A., Li, W.H., 2008. Gene number expansion and contraction in vertebrate genomes with respect to invertebrate genomes. Genome Res. 18, 221–232.

Román-Padilla, J., Rodriguez-Rua, A., Claros, M.G., Hachero-Cruzado, I., Manchado, M., 2016. Genomic characterization and expression analysis of four apolipoprotein A-IV paralogs in Senegalese sole (*Solea senegalensis* Kaup). Comp. Biochem. Physiol. B 191, 84–98.

Sanchez-Ramos, I., Cross, I., Macha, J., Martinez-Rodriguez, G., Krylov, V., Rebordinos, L., 2012. Assessment of tools for marker-assisted selection in a marine commercial species: significant association between MSTN-1 gene polymorphism and growth traits. Sci. World J. 2012, 369802.

Sarropoulou, E., Franch, R., Louro, B., Power, D.M., Bargelloni, L., Magoulas, A., Senger, F., Tsalavouta, M., Patarnello, T., Galibert, F., Kotoulas, G., Geisler, R., 2007. A gene-based radiation hybrid map of the gilthead sea bream *Sparus aurata* refines and exploits conserved synteny with *Tetraodon nigroviridis*. BMC Genomics 8, 44.

Sarropoulou, E., Kotoulas, G., Power, D.M., Geisler, R., 2005. Gene expression profiling of gilthead sea bream during early development and detection of stress-related genes by the application of cDNA microarray technology. Physiol. Genomics 23, 182–191.

Senger, F., Priat, C., Hitte, C., Sarropoulou, E., Franch, R., Geisler, R., Bargelloni, L., Power, D., Galibert, F., 2006. The first radiation hybrid map of a perch-like fish: the gilthead seabream (*Sparus aurata* L). Genomics 87, 793–800.

Teles, M., Boltana, S., Reyes-Lopez, F., Santos, M.A., Mackenzie, S., Tort, L., 2013. Effects of chronic cortisol administration on global expression of GR and the liver transcriptome in *Sparus aurata*. Mar. Biotechnol. (NY) 15, 104–114.

Tsigenopoulos, C.S., Louro, B., Chatziplis, D., Lagnel, J., Vogiatzi, E., Loukovitis, D., Franch, R., Sarropoulou, E., Power, D.M., Patarnello, T., Mylonas, C.C., Magoulas, A., Bargelloni, L., Canario, A., Kotoulas, G., 2014. Second generation genetic linkage map for the gilthead sea bream *Sparus aurata* L. Mar. Genomics 18 Pt A, 77–82.

Vieira, F.A., Thorne, M.A., Stueber, K., Darias, M., Reinhardt, R., Clark, M.S., Gisbert, E., Power, D.M., 2013. Comparative analysis of a teleost skeleton transcriptome provides insight into its regulation. Gen. Comp. Endocrinol. 191, 45–58.

Yufera, M., Halm, S., Beltran, S., Fuste, B., Planas, J.V., Martinez-Rodriguez, G., 2012. Transcriptomic characterization of the larval stage in gilthead seabream (*Sparus aurata*) by 454 pyrosequencing. Mar. Biotechnol. (NY) 14, 423–435.

Molluscs

9

Rebeca Moreira Sanmartín, Steven Roberts**, Antonio Figueras**
*Marine Research Institute (IIM), Immunology & Genomics group, Spanish National Research Council (CSIC), Vigo, Spain; **School of Aquatic and Fishery Sciences University of Washington, Seattle, WA, United Sates

Life history and biology

Molluscs represent a phylum with approximately 50,000 species that is marked by tremendous diversity from cockles (genus *Cerastoderma*) to giant squids (genus *Architeuthi).* Taxonomic classes include bivalves, scaphopods, gastropods, cephalopods, monoplacophorans, polyplacophorans, and aplacophorans (including solenogastres and caudofoveata groups). Despite this diversity, key characters of this group are an epithelial derived mantle that can secrete shell(s) or spicules, a foot that develops from body wall muscles, and a radula, a feeding structure that is not present in bivalves (Fig. 9.1).

Since prehistory, ponds were used for algae culture and for the farming and maintenance of aquatic animals (Farber, 1997). Oysters were cultured in Japan, Greece, or Italy more than 2000 years ago (Pillay, 1997). Nowadays, molluscan aquaculture contributes a significant proportion of global aquaculture with 15 million tons produced in 2012, FAO, 2014. In this chapter we set out to discuss the current state of genomics in molluscs with a particular focus on those species that make up a significant portion of this global market.

In considering the role of genomics in molluscan aquaculture it is important to consider basic hatchery and production techniques. First one must consider the general life history. A majority of species that are farmed are broadcast spawners, meaning the gametes are commonly released into the water where external fertilization occurs. This is the case for most bivalves including mussels, scallops, oysters, and clams as well as for gastropods such as abalone. Cephalopods such as octopus are an emerging aquacultured group (Vidal et al., 2014) and are quite different in this respect as internal fertilization is accomplished with a copulatory organ, often a modified arm referred to as a hectocotylus. While there are no known hermaphroditic cephalopod species (Pechenik, 2010), many of the other cultured species are, including examples of protandric hermaphrodites (oysters i.e., *Crassostrea*), simultaneous hermaphrodites (scallops i.e., *Argopecten*), and rhythmical consecutive hermaphrodites (oysters i.e., *Ostrea*).

Larvae that develop from external fertilization are consequently free-living, and remain in the water in the order of days to weeks where they undergo metamorphosis. The length of time is dependent upon the taxa as well as environmental conditions. During this larval period development from an embryo to a trochophore occurs.

Genomics in Aquaculture. http://dx.doi.org/10.1016/B978-0-12-801418-9.00009-3

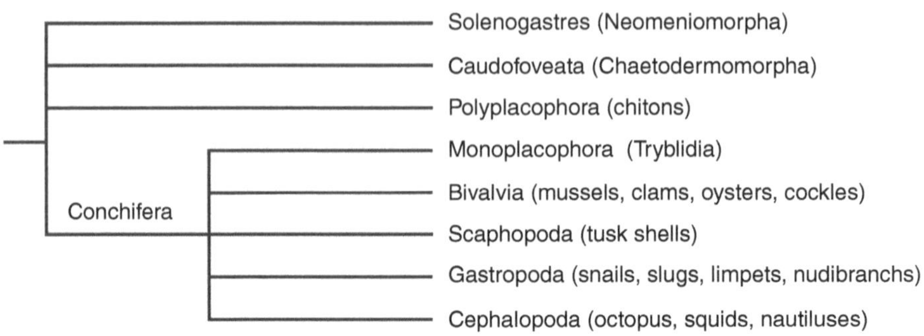

Figure 9.1 ***Mollusca*** **phylum.**
Source: Own work from Tree of life web project, http://tolweb.org/Mollusca/2488.

A velum soon appears, which is a ciliated organ used for locomotion, gas exchange, and food collection. At this point the larva is considered a veliger. In gastropods, such as abalone, larvae undergo a process known as torsion, a 180 degree twisting of the nervous system and digestive system. Once metamorphosis is complete the body plan is more similar to that of the adult form and the larvae will settle out of the water column where they might attach to a substrate (e.g., mussels and oysters) or have limited locomotory capabilities (e.g., abalone, clams, and scallops). The primary diet of these molluscs consists of plankton (bivalves) and macroalgae (gastropods) (Fig. 9.2).

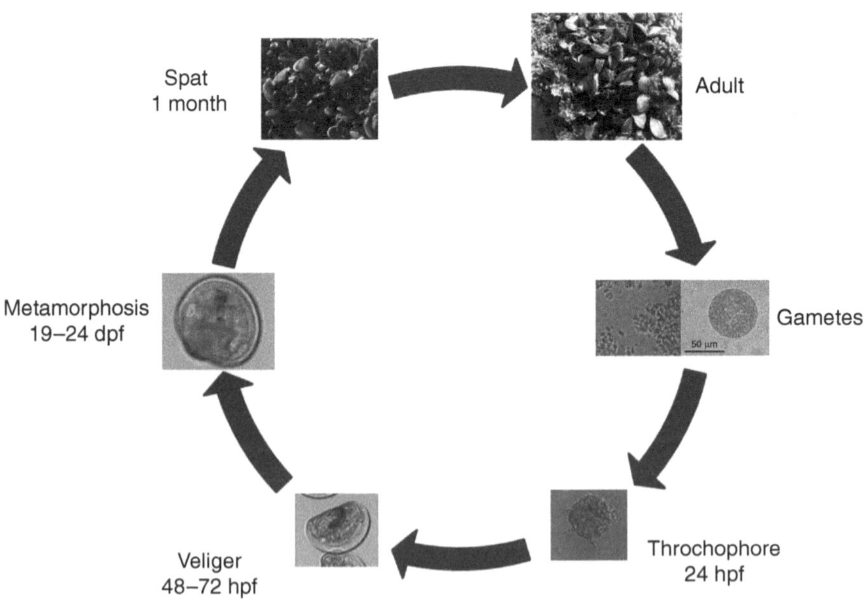

Figure 9.2 Mussel (*Mytilus galloprovincialis*) life circle. Gametes image shows sperm (*left*) and an oocyte (*right*).
Source: Own work. Photographs of larvae courtesy of Antonio Figueras lab.

Mollusc aquaculture is different from most finfish aquaculture in that most adults are commonly placed back into open water with intensive culture practices focusing on larval rearing. Often the culture process will start with the "conditioning" of broodstock with the feeding of nutrient rich microalgae usually grown on premises. It is worth pointing out that a significant portion of mollusc aquaculture is in fact algae culture. Once selected broodstock are reproductively mature, spawning and fertilization is initiated by either manipulating environmental conditions (i.e., temperature, food availability) or in some cases gametes are harvested and controlled crosses carried out. In some instances juveniles that have recently metamorphosed and settled, referred to as seed, are collected from the wild. As embryos develop larvae are grown up through metamorphosis until the juveniles are of sufficient size to move out of the hatchery to either a nursery setting of open water to take advantage of the natural food supply. Adult culture technology varies and includes rope culture (i.e., mussels), cages/bags (i.e., oysters), and "planting" in sediment (i.e., clams).

Genomes: diversity, structure, and organization

Genomic advances are occurring in an exponential manner with sequencing technology allowing for us to characterize genomes at a rate not even considered a few years ago. In the coming years the amount of available information will greatly expand our understanding of mollusc biology as it pertains to aquaculture. Ten years ago the information on molluscs in the genomic databases was about 25,000 nucleotide sequences, nowadays more than 3 million sequences are publicly available in Genbank and 2359 SRA projects are accessible. Regarding bivalve nucleotide sequences, Genbank has more than 720,000 results, and the SRA repository has 771 entries.

There are currently a handful of publicly available draft genomes and numerous transcriptomes and proteomes from cultured molluscs. The California sea hare, *Aplysia californica*, was the first mollusc to be sequenced. Its genome sequence will be useful in the study of many scientific areas, but it will be best used in the study of the sea hare's remarkable nervous system. The whole sea hare genome shotgun assembly was produced by the Broad Institute using the Arachne assembler. The Broad Institute has sequenced to 11× coverage resulting in a total genome assembly size of approximately 712 Mb.

The next mollusc genome to be sequenced was the *Lottia gigantea* genome (Simakov et al., 2013). This gastropod was chosen because the species is an emerging model in evolution and development, ecology, and conservation. In fact, its publication described the comparison among other animal genomes to investigate the origin and diversification of bilaterians. The size of its genome is 348 Mb, and the repetitive content represents the 21% of the assembled genome. There were approximately 23,800 genes predicted from its genome. The browser of this genome is available at http://genome.jgi.doe.gov.

With respect to the Pacific oyster (*Crassostrea gigas*), a mollusc that constitutes a significant economic role in global aquaculture production, it has numerous genomic resources including a draft genome. One of the first integrative resources for this species was the GigasDatabase (Fleury et al., 2009) where publically available transcriptomic data such as expressed sequence tags (ESTs) were assembled into contigs and annotated. This resource greatly increased the rate of scientific discovery in the field of mollusc genomics.

A draft genome was published in 2012 for *C. gigas* (Zhang et al., 2012). This effort used a short-read sequencing and fosmid pooling strategy that resulted in an assembly with 559 Mb and a scaffold N50 size of 401 kb (Zhang et al., 2012). A total of 11969 scaffolds were assembled with more than 90% of the assembly covered by the longest 1670 scaffolds (Zhang et al., 2012). A total of 28,027 genes were predicted in this effort with 49 RNA-seq libraries from different developmental stages and different adult organs and 59 libraries sequenced from oysters subjected to various stressors (Zhang et al., 2012). Other general features described with the *C. gigas* genome include that 36% of the genome was repetitive sequence as well as evidence for active transposable elements. These oyster genome datasets were subsequently incorporated into the EnsemblMetazoa (http://metazoa.ensembl.org/) where the assembly is continually annotated. Only 7658 of the scaffolds have been used in the Ensembl platform, corresponding to 26,101 coding genes. EnzemblMetazoa provides novel genome features compared to the source data including coordinate information for noncoding RNA and sequence repeats.

A pearl oyster (*Pinctada fuctata*) draft genome has also been sequenced representing approximately 1150 Mb and 23,257 complete gene models (Takeuchi et al., 2012). Over 800,000 scaffolds were assembled 629 at least 100,000 bp in length. The draft genome assembly and predicted gene products are available at http://marinegenomics.oist.jp/.

The octopus (*Octopus bimaculoides*) genome has recently been released (Albertin et al., 2015). With more than 300,000 scaffolds and a scaffold N50-length of 470 kb, the genome assembly captures more than 97% of expressed protein-coding genes and 83% of the estimated 2.7 Gb genome size. There are 33,638 protein-coding genes predicted. The unassembled fraction is dominated by high-copy repetitive sequences. Nearly 45% of the assembled genome is composed of repetitive elements. A browser of this genome assembly is available at http://octopus.metazome.net/.

The *M. galloprovincialis* genome is still in progress. The sequencing strategy was similar to that of *C. gigas*, fosmid pools and short-read sequencing of a wild female sampled in the Ría de Vigo. The actual assembly comprises 22,279 scaffolds and the estimated the genome size is 1.42 Gb (personal communication). There are however a wealth of transcriptomic resources for mussels including MytiBase (http://mussel.cribi.unipd.it) (Venier et al., 2009). Mytibase is a compilation of annotated ESTs obtained from different origins gathered in a platform for expression studies, marker validation and genetic linkage analysis.

To date there is yet to be a complete mollusc genome (i.e., chromosome organization) although there are other mollusc species with significant resources that allow for comparative analysis. Regarding the *Mytilus* and *Crassostrea* genus, there are some studies about their karyotypes showing that 4 different species of mussel have 14 chromosome pairs (Pérez-García et al., 2014), meanwhile 5 different oyster species have 10 pairs (Wang et al., 2004). These studies could help to solve the

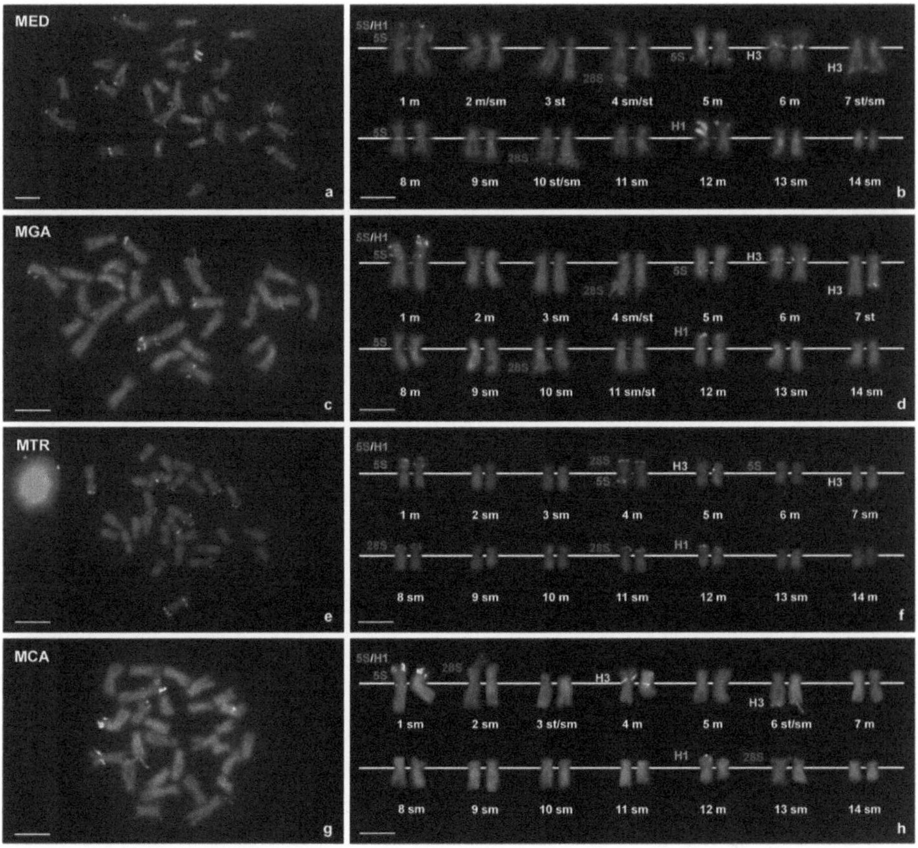

Figure 9.3 ***Mytilus*** **kariotypes.**
Source: Pérez-García *et al.* (2014).

genomic architecture for these species as more sequence information becomes available (Fig. 9.3).

With the ever-increasing advances in DNA sequencing technology it is certain there will be a number of other draft genomes completed (undoubtedly before the publication of this chapter). For example, new genomic resources are expected for *Crassostrea virginica* (Gómez-Chiarri et al., 2015), *M. galloprovincialis* (Maria et al., 2014) and *Patinopecten yessoensis* (PRJNA253231; PRJNA259405), as shown in Table 9.1. In this table it is also described recent genomic research regarding the most valuable molluscs in terms of aquaculture interest: oysters, mussels, clams, scallops, and octopus.

As highlighted by Gómez-Chiarri et al. and Murgarella et al. there are several challenges to prepare a complete assembly of these genomes including the highly polymorphic nature and high percentage of repetitive elements. These features are likely associated with the life history characteristics of marine bivalves, the predominantly

Table 9.1 **Summary of genomic studies in molluscs with an economic importance in aquaculture**

Species	Resource	Accession no.°	Tissue	References
Crassostrea gigas	Transcriptome		Hemocytes	Gueguen et al. (2003)
	Transcriptome	PRJNA149121, GSM667902, GSM667901	Hemocytes	Rosa et al. (2012)
	Transcriptome	PRJNA167099	Gill	Zhao et al. (2012)
	Transcriptome	PRJNA198468	Whole body section	Jouaux et al. (2013)
	BAC library		Sperm	Cunningham et al. (2006)
	Genome	PRJNA70283	Adductor muscle, gill, mantle, gonad	Zhang et al. (2012)
	Epigenome		Gill	Gavery and Roberts (2013)
	Proteome		Gill	Zhang et al. (2014b)
Crassostrea virginica	Transcriptome	PRJNA82611	Hemocytes, gill, digestive gland, adductor muscle, mantle	Zhang et al. (2014a)
	Transcriptome	PRJNA248114	Whole juvenile	McDowell et al. (2014)
	Transcriptome	PRJNA227565	Gill, adductor muscle, mantle	Eierman and Hare (2014)
	BAC Library		Sperm	Cunningham et al. (2006)
	Genome (in progress)			Gómez-Chiarri et al. (2015)
Mytilus edulis	Transcriptome	PRJEB2700	Digestive gland, foot, adductor muscle	Philipp et al. (2012)
	Transcriptome	PRNJNA182066	Hemocytes	Tanguy et al. (2013)
	Transcriptome	PRJNA252953	Larvae	Bassim et al. (2014a)
	Transcriptome	PRJEB4516	Mantle	Freer et al. (2014)
Mytilus galloprovincialis	Transcriptome	PRJNA80091	Hemocytes	Rosani et al. (2011)
	Transcriptome	PRJNA230138	Hemocytes, gill, muscle, mantle	Moreira et al., 2015
	Genome (in progress)	PRJNA178783	Mantle	Nguyen et al. (2011)
	Genome (in progress)		Mantle	Murgarella et al. (2015)
	Proteome		Hemolymph	Oliveri et al. (2014)

Patinopecten yessoensis	Transcriptome	PRJNA79873	Larvae; adult adductor muscle, digestive gland, gonad	Hou et al. (2011)
	Transcriptome	PRJNA186890	Gill, digestive gland	Meng et al. (2013)
	Genome (in progress)	PRJNA253231, PRJNA259405		
	Metagenome	PRJNA242688		
Pecten maximus	Transcriptomes	PRJNA222492	Hemocytes	Pauletto et al. (2014)
Pinctada fucata	Transcriptome	PRJDA63487	Pearl sac, mantle	Kinoshita et al. (2011)
	Genome (draft)	PRJDB2628	Sperm	Takeuchi et al. (2012)
Pinctata maxima	Transcriptome	PRJNA187136		Jones et al. (2011)
	Transcriptome	PRJNA114555	Mantle	Gardner et al. (2011)
Ruditapes philipinarum	Transcriptome	SRX100159	Hemocytes	Moreira et al. (2012)
	Transcriptome	PRJNA234093, PRJNA234077, PRJNA2329		
Octopus bimaculoides	Genome	PRJNA270931, PRJNA285380		Albertin et al. (2015)

cultured mollusc group. Several strategies have been tried to resolve these challenges. From the biological point of view, the development of highly inbred lines for these species is a strategy that could be of help. In the *C. gigas* genome, the use of inbred individuals reduced the polymorphism rate by 44%. From the bioinformatic point of view sequencing long reads or fosmid libraries and novel assembly algorithms for highly polymorphic genomes are the proposed solutions. But even with this support the issue of polymorphisms and repetitions persist and are difficult to solve.

Genome sequencing of molluscan pathogens is also of interest in aquaculture. These efforts can be of help in aquaculture, especially for larval survival and adult fitness. Information regarding the genomes of bacterial (i.e., *Vibrio*, PRJNA256191, Schreier and Schott, 2014) and protozoan (i.e., *Perkinsus*, PRJNA46451 and PRJNA237117) pathogens is increasing each year. Viruses are still a challenge in this field however the genome of the Ostreid herpesvirus-1 is available (PRJNA14552, Davison et al., 2005). Pathogen genomes would allow the identification of molecular variants and mechanisms of virulence, resulting in an increase of the performance in hatcheries and nurseries through specific management tools.

Functional and applied aspects

Much of what we have learned with regard to genome activity (gene and protein expression) in aquacultured molluscs has centered upon fundamental physiological processes that are critical for improving aquaculture production. In this section we will address key aspects of their biology with first reviewing how genomic approaches have improved our fundamental understanding, followed by direct comparison of the genomic underpinnings of beneficial traits. Note that here we will be focusing primarily on the expressed part of the genome and not as much on genomic approaches.

Some of the first functional genomic studies of aquacultured molluscs using bivalves were conducted using homology cloning. Often this approach was not successful because of limited information on bivalve genes in databases. The next step using functional genomic approaches was to describe differentially expressed genes using expressed sequence tag (ESTs) collections from cDNA or subtractive hybridization libraries (SSH). The major advance in recent genomic research occurred with the development of next-generation sequencing technologies (NGS). In bivalves, the NGS and the microarray technologies have been applied to investigate numerous different physiological processes (i.e., reviewed in Romero et al., 2012 and in Suárez-Ulloa et al., 2013), including physiological processes fundamentally linked to aquaculture such as growth, immune function, and reproduction.

Growth

Like any agricultural product, growth is an important factor in terms of reaching market size with minimum expenditure on resources. There are numerous aspects in regard to factors contributing to growth with functional genomic studies providing important insight into underlying mechanisms.

In bivalves and in particular *C. gigas*, growth heterosis is observed whereby hybrids often grow faster than respective parental lines (Pace et al., 2006). In one of the earliest high-throughput sequencing efforts in bivalves, (Hedgecock et al., 2007) used massively parallel signature sequencing (MPSS) to examine gene expression patterns in two partially inbred and two hybrid larval populations. Previous work had shown that hybrid oysters have higher feeding rates and feeding efficiencies than inbred oysters (Bayne et al., 1999; Pace et al., 2006) but nothing was known about underlying transcriptomic differences. Hedgecock et al., 2007 found that approximately 1.5% of genes control the phenotypes associated with growth heterosis and that protein metabolism was involved based upon gene annotation results. Later work (Meyer and Manahan, 2010) also found protein metabolism drives growth variation in a study where using the same genetic families as (Hedgecock et al., 2007), they examined 188 candidate genes. In this study, the majority of candidate genes were ribosomal proteins and the authors indicate that given the high metabolic cost of protein synthesis and degradation, inefficiencies in this process could certainly impact energy available for growth (Meyer and Manahan, 2010).

Immune function

According to the FAO, the future of the molluscan production needs to be planned as an integral program that includes aspects such as the creation of a special research plan on molluscan pathologies. The high overall mortality of bivalves at multiple life stages is the main problem in mollusc aquaculture such mortalities causes dramatic economic losses in the industry. Nevertheless, the knowledge of the immune response in these species is still limited and the fight against the pathologies is based in preventive strategies and the removal of diseased individuals. The molecular basis of the immune response in bivalves has been of great interest in the last decade and many studies have been published about bivalve genomics, reviewed in Saavedra and Bachère (2006) and Romero et al. (2012).

Vertebrate immune system is characterized by a nonspecific innate response and an acquired response, with memory after a previous contact with the pathogen. Bivalves lack the acquired response in a narrow sense, but in addition of physical barriers such as the shell and the mucus, they have a potent and efficient cellular and humoral innate immune system.

Defense cells in bivalve mollusks are the hemocytes. These cells are traditionally classified depending on their morphologic and functional characteristics in granulocytes and hyalinocytes. Hemocytes trigger their defense mechanisms after pathogen recognition and eliminate these cells or foreign particles by phagocytosing and destroying them by lysosomal enzymes, reactive oxygen species (ROS) synthesis or nitric oxide (NO) production (Fig. 9.4).

Humoral immunity is mediated by the proteins found in the hemolymph, the hemocytes or both. These proteins are the focus of the omics studies in mollusc immunology.

Traditionally, research in mollusc pathology was mainly based on functional and observational studies such as phagocytosis, ROS/NO production or mortality and histology assays. Then, studies focused in a molecular point of view, and characterization

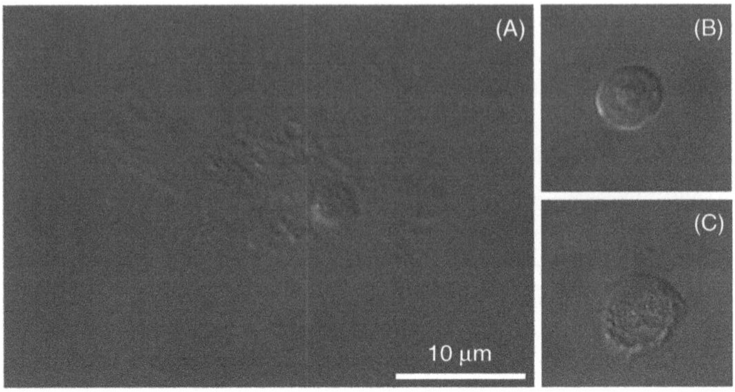

Figure 9.4 Granulocyte (A) and hyalinocyte (B and C) of carpet shell clam (*Ruditapes decussatus*).
Source: Own work, phase contrast microscopy images courtesy of Antonio Figueras lab; Carballal, M.J., Lopez, M.C., Azevedo, C., Villalba, A., 1997. Hemolymph cell types of the mussel *Mytilus galloprovincialis*. Dis. Aquat. Org. 29, 127–135.

of specific immune-related molecules were performed with techniques such as homology cloning, BAC libraries or Expressed Sequence Tags (EST) collections from Suppression Subtractive Hybridization (SSH). These methods were especially useful in organisms where genomic data were scarce or not available. They facilitated the identification of new genes through the study of differential gene expression after the interaction between the host immune system and different pathogens.

In molluscs, the first published library in an immune framework was constructed in *C. virginica* (Jenny et al., 2002). After that similar studies were performed in oyster, clam, or mussel to identify genes related to pathogen infections (i.e., *Vibrio* sp. and *Perkinsus* sp.) or focused on differential susceptibility to disease (reviewed in Romero et al., 2012).

The SSH technology and the information available in the MytiBase led to the detection of a considerable number of antimicrobial peptides (AMPs) in different mollusc species. AMPs are small amphipathic proteins highly conserved throughout evolution. Their antimicrobial function relies principally on their integration on the pathogen's membrane, unstabilizing it and generating pores, which lyses the cells, but AMPs can also act intracellularly avoiding DNA, protein or cell wall synthesis.

AMPs were originally classified in four main groups: defensins, mytilins, myticins, and mytimycins (Mitta et al., 2000). Their importance in bivalve immune response lead to a large number of recent publications, such as the various studies carried on myticin C to explain its high diversity (Costa et al., 2009), its antiviral and immunoregulatory properties (Balseiro et al., 2011) and its genomic organization, molecular diversification, and evolution (Vera et al., 2011). There is also a recent study on the variability of the AMPs (Rosani et al., 2011). Even in a group such as AMPs, very studied in the last 15 years, there have been recent discoveries, like the characterization of big defensins and mytimacins (Gerdol et al., 2012); myticusins (Liao et al., 2013), and mytichitins (Qin et al., 2014).

More recently, the NGSs technology were the boost that the genomics research in mollusc immunology needed to enrich the databases and explore new tools, such as microarrays and RNAseq to study host pathogen interactions. The European project ReProSeed (http://www.reproseed.eu/), finished in 2013, made an important effort to enrich the databases in sequences of several species of cultured bivalves: oyster (*C. gigas*), mussels (*M. galloprovincialis* and *Mytilus edulis*), clams (*Ruditapes decussatus*) and scallop (*Pecten maximus*). This techniques lead to the identification and characterization of relevant immune-related genes in molluscs. Several immune processes such as apoptosis, the Toll-like signaling pathway and the complement cascade were described for *Ruditapes philippinarum*, together with a large number of other immune related genes (Moreira et al., 2012). Valuable information about the apoptotic process in *M. galloprovincialis* was obtained from the MytiBase database by Romero et al. (2011). They were able to identify for the first time the most important molecules involved in apoptosis, as a result six different caspase genes were characterized.

Transcriptomic studies lead to many publications with microarray technology. Microarrays were principally used to study the response to several pathogens of protozoan, bacterial, and viral origin which infect a wide variety of marine molluscs. *Perkinsus* spp. belongs to a family of protozoan parasites that are associated with mass mortalities. It is the causal agent of the dermo disease. This parasite was used as a challenge in several microarray studies in *C. virginica* (Wang et al., 2010), *Ruditapes decusattus* (Leite et al., 2013), and *R. philippinarum* (Romero et al., 2015). All of them used different approaches to study this chronic condition: an end point infection, natural infection, and a time course but all found that pathogen recognition (as *Perkinsus* is an intracellular pathogen these are important genes in the initial stages of the infection for the parasite to be internalized), antimicrobial activity, redox processes, and especially, apoptosis were important processes to fight the infection.

Regarding bacterial infections, *Vibrio* spp. is a universal marine pathogen that has been studied in mollusc hosts such as *M. galloprovincialis* (Venier et al., 2011), *R. philippinarum* (Moreira et al., 2014; Allam et al., 2014) and the disk abalone *Haliotis discus discus*, a species with small and recent representation in aquaculture (De Zoysa et al., 2011). Although the pathogen species was different and the infection timing differs among studies a general tendency can be followed to fight the acute infection: apoptosis, as in the case of *Perkinsus* infections, is an important process activated in the first hours after the infection. Recognition molecules interacting with bacterial PAMPs, genes related to signaling pathways and transcription factors are present in all the studied molluscs, as well as cytokines such as LITAF, MIF, or IL-17. It is also observed oxidative burst-related genes. Chemotactic and phagocytic behavior of hemocytes is also important in the critical phase of the infection. In the phase of overcoming the infection a predominant downregulation of gene expression, as well as activation of processes related to wound healing such as biomineralization are observed.

Only one study was performed to date to study the interaction between viruses and molluscs, Ostreid herpes virus type 1 (OsHV-1) in Pacific oyster *C. gigas* (Jouaux et al., 2013). As well as against Perkinsus and Vibrio infections apoptosis plays an important role to overcome the infection. Immune system has a more limited activity to fight viral infections but cell signaling in the host and its ability to regulate the

OsHV-1 genome activity seems to be the key to survival. Indeed, another study with resistant and susceptible *C. gigas* to summer mortality (a phenomenon resulting from viral and bacterial infection of individuals weakened by abiotic stress and reproduction) (Fleury and Huvet, 2012) shows that the regulation of the NF–κB signaling pathway is key to determine the susceptibility or resistance to summer mortality. These genes are possible candidates to allow marker-aided selection to improve oyster and probably other molluscs culture.

RNA-seq studies in molluscs are still scarce but several are available or in progress. The first SOLiD RNA sequencing work was performed in *C. gigas* (Gavery and Roberts, 2012) and it demonstrated that this technology was able to generate novel information and identify differentially expressed genes, which is very useful in non-model species. In the next years other approaches in *M. chilensis*, *Octopus vulgaris*, *P. maximus*, and *M. galloprovincialis* were released. These studies covered different areas like SNP identification, yielding 20,306 polymorphisms associated to immune-related genes in *M. chilensis* (Núñez-Acuña and Gallardo-Escárate, 2013); or tissue characterization in *M. galloprovincialis* (Moreira et al., 2015) showing the great importance of the immune system in all the tissues and finding new functions like the hematopoietic, antifungal, and sensorial functions of mantle. Of course, it was studied the interaction with PAMPs and pathogens. There was generated new information about coccidiosis in octopus, which revealed genes involved in NF-kB and TLR pathways and in the complement cascade (Castellanos-Martínez et al., 2014), there were identified at least four TLRs in scallop hemocyte transcriptome (Pauletto et al., 2014), and new virulence factors in *Perkinsus* (Pales et al., 2014).

Reproduction

As indicated earlier the aquaculture production cycle starts with the reproductive maturation of broodstock which will lay the foundation for larval production and subsequent adult harvest. From a functional genomics perspective, there has been deserved attention towards understanding processes associated with reproductive maturation and spawning. Having a better understanding of reproductive biology has the potential and promise to increase efficiency and output. While there have been numerous efforts targeting candidate gene function during reproduction, one example of a global approach to examining gametogenesis in an aquaculture relevant species was Dheilly et al. (2012) study using a microarray platform in *C. gigas*. This oligonucleotide microarray was composed of 31,918 expressed sequence tags (ESTs) (Dheilly et al., 2011). In comparing males and females Dhielly et al identified 77 genes that were sex specific, which could have practical application in identifying broodstock. For example, there are several aquacultured mollusc where sexual dimorphism is lacking. Gene or possibly protein expression could be used to distinguish sexes as well as developmental status. More recently there have been numerous studies that examine reproductive genes other molluscs (i.e., De Sousa et al., 2014; Teaniniuraitemoana et al., 2014, 2015; Valenzuela-Muñoz et al., 2014).

Taken together we now have a more complete understanding gametogenesis in commercially important species. Primordial germ cells are the cells, which develop into gametes, which go onto to create a new individual through sexual reproduction. The establishment of the germline varies across taxa with a general difference of either being specified via inherited factors (preformation) or by inductive signals (epigenesis) (see Extavour and Akam, 2003 for review). In several molluscs it is not clear (or varies) on the origin of the germline. Using a vasa-like gene as a marker in *C. gigas*, (Fabioux et al., 2004) concluded that in *C. gigas* that the germline is specified by maternal cytoplasmic determinants (preformation) through the 4d cell lineage and larvae possess putative PGCs. Also using vasa gene expression (Kranz et al., 2010) inferred that in the Abalone (*Haliotis asinina*) that PGCs are not determined completely by maternal determinants, but also by inductive signals. The germ cells eventually become sexually differentiated and enter gametogenesis. As indicated in the previous section, sexual reproduction can be quite complex in this group due to different temporal degrees of hermaphroditism and sex change. A recent study by (Zhang et al., 2014) used transcriptome sequencing to identify genes in sex-determining pathways.

These gonia then go through oogenesis and spermatogenesis and ripen into mature ova and spermatozoa. Gametogenesis in many marine invertebrates, is a substantial energy demanding process as these organisms have high fecundity. Fecundity and subsequent larval survival is key concern in aquaculture as this serves as the base of production. As indicated the lability of sex in molluscs is complex (and fascinating) and functional genomic approaches are actively being used in the better understand this. As with many aspects of physiology this includes attention to how large scale environmental change might influence reproductive biology.

The first stage of gametogenesis (Stage 0) is the initial differentiation of primordial germ cells where the sex cannot be determined. The next stage (Stage 1) is when germ cells undergo mitosis and a large number of gonia are produced. Stage 2 gonads contain maturing cells that depend on energy from surrounding tissues. Stage 3 gonads are considered fully mature. Using a microarray platform to study the gonadal maturation process in males (spermatogenesis) and females (oogenesis) (Dheilly et al., 2012) identified a suite of genes indicative of the respective processes. In mature male gonads, there was a number of genes involved in ubiquitination and proteasomal degradation of ubiquitinated proteins. In females, researchers found evidence of elevated cell cycle activity in maturing gonads and interestingly found Methyl-CpG binding domain protein 2 to be highly expressed in mature female gonads. Methyl-CpG binding domain protein 2 is involved in binding methylated DNA and thus presumably involved in gene regulatory activity. (Dheilly et al., 2012) suggested that the high expression of methyl-CpG binding domain protein indicated a epigenetic transfer of information. This process could have important implications for aquaculture production as it would indicate parental factors (i.e., maternal RNA) have direct impact on larval development, thus broodstock physiology is important to consider. Maternal factors are considered important for fertilization and embryonic genome activation (Li et al., 2010). The maternal contribution is also receiving attention in the aquaculture production of teleost fish (i.e., Sullivan et al., 2015). (Dheilly et al., 2012) also found a number of genes associated with early embryological development (i.e., Forkhead box Q2, Frizzled).

Focusing on another important aquaculture species, the European clam (*R. decussatus*), (De Sousa et al., 2015) used a microarray to identify potential biomarkers of oocyte quality, also using D-larval yield as a metric. From this work it was determined that chaperone molecules are major determinants of good quality oocytes and genes involved in stress response, including apoptosis, are associated with poor oocyte quality.

Proteomic approaches are increasingly becoming important tools to assess gamete and egg quality, particularly for species with abundant transcriptomic resources necessary for annotation. Using a 2-DE proteomic approach (Corporeau et al., 2012) identified factors indicative of oocyte quality. Specifically oocyte quality was determined by D-larva yields. In low quality oocytes, there was a higher expression of 10 proteins, including 5 vitellogenins proteins some of which were breakdown products. It was suggested that cleavage products could be an indication of resorbtion and or oocyte aging. In high quality oocytes, again vitellogenin was identified as well as molecular chaperones and protein involved in cell cycle control.

In an analogous study that set out to identify proteins associated with mature spermatozoa (Kingtong et al., 2013) used 2-DE proteomics to compare mature sperm with germ/less mature cells. Here researchers identified 31 differentially expressed proteins with mature spermatozoa having increased expression of proteins associated with flagellum, energy production, and acrosome reaction. Researchers proposed that the suite of proteins identified would be good biomarkers for spermatozoa in Pacific oysters. This use of biomarkers would be particularly useful in this species as the Pacific oyster is one organism where gonad stripping (dissection) is commonly used for direct, controlled breeding.

Future directions

It is evident that molluscan aquaculture faces significant challenges associated with an increasing human population. This will include both providing nourishment to the growing population as well as continuing to sustainably produce under changing environmental conditions. Functional genomic research will play an important role on both of these fronts. There is still a considerable amount of fundamental biology that we do not fully understand regarding the ability of molluscs to adapt to changing environment. A majority of studies to date have focused on acute impacts of environmental stress with a general trend that the outcome will have an unfavorable impact on shellfish aquaculture. In order to gain a better idea of how commercially important species will respond to a changing environment, more attention should be paid to the capacity for an evolutionary response as opposed to a within generation response. From a practical perspective this refers to the possibility that an acute exposure to predicted future environmental conditions might result in significant mortality, while from an evolutionary perspective those individuals that survive could contribute to more robust broodstock. Functional genomic research efforts will play a central role in evaluating population responses, determining mechanisms, and providing information for modeling responses under different scenarios. Specifically this may include the linking genotypes and epigenotypes to phenotype (including interactive effects) and measuring the inherent genomic variation that occurs across generations. The latter is something that is not necessarily considered in the context of aquaculture production as their livestock counterparts (i.e., cattle, pigs, sheep) have considerably different life history strategies. Aquacultured molluscs can produce millions of offspring and whether an advantageous trait and/or a result of excessive gametogenesis, there is evidence that genomic variation occurs at the nucleotide and transposable element level. Given the number of offspring produced at each spawning event and the propensity for variation, genomic tools that can quantify variation, selection, and mutations will be critical to understanding how to optimize aquaculture practices and predict yield.

Although genomes, transcriptomes, and proteomes for marine molluscs are scarce compared to those for vertebrates, especially model organisms, high-throughput sequencing technologies and the associated economics is continuing to expand our understanding of mollusc biology. A few examples of studies that have leveraged these molecular technologies and have impacts in variety of disciplines include works focuses on mollusc biology in the context of gene characterization (Saranya et al., 2012; Maldonado-Aguayo et al., 2014; Niu et al., 2014), environmental stress (Suárez-Ulloa et al., 2013; Menike et al., 2014), and ontology (Bassim et al., 2014a,b; Balseiro et al., 2013). Likewise the enrichment of the mollusc sequences in databases has lead to five functional categories specific to molluscan biology to be added to the Gene Ontology database, including "Byssus Formation" and "Shell Formation" (Kawashima et al., 2013).

Currently, our knowledge surrounding the molecular and cellular mechanisms involved in the physiological processes of interest in aquaculture (reproduction, growth, and immunity) is rapidly increasing thanks to genomic technologies. The next challenge is to find the specific functional roles new genes and proteins, and to curate the possible issues in sequencing, mapping and assembly or annotation. The number of genomic studies is rising every year but, in contrast, the identification and characterization of proteins of interesting pathways is not accompanying these findings. A strong effort focused on the characterization of genes and proteins is needed, and will be facilitated, now that many large transcriptome databases and some genomes are available. On the other hand, the "big data" obtained by NGS tends to generate large numbers of errors, requiring monitoring for methodological biases and strategies for replication and validation (Goldman et al., 2014). Despite the recent progress omic tools have enabled in the field of mollusc biology, there is much more work to be done. To date, only a few mollusk genomes have been sequenced completely, but the inaccuracy in assembly is still a problem difficult to solve. The Pacific oyster genome assembly is still ongoing (Hedgecock, 2015) after linkage maps detected errors in the genome. As other researchers have pointed before, more genomes need to be sequenced. Model species studies have proven that comparisons between related species provide many times more inferential power than that which is possible from the analysis of a single genome.

Exciting new directions for mollusc research in the context of aquaculture production are emerging "postgenomic" fields, including epigenetics and ribosome profiling. Epigenetics refers to phenomenon that can influence phenotype that do not alter the nucleotide sequence. Commonly studied mechanisms include short RNAs, histone modification, and DNA methylation. To date, most of this research on aquacultured species has centered around *C. gigas* due to the availability of a draft genome, which greatly facilitates approaches that rely on mapping short read sequences (i.e., high-throughput bisulfite sequencing). Recent work has included the role of immune related microRNAs (Zhou et al., 2014), temperature influences on histone methylation (Fellous et al., 2015), and the functional role of DNA methylation (Gavery and Roberts, 2013; Olson and Roberts, 2014; Saint-Carlier and Riviere, 2015; Rivière, 2014). While there have been a number of studies on DNA methylation in oysters, there is not yet a clear picture of the precise mechanisms involved and questions regarding inheritance and plasticity remain (Gavery and Roberts, 2014). There is also increasing interest in epigenetic modification in molluscs, particularly in the context of environmental contaminants (Suarez-Ulloa et al., 2015).

Ribosome profiling (Ingolia et al., 2009) is a method that reflects the translation process of specific tissues and developmental stages or conditions. As a result, it was suggested to estimate the translation efficiency of genes as well as the interaction between translation initiation, elongation, and termination. Expression and sequencing tools alone cannot explain critical aspects of gene expression regulation. This technique fills the gap between gene expression and protein quantification. In combination with methods such as RNA immunoprecipitation, miRNA profiling or proteomics, it is possible to get a new point of view of posttranscriptional and translational gene regulation. In addition, these techniques also provide new insight into new regulatory elements, such as alternative open reading frames and translational regulation under different conditions.

A constant effort going forward will be to improve our knowledge of what connects the diverse expression of genomic and epigenomic information and most importantly how this translates in phenotypes and traits of economic importance. This will result in a better understanding of mollusc physiology, and eventually that can be translated into a better management of these commercially and ecologically important species.

References

Albertin, C.B., Simakov, O., Mitros, T., Wang, Z.Y., Pungor, J.R., Edsinger-Gonzales, E., Brenner, S., Ragsdale, C.W., Rokhsar, D.S., 2015. The octopus genome and the evolution of cephalopod neural and morphological novelties. Nature 524, 220–224.

Allam, B., Pales, Espinosa E., Tanguy, A., Jeffroy, F., Le Bris, C., Paillard, 2014 Nov. Transcriptional changes in Manila clam (*Ruditapes philippinarum*) in response to Brown Ring Disease. Fish Shellfish Immunol. 41 (1), 2–11.

Balseiro, P., Falcó, A., Romero, A., Dios, S., Martínez- López, A., Figueras, A., et al., 2011. *Mytilus galloprovincialis* myticin C: a chemotactic molecule with antiviral activity and immunoregulatory properties. PLoS One 6, e23140.

Balseiro, P., Moreira, R., Chamorro, R., Figueras, A., Novoa, B., 2013. Immune responses during the larval stages of *Mytilus galloprovincialis*: metamorphosis alters immunocompetence, body shape and behavior. Fish Shellfish Immunol. 35 (2), 438–447.

Bassim, S., Genard, B., Gauthier-Clerc, S., Moraga, D., Tremblay, R., 2014a. Ontogeny of bivalve immunity: assessing the potential of next-generation sequencing techniques. Rev. Aquacul. 6, 1–21.

Bassim, S., Tanguy, A., Genard, B., Moraga, D., Tremblay, R., 2014b. Identification of *Mytilus edulis* genetic regulators during early development. Gene 551, 65–78.

Bayne, B.L., Hedgecock, D., McGoldrick, D., Rees, R., 1999. Feeding behaviour and metabolic efficiency contribute to growth heterosis in Pacific oysters [*Crassostrea gigas* (Thunberg)]. J. Exp. Mar. Biol. Ecol. 233 (1), 115–130.

Bettencourt, R., Pinheiro, M., Egas, C., Gomes, P., Afonso, M., Shank, T., Santos, R.S., 2010. High-throughput sequencing and analysis of the gill tissue transcriptome from the deep-sea hydrothermal vent mussel *Bathymodiolus azoricus*. BMC Genomics 11, 559.

Carballal, M.J., Lopez, M.C., Azevedo, C., Villalba, A., 1997. Hemolymph cell types of the mussel *Mytilus galloprovincialis*. Dis. Aquat. Org. 29, 127–135.

Castellanos-Martínez, S., Arteta, D., Catarino, S., Gestal, C., 2014. De novo transcriptome sequencing of the *Octopus vulgaris* hemocytes using Illumina RNA-Seq technology: response to the infection by the gastrointestinal parasite *Aggregata octopiana*. PLoS One 9 (10), e107873.

Corporeau, C., Vanderplancke, G., Boulais, M., Suquet, M., Quéré, C., Boudry, P., Huvet, A., Madec, S., 2012. Proteomic identification of quality factors for oocytes in the Pacific oyster *Crassostrea gigas*. J. Proteomics 75 (18), 5554–5563.

Costa, M.M., Dios, S., Alonso-Gutierrez, J., Romero, A., Novoa, B., Figueras, A., 2009. Evidence of high individual diversity on myticin C in mussel (*Mytilus galloprovincialis*). Dev. Comp. Immunol. 33, 162–170.

Cunningham, C., Hikima, J., Jenny, M.J., Chapman, R.W., Fang, G.-C., Saski, C., Lundqvist, M.L., Wing, R.A., Cupit, P.M., Gross, P.S., Warr, G.W., Tomkins, J.P., 2006. New resources for marine genomics: bacterial artificial chromosome libraries for the Eastern and Pacific oysters (*Crassostrea virginica* and *C. gigas*). Mar. Biotechnol. 8, 521–533.

Davison, A.J., Trus, B.L., Cheng, N., Steven, A.C., Watson, M.S., Cunningham, C., Deuff, R.-M.L., Renault, T., 2005. A novel class of herpesvirus with bivalve hosts. J. Gen. Virol. 86, 41–53.

De Sousa, J.T., Milan, M., Bargelloni, L., Pauletto, M., Matias, D., Joaquim, S., Matias, A.M., Quillien, V., Leitão, A., Huvet, A., 2014. A microarray-based analysis of gametogenesis in two Portuguese populations of the European clam *Ruditapes decussatus*. PLoS One 9 (3), e92202.

De Sousa, J.T., Milan, M., Pauletto, M., Bargelloni, L., Joaquim, S., Matias, D., Matias, A.M., Quillien, V., Leitão, A., Huvet, A., 2015. A microarray-based analysis of oocyte quality in the European clam *Ruditapes decussatus*. Aquaculture 446, 17–24.

De Zoysa, M., Nikapitiya, C., Oh, C., Lee, Y., Whang, I., Lee, J.S., Choi, C.Y., Lee, J., 2011 Feb. Microarray analysis of gene expression in disk abalone *Haliotis discus discus* after bacterial challenge. Fish Shellfish Immunol. 30 (2), 661–673.

Dheilly, N.M., Lelong, C., Huvet, A., Favrel, P., 2011. Development of a Pacific oyster (*Crassostrea gigas*) 31,918-feature microarray: identification of reference genes and

Dheilly, N.M., Lelong, C., Huvet, A., Kellner, K., Dubos, M.-P., Riviere, G., Boudry, P., Favrel, P., 2012. Gametogenesis in the Pacific oyster *Crassostrea gigas*: a microarrays-based analysis identifies sex and stage specific genes. PLoS One 7 (5), e36353.

Eierman, L.E., Hare, M.P., 2014. Transcriptomic analysis of candidate osmoregulatory genes in the eastern oyster *Crassostrea virginica*. BMC Genomics 15, 503.

Extavour, C.G., Akam, M., 2003. Mechanisms of germ cell specification across the metazoans: epigenesis and preformation. Development 130 (24), 5869–5884.

Fabioux, C., Huvet, A., Lelong, C., Robert, R., Pouvreau, S., Daniel, J.Y., Minguant, C., Le Pennec, M., 2004. Oyster vasa-like gene as a marker of the germline cell development in *Crassostrea gigas*. Biochem. Biophys. Res. Commun. 320 (2), 592–598.

Farber, J.M., 1997. Ancient Hawaiian fishponds: can restoration succeed on Moloka'i? Cornell University, first ed. Neptune House Publications, California.

Fellous, A., Favrel, P., Riviere, G., 2015. Temperature influences histone methylation and mRNA expression of the Jmj-C histone-demethylase orthologues during the early development of the oyster *Crassostrea gigas*. Mar. Genomics 19, 23–30.

Fleury, E., Huvet, A., 2012. Microarray analysis highlights immune response of pacific oysters as a determinant of resistance to summer mortality. Mar. Biotechnol. 14 (2), 203–217.

Fleury, E., Huvet, A., Lelong, C., de Lorgeril, J., Boulo, V., Gueguen, Y., Bachere, E., Tanguy, A., Moraga, D., Fabioux, C., Lindeque, P., Shaw, J., Reinhardt, R., Prunet, R., Davey, G., Lapegue, S., Sauvage, C., Corporeau, C., Moal, J., Gavory, F., Wincker, P., Moreews, F., Klopp, C., Mathieu, M., Boudry, P., Favrel, B., 2009. Generation and analysis of a 29,745 unique Expressed Sequence Tags from the Pacific oyster (*Crassostrea gigas*) assembled into a publicly accessible database: the GigasDatabase. BMC Genomics 10, 341.

Food and Agriculture Organization of the United States and, Food and Agriculture Organization of the United Nations, 2014. State of the World Fisheries and Aquaculture 2014, Rome, FAO.

Freer, A., Bridgett, S., Jiang, J., Cusack, M., 2014. Biomineral proteins from *Mytilus edulis* mantle tissue transcriptome. Mar. Biotechnol. 16, 34–35.

Gardner, L., Mills, D., Wiegand, A., Leavesley, D., Elizur, A., 2011. Spatial analysis of biomineralization associated gene expression from the mantle organ of the pearl oyster *Pinctada maxima*. BMC Genomics 12, 455.

Gavery, M.R., Roberts, S.B., 2012. Characterizing short read sequencing for gene discovery and RNA-Seq analysis in *Crassostrea gigas*. Comp. Biochem. Physiol. Part D Genomics Proteomics 7, 94–99.

Gavery, M.R., Roberts, S.B., 2013. Predominant intragenic methylation is associated with gene expression characteristics in a bivalve mollusc. PeerJ. 1, e215.

Gavery, M.R., Roberts, S.B., 2014. A context dependent role for DNA methylation in bivalves. Brief Funct. Genomics 13, 217–222.

Gerdol, M., De Moro, G., Manfrin, C., Venier, P., Pallavicini, A., 2012. Big defensins and mytimacins, new AMP families of the Mediterranean mussel *Mytilus galloprovincialis*. Dev. Comp. Immunol. 36, 390–399.

Goldman, D., Domschke, K., 2014. Making sense of deep sequencing. Int. J. Neuropsychopharmacol. 17, 1717–1725.

Gómez-Chiarri, M., Marta, G.-C., Warren, W.C., Ximing, G., Dina, P., 2015. Developing tools for the study of molluscan immunity: The sequencing of the genome of the eastern oyster, *Crassostrea virginica*. Fish Shellfish Immunol. 46 (1), 2–4.

Gueguen, Y., Cadoret, J.-P., Flament, D., Barreau-Roumiguière, C., Girardot, A.-L., Garnier, J., Hoareau, A., Bachère, E., Escoubas, J.-M., 2003. Immune gene discovery by expressed sequence tags generated from hemocytes of the bacteria-challenged oyster, *Crassostrea gigas*. Gene 303, 139–145.

Hedgecock, D., 2015. Second-generation linkage maps reveal errors in the assembly of the pacific oyster (*Crassostrea gigas*) genome and factors affecting map lengths and marker orders. ISGA XII abstract book, O36, p. 82. http://www.isga2015.com/isga-2015-Abstract-Book.pdf

Hedgecock, D., Lin, J.-Z., DeCola, S., Haudenschild, C.D., Meyer, E., Manahan, D.T., Bowen, B., 2007. Transcriptomic analysis of growth heterosis in larval Pacific oysters (*Crassostrea gigas*). Proc. Natl. Acad. Sci. USA 104 (7), 2313–2318.

Hou, R., Bao, Z., Wang, S., Su, H., Li, Y., Du, H., et al., 2011. Transcriptome sequencing and de novo analysis for Yesso scallop (*Patinopecten yessoensis*) using 454 GS FLX. PLoS One 6, e21560.

Ingolia, N.T., Ghaemmaghami, S., Newman, J.R., Weissman, J.S., 2009. Genome-wide analysis in vivo of translation with nucleotide resolution using ribosome profiling. Science 324, 218–223.

Jenny, M.J., Ringwood, A.H., Lacy, E.R., Lewitus, A.J., Kempton, J.W., Gross, P.S., et al., 2002. Potential indicators of stress response identified by expressed sequence tag analysis of hemocytes and embryos from the American oyster, *Crassostrea virginica*. Mar. Biotechnol. 4, 81–93.

Jones, D., Zenger, K., Jerry, D., 2011. In silico whole genome EST analysis reveals 2322 novel microsatellites for the silver-lipped pearl oyster, *Pinctada maxima*. Mar. Genomics 4, 287–290.

Jouaux, A., Lafont, M., Blin, J.L., Houssin, M., Mathieu, M., Lelong, C., 2013. Physiological change under OsHV-1 contamination in Pacific oyster *Crassostrea gigas* through massive mortality events on fields. BMC Genomics 14, 590.

Kawashima, T., Takeuchi, T., Koyanagi, R., Kinoshita, S., Endo, H., Endo, K., 2013. Initiating the mollusk genomics annotation community: toward creating the complete curated gene-set of the Japanese Pearl Oyster, *Pinctada fucata*. Zool. Sci. 30, 794–796.

Kingtong, S., Kellner, K., Bernay, B., Goux, D., Sourdaine, P., Berthelin, C.H., 2013. Proteomic identification of protein associated to mature spermatozoa in the Pacific oyster *Crassostrea gigas*. J. Proteomics 82, 81–91.

Kinoshita, S., Wang, N., Inoue, H., Maeyama, K., Okamoto, K., Nagai, K., Kondo, H., Hirono, I., Asakawa, S., 2011. Deep sequencing of ESTs from nacreous and prismatic layer producing-tissues and a screen for novel shell formation related genes in the Pearl Oyster. PLoS One 6, e21238.

Kranz, A.M., Tollenaere, A., Norris, B.J., Degnan, B.M., Degnan, S.M., 2010. Identifying the germline in an equally cleaving mollusc: Vasa and Nanos expression during embryonic and larval development of the vetigastropod *Haliotis asinina*. J. Exp. Zool. Part B Mol. Dev. Evol. 314 (4), 267–279.

Leite, R.B., Milan, M., Coppe, A., Bortoluzzi, S., dos Anjos, A., Reinhardt, R., et al., 2013. mRNA-Seq and microarray development for the Grooved Carpet shell clam, *Ruditapes decussatus*: a functional approach to unravel host-parasite interaction. BMC Genomics 14, 741.

Li, L., Zheng, P., Dean, J., 2010. Maternal control of early mouse development. Development 137 (6), 859–870.

Liao, Z1, Wang, X.C., Liu, H.H., Fan, M.H., Sun, J.J., Shen, W., 2013. Molecular characterization of a novel antimicrobial peptide from *Mytilus coruscus*. Fish Shellfish Immunol. 34 (2), 610–616.

Maldonado-Aguayo, W., Teneb, J., Gallardo-Escárate, C., 2014. A galectin with quadruple-domain from red abalone *Haliotis rufescens* involved in the immune innate response against to *Vibrio anguillarum*. Fish Shellfish Immunol. 40, 1–8.

Maria, M., André, C., Tyler, A., Beatriz, N., Antonio, F., David, P., Carlos, C., 2014. Genomic characterization of the aquaculture resource *Mytilus galloprovincialis*. Front. Mar. Sci., 1, 113. Available from: http://www.frontiersin.org/Journal/FullText.aspx?f=45&name=marine_science&ART_DOI=10.3389/conf.fmars.

McDowell, I.C., Nikapitiya, C., Aguiar, D., Lane, C.E., Istrail, S., Gomez-Chiarri, M., 2014. Transcriptome of American oysters, *Crassostrea virginica*, in response to bacterial challenge: insights into potential mechanisms of disease resistance. PLoS One 9, e105097.

Meng, X., Lui, M., Jiang, K., Wang, B., Tian, X., Sun, S., Luo, Z., Qiu, C., Wang, L., 2013. De novo characterization of Japanese Scallop, *Mizuhopecten yessoensis* transcriptome and analysis of its gene expression following cadmium exposure. PLoS One 8, e6485.

Menike, U., Lee, Y., Oh, C., Wickramaarachchi, W.D., Premachandra, H.K., Park, S.C., Lee, J., De Zoysa, M., 2014. Oligo-microarray analysis and identification of stress-immune response genes from manila clam (*Ruditapes philippinarum*) exposure to heat and cold stresses. Mol. Biol. Rep. 41, 6457–6473.

Meyer, E., Manahan, D.T., 2010. Gene expression profiling of genetically determined growth variation in bivalve larvae (*Crassostrea gigas*). J. Exp. Biol. 213 (5), 749–758.

Mitta, G., Vandenbulcke, F., Roch, P., 2000. Original involvement of antimicrobial peptides in mussel innate immunity. FEBS Lett. 486, 185–190.

Moreira, R., Balseiro, P., Planas, J.V., Fuste, B., Beltran, S., Novoa, B., Figueras, A., 2012. Transcriptomics of in vitro immune-stimulated hemocytes from the Manila clam *Ruditapes philippinarum* using high-throughput sequencing. PLoS One 7 (4), e35009.

Moreira, R., Milan, M., Balseiro, P., Romero, A., Babbucci, M., Figueras, A., Bargelloni, L., Novoa, B., 2014. Gene expression profile analysis of Manila clam (*Ruditapes philippinarum*) hemocytes after a *Vibrio alginolyticus* challenge using an immune-enriched oligo-microarray. BMC Genomics 15, 267.

Moreira, R., Pereiro, P., Canchaya, C., Posada, D., Figueras, A., Novoa, B., 2015. RNA-Seq in *Mytilus galloprovincialis*: comparative transcriptomics and expression profiles among different tissues. BMC Genomics 16, 728.

Murgarella, M., Corvelo, A., Alioto, T., Novoa, B., Figueras, A., Posada, D., Canchaya, C.A., 2015. Genomic characterization of the aquaculture resource *Mytilus galloprovincialis*. Front. Mar. Sci. Conference Abstract: IMMR International Meeting on Marine Research 2014.

Nguyen, T.T., Hayes, B.J., Guthridge, K., Ab Rahim, E.S., Ingram, B.A., 2011. Use of a microsatellite-based pedigree in estimation of heritabilities for economic traits in Australian blue mussel, *Mytilus galloprovincialis*. J. Anim. Breed Genet. 128 (6), 482–490.

Niu, D., Xie, S., Bai, Z., Wang, L., Jin, K., Li, J., 2014. Identification, expression, and responses to bacterial challenge of the cathepsin C gene from the razor clam *Sinonovacula constricta*. Dev. Comp. Immunol. 46, 241–245.

Núñez-Acuña, G., Gallardo-Escárate, C., 2013. Identification of immune-related SNPs in the transcriptome of *Mytilus chilensis* through high-throughput sequencing. Fish Shellfish Immunol. 35 (6), 1899–1905.

Oliveri, C., Peric, L., Sforzini, S., Banni, M., Viarengo, A., Cavaletto, M., Marsano, F., 2014. Biochemical and proteomic characterisation of haemolymph serum reveals the origin of the alkali-labile phosphate (ALP) in mussel (*Mytilus galloprovincialis*). Comp. Biochem. Physiol. D Genomics Proteomics 11, 29–36.

Olson, C.E., Roberts, S.B., 2014. Genome-wide profiling of DNA methylation and gene expression in *Crassostrea gigas* male gametes. Front. Physiol. 5, 224.

Pace, D.A., Marsh, A.G., Leong, P.K., Green, A.J., Hedgecock, D., Manahan, D.T., 2006. Physiological bases of genetically determined variation in growth of marine invertebrate larvae: A study of growth heterosis in the bivalve *Crassostrea gigas*. J. Exp. Mar. Biol. Ecol. 335 (2), 188–209.

Pales, Espinosa E., Corre, E., Allam, B., 2014. Pallial mucus of the oyster *Crassostrea virginica* regulates the expression of putative virulence genes of its pathogen *Perkinsus marinus*. Int. J. Parasitol. 44 (5), 305–317.

Pauletto, M., Milan, M., Moreira, R., Novoa, B., Figueras, A., Babbucci, M., Patarnello, T., Bargelloni, L., 2014. Deep transcriptome sequencing of *Pecten maximus* hemocytes: a genomic resource for bivalve immunology. Fish Shellfish Immunol. 37 (1), 154–165.

Pechenik, J.A., 2010. Biology of the Invertebrates. McGraw-Hill Higher Education, New York.

Pérez-García, C., Morán, P., Pasantes, J.J., 2014. Karyotypic diversification in *Mytilus* mussels (*Bivalvia: Mytilidae*) inferred from chromosomal mapping of rRNA and histone gene clusters. BMC Genet. 15, 84.

Philipp, E.E.R., Kraemer, L., Melzner, F., Poustka, A.J., Thieme, S., Findeisen, U., Schreiber, S., Rosenstiel, P., 2012. Massively parallel RNA sequencing identifies a complex immune gene repertoire in the lophotrochozoan *Mytilus edulis*. PLoS One 7, e33091.

Pillay, T.V.R., 1997. Economic and social dimensions of aquaculture management. Aquacult. Econ. Manag. 1, 3–11.

Qin, C.L., Huang, W., Zhou, S.Q., Wang, X.C., Liu, H.H., Fan, M.H., Wang, R.X., Gao, P., Liao, Z., 2014. Characterization of a novel antimicrobial peptide with chitin-biding domain from *Mytilus coruscus*. Fish Shellfish Immunol. 41 (2), 362–370.

Rivière, G., 2014. Epigenetic features in the oyster *Crassostrea gigas* suggestive of functionally relevant promoter DNA methylation in invertebrates. Front. Physiol., 5 APR. Available from: http://dx.doi.org/10.3389/fphys.2014.00129

Romero, A., Estévez-Calvar, N., Dios, S., Figueras, A., Novoa, B., 2011. New insights into the apoptotic process in mollusks: characterization of caspase genes in *Mytilus galloprovincialis*. PLoS One 6 (2), e17003.

Romero, A., Novoa, B., Figueras, A., 2012. Genomics, immune studies and diseases in bivalve aquaculture. Invertebr. Surviv. J. 9, 110–121.

Romero, A., Forn-Cuní, G., Moreira, R., Milan, M., Bargelloni, L., Figueras, A., Novoa, B., 2015. An immune-enriched oligo-microarray analysis of gene expression in Manila clam (*Venerupis philippinarum*) haemocytes after a Perkinsus olseni challenge. Fish Shellfish Immunol. 43 (1), 275–286.

Rosa, R.D., de Lorgeril, J., Tailliez, P., Bruno, R., Piquemal, D., Bachère, E., 2012. A hemocyte gene expression signature correlated with predictive capacity of oysters to survive Vibrio infections. BMC Genomics 13, 252.

Rosani, U., Varotto, L., Rossi, A., Roch, P., Novoa, B., Figueras, A., et al., 2011. Massively parallel amplicon sequencing reveals isotype-specific variability of antimicrobial peptide transcripts in *Mytilus galloprovincialis*. PLoS One 6, e26680.

Saavedra, C., Bachère, E., 2006. Bivalve genomics. Aquaculture 256, 1–14.

Saint-Carlier, E., Riviere, G., 2015. Regulation of Hox orthologues in the oyster *Crassostrea gigas* evidences a functional role for promoter DNA methylation in an invertebrate. FEBS lett. 589 (13), 1459–1466.

Saranya, Revathy K., Umasuthan, N., Lee, Y., Choi, C.Y., Whang, I., Lee, J., 2012. First molluscan theta-class Glutathione S-Transferase: identification, cloning, characterization and transcriptional analysis post immune challenges. Comp. Biochem. Physiol. B Biochem. Mol. Biol. 162, 10–23.

Schreier, H.J., Schott, E.J., 2014. Draft genome sequence of the shellfish bacterial pathogen Vibrio sp. strain B183. Genome Announc. 2 (5), , e00914–14.

Simakov, O., Marletaz, F., Cho, S.J., Edsinger-Gonzales, E., Havlak, P., Hellsten, U., Kuo, D.H., Larsson, T., Lv, J., Arendt, D., Savage, R., Osoegawa, K., de Jong, P., Grimwood, J., Chapman, J.A., Shapiro, H., Aerts, A., Otillar, R.P., Terry, A.Y., Boore, J.L., Grigoriev, I.V., Lindberg, D.R., Seaver, E.C., Weisblat, D.A., Putnam, N.H., Rokhsar, D.S., 2013 Jan 24. Insights into bilaterian evolution from three spiralian genomes. Nature. 493 (7433), 526–531.

Suarez-Ulloa, V., Gonzalez-Romero, R., Eirin-Lopez, J.M., 2015. Environmental epigenetics: A promising venue for developing next-generation pollution biomonitoring tools in marine invertebrates. Mar. Pollut. Bull. 98 (1–2), 5–13.

Suárez-Ulloa, V., Fernández-Tajes, J., Manfrin, C., Gerdol, M., Venier, P., Eirín-López, J.M., 2013. Bivalve omics: state of the art and potential applications for the biomonitoring of harmful marine compounds. Mar. Drugs 11, 4370–4389.

Sullivan, C.V., Chapman, R.W., Reading, B.J., Anderson, P.E., 2015. Transcriptomics of mRNA and egg quality in farmed fish: Some recent developments and future directions. Gen. Comp. Endocrinol. 221, 23–30.

Takeuchi, T., Kawashima, T., Koyanagi, R., Gyoja, F., Tanaka, M., Ikuta, T., Shoguchi, E., Fujiwara, M., Shinzato, C., Hisata, K., Fujie, M., Usami, T., Nagai, K., Maeyama, K., Okamoto, K., Aoki, H., Ishikawa, T., Masaoka, T., Fujiwara, A., Endo, K., Endo, H., Nagasawa, H., Kinoshita, S., Asakawa, S., Watabe, S., Satoh, N., 2012. Draft genome of the pearl oyster Pinctada fucata: a platform for understanding bivalve biology. DNA Res. 19, 117–130.

Tanguy, M., McKenna, P., Gauthier-Clerc, S., Pellerin, J., Danger, J.-M., Siah, A., 2013. Sequence analysis of a normalized cDNA library of *Mytilus edulis* hemocytes exposed to *Vibrio splendidus* LGP32 strain. Results Immunol. 3, 40–50.

Teaniniuraitemoana, V., Huvet, A., Levy, P., Klopp, C., Lhuillier, E., Gaertner-Mazouni, N., Gueguen, Y., Le Moullac, G., 2014. Gonad transcriptome analysis of pearl oyster *Pinctada margaritifera*: identification of potential sex differentiation and sex determining genes. BMC Genomics 15, 491.

Teaniniuraitemoana, V., Huvet, A., Levy, P., Gaertner-Mazouni, N., Gueguen, Y., Le Moullac, G., 2015. Molecular signatures discriminating the male and the female sexual pathways in the pearl oyster *Pinctada margaritifera*. PLoS One 10 (3), e0122819.

Valenzuela-Muñoz, V., Bueno-Ibarra, M.A., Escárate, C.G., 2014. Characterization of the transcriptomes of *Haliotis rufescens* reproductive tissues. Aquacult. Res. 45 (6), 1026–1040.

Venier, P., De Pittà, C., Bernante, F., Varotto, L., De Nardi, B., Bovo, G., et al., 2009. MytiBase: a knowledgebase of mussel (*M. galloprovincialis*) transcribed sequences. BMC Genomics 9, 10–72.

Venier, P., Varotto, L., Rosani, U., Millino, C., Celegato, B., Bernante, F., Lanfranchi, G., Novoa, B., Roch, P., Figueras, A., Pallavicini, A., 2011. Insights into the innate immunity of the Mediterranean mussel *Mytilus galloprovincialis*. BMC Genomics 12, 69.

Vera, M., Martínez, P., Poisa-Beiro, L., Figueras, A., Novoa, B., 2011. Genomic organization, molecular diversification, and evolution of antimicrobial peptide myticin-C genes in the mussel (*Mytilus galloprovincialis*). PLoS One 6, e24041.

Vidal, E.A.G., Villanueva, R., Andrade, J.P., Gleadall, I.G., Iglesias, J., Koueta, N., Rosas, C., Segawa, S., Grasse, B., Franco-Santos, R.M., Albertin, C.B., Caamal-Monsreal, C., Chimal, M.E., Edsinger-Gonzales, E., Gallardo, P., Le Pabic, C., Pascual, C., Roumbedakis, K., Wood, J., 2014. Cephalopod culture: current status of main biological models and research priorities. Adv. Mar. Biol. 67, 1–98.

Wang, Y., Xu, Z., Guo, X., 2004. Differences in the rDNA-bearing chromosome divide the Asian-Pacific and Atlantic species of *Crassostrea* (*Bivalvia, Mollusca*). Biol. Bull. 206, 46–54.

Wang, S., Peatman, E., Liu, H., Bushek, D., Ford, S.E., Kucuktas, H., Quilang, J., Li, P., Wallace, R., Wang, Y., Guo, X., Liu, Z., 2010. Microarray analysis of gene expression in eastern oyster (*Crassostrea virginica*) reveals a novel combination of antimicrobial and oxidative stress host responses after dermo (*Perkinsus marinus*) challenge. Fish Shellfish Immunol. 29, 921–929.

Zhang, G., Fang, X., Guo, X., Li, L., Luo, R., Xu, F., Yang, P., Zhang, L., Wang, X., Qi, H., Xiong, Z., Que, H., Xie, Y., Holland, P.W., Paps, J., Zhu, Y., Wu, F., Chen, Y., Wang, J., Peng, C., Meng, J., Yang, L., Liu, J., Wen, B., Zhang, N., Huang, Z., Zhu, Q., Feng, Y., Mount, A., Hedgecock, D., Xu, Z., Liu, Y., Domazet-Lošo, T., Du, Y., Sun, X., Zhang, S., Liu, B., Cheng, P., Jiang, X., Li, J., Fan, D., Wang, W., Fu, W., Wang, T., Wang, B., Zhang, J., Peng, Z., Li, Y., Li, N., Wang, J., Chen, M., He, Y., Tan, F., Song, X., Zheng, Q., Huang, R., Yang, H., Du, X., Chen, L., Yang, M., Gaffney, P.M., Wang, S., Luo, L., She, Z., Ming, Y., Huang, W., Zhang, S., Huang, B., Zhang, Y., Qu, T., Ni, P., Miao, G., Wang, J., Wang, Q., Steinberg, C.E., Wang, H., Li, N., Qian, L., Zhang, G., Li, Y., Yang, H., Liu, X., Wang, J., Yin, Y., Wang, J., 2012. The oyster genome reveals stress adaptation and complexity of shell formation. Nature 490, m49–m54.

Zhang, N., Xu, F., Guo, X., 2014. Genomic analysis of the Pacific oyster (*Crassostrea gigas*) reveals possible conservation of vertebrate sex determination in a mollusc. G3 4 (11), 2207–2217.

Zhang, L., Li, L., Zhu, Y., Zhang, G., Guo, X., 2014a. Transcriptome analysis reveals a rich gene set related to innate immunity in the eastern oyster (*Crassostrea virginica*). Mar. Biotechnol. 16, 17–33.

Zhang, Y., Sun, J., Mu, H., Li, Jun., Hang, Y., Qian, P., Qiu, J., Yu, Z., 2014b. Proteomic basis of stress responses in the gills of the pacific oyster *Crassostrea gigas*. J. Proteome Res. 14 (1), 304–317.

Zhao, X., Yu, H., Kong, L., Li, Q., 2012. Transcriptomic responses to salinity stress in the pacific oyster *Crassostrea gigas*. PLoS One 7, e46244.

Zhou, Z., et al., 2014. The identification and characteristics of immune-related microRNAs in haemocytes of oyster *Crassostrea gigas*. PLoS One 9 (2), e88397.

Genomics in the common carp

10

*Peng Xu****, *Yanliang Jiang**, *Jian Xu**, *Jiongtang Li**, *Xiaowen Sun*[†]
*Beijing Key Laboratory of Fishery Biotechnology, Centre for Applied Aquatic Genomics, Chinese Academy of Fishery Sciences, Beijing, China; **State Key Laboratory of Marine Environmental Science, College of Ocean & Earth Science, Xiamen University, Xiamen, China; [†]Heilongjiang Fishery Research Institute, Chinese Academy of Fisheries Sciences, Harbin, China

Common carp biology and aquaculture

The common carp (*Cyprinus carpio*) belongs to the class Osteichthyes (the bony fishes), the order Cypriniformes, and the family Cyprinidae. According to previous taxonomic and phylogenetic studies (Balon, 1995; Jhingran and Pullin, 1985; Wang and Li, 2004; Zhou et al., 2004), four subspecies of the common carp have been identified. These subspecies include *Cyprinus carpio carpio*, which is distributed from the Danube River basin in the natural waters up to the Ural Mountain range, *Cyprinus carpio aralensis*, which is distributed in the Central Asia area, *Cyprinus carpio haematopterus*, which is distributed from the Amur River (also known as the Heilongjiang River) basin to southern China, and *Cyprinus carpio viridiviolaceus*, which is distributed in Southeast Asia. Other researchers have suggested that there are only two subspecies: *C. carpio carpio* and *C. carpio haematopterus*, representing the European subspecies and East Asia subspecies, respectively (Balon, 1995; Dong et al., 2015).

The common carp has been introduced into all continents except Antarctica over the past two centuries. In many of the natural waters where it has been introduced, the common carp is considered to be an invasive species. The common carp is one of the most widely cultured freshwater fish species in the world (Hasan et al., 2007; Welcomme, 1988). Aquaculture production of the common carp increases in parallel with the increase in global aquaculture production of freshwater fishes. According to FAO (FIGIS, 2013), the common carp is cultured in over 100 countries worldwide and accounts for over 3.4 million metric tons of annual production throughout the world (Bostock et al., 2010). In addition to its value as a food source, the common carp is also an important ornamental fish species. One of its variants (koi) is the most popular outdoor ornamental fish due to its distinctive color and scale patterns. There are several renowned food/ornamental dual-purpose strains in China, including the Oujiang color carp, Hebao red carp, and Xingguo red carp, which have been selected and cultured in certain geographic regions in China and have great economic value for the domestic people.

Fishes exhibit the most extensive polyploidy among the vertebrates. The most well-characterized and recognized polyploid fishes include *Salmonidae* (Allendorf and Thorgaard, 1984; Phillips and Rab, 2001) and *Cyprinidae* (Ohno et al., 1967; Wolf et al., 1969). The polyploidized salmonid genomes resulted from an autopolyploidization event (genome doubling), while the tetraploidized common carp and crucian carp

Genomics in Aquaculture. http://dx.doi.org/10.1016/B978-0-12-801418-9.00010-X

genomes likely resulted from an allopolyploidization event (species hybridization). These polyploidization events are also called the fourth round of whole genome duplication (WGD) and introduce more complexity into the genomes. Cytogenetic evidence of *C. carpio*'s allotetraploidization has suggested that 50 bivalents rather than 25 quadrivalents are formed during meiosis (Ohno et al., 1967), implying that the fish retain two sets of ancestral genomes from the parental species. Molecular evidence collected over the past two decades suggested that the polyploidization event could have occurred from 58 to 12 MYA (David et al., 2003; Larhammar and Risinger, 1994; Wang et al., 2012a; Xu et al., 2014a); thus, it is considered quite a new event and has attracted increasing attention from evolutionary biologists. Unlike diploid genomes, polyploid genomes involve multiple rounds of WGD and segmental duplications and produce redundant genes that provide an important genetic material basis for phenotypic complexity, which would potentially benefit an organism in terms of its adaptation to environmental changes (Ohno, 1970).

Owing to the importance of the aquaculture and biology of the common carp, scientists performed various genomic and genetic studies on this species over the past few decades. Over time, researchers have collected abundant genomic resources and data, including the completely sequenced and annotated genome. In this chapter, we will summarize the genomics and related genetic studies on the common carp over past few decades and applications dealing with various aquaculture and biological targets.

The genome: diversity, structure, organization, and evolution

BAC-based genomics

A variety of genome resources and tools were developed prior to the early 2000s, including a large number of genetic markers (Ji et al., 2012b; Xu et al., 2012) and over 30,000 expressed sequence tags (ESTs) (Christoffels et al., 2006). However, few sufficient genome sequences were available to assess the genomic complexity of the common carp, which is an essential step for designing strategies to ultimately unveil the genome sequence and structure. Considering the complexity of the tetraploid genome of the common carp, large insert genome libraries are essential resources for genome characterization, sequence contig assembly and scaffolding, and target sequence screening and cloning. Therefore, a common carp bacterial artificial chromosome (BAC) library was constructed in 2009 that contained a total of 92,160 BAC clones with an average insert of 141 kb covering 7.7X haploid genome equivalents (Li et al., 2011). Thereafter, a large set of BAC clones were sequenced from both ends, generating 65,720 clean BAC-end sequences (BESs) that represented approximately 2.5% of the common carp genome (Xu et al., 2011a). Hence, the first survey of the common carp genome was performed based on these BESs and provided a useful evaluation for the whole genome sequencing project. The survey revealed that the proportion of repetitive elements in the common carp genome was approximately 27.9% of the surveyed genome regions. The most abundant type of repetitive element was DNA transposons, followed by retroelements including LINEs, LTR elements, and SINEs. The repeats divergence rate of DNA transposons (percentage of substitutions in the matching region compared with consensus repeats in the constructed libraries) showed a nearly normal distribution with a peak at 24%. A fraction of the LTR retrotransposons, LINEs and SINEs had nearly the same divergence rates as the DNA transposons (peaks at 30, 28, and 22%, respectively), indicating relatively old origins. The repetitive sequences in the common carp genome were much higher compared to the genomes of the Takifugu (7.1%), Tetroadon (5.7%), and stickleback (13.48%), but lower compared to the Atlantic salmon genome (30–35%) (Davidson et al., 2010) and zebrafish genome (52.2%) (Howe et al., 2013). Although this estimate of repetitive sequences is significantly lower than the repeat calculation based on the complete genome that revealed that 37% of the common carp genome was repeat sequences, it represented the first comprehensive repetitive element evaluation of the common carp genome and provided useful information for further genomic studies.

The BESs were aligned onto the zebrafish genome (one of the sequenced cyprinid species) to evaluate their similarity. A total of 39,335 BESs from the common carp had conserved homologs on the zebrafish genome, demonstrating the high similarity between the zebrafish and common carp genomes and indicating the feasibility of using the zebrafish reference genome to assess the architecture of the common carp genome. Therefore, a comparative mapping approach was employed to assess and characterize the tetraploidized common carp genome. A large number of BAC-anchored microsatellite and SNP markers were genotyped and a high density genetic map with 1209 genetic markers was constructed with 50 linkage groups (Zhao et al., 2013) to provide the dense framework for a comparative genomics study between the zebrafish and common carp. Next, the BAC-based physical map and BAC sequences were anchored and integrated onto the genetic map (the genome frame). This was the first chromosome-scale integration map for the common carp genome prior to the whole genome sequencing project. Then, the anchored BESs were anchored onto the zebrafish reference genome and the first comparative map was constructed between the two cyprinid genomes. A perfect "2 versus 1" homologous relationship was revealed between 50 common carp linkage groups and 25 zebrafish chromosomes, providing the first genome-scale evidence for the tetraploidization of the common carp genome. To better understand the genome structure and facilitate future genetic studies, we decided to number the common carp chromosomes following zebrafish chromosome numbering. For instance, the two chromosomes that homologize to zebrafish chromosome 1 were designated as chromosome 1 and 2, respectively.

Thus, we evaluated the architecture of the common carp genome based on the limited genome sequence data and genetic maps derived from the BAC libraries and substantiated the tetraploidization manner of the common carp genome for the first time. These results provided a large set of mate-paired BESs, physical maps, and genetic maps for whole genome sequencing, assembly, and scaffolding.

Tetraploidized genome: genome structure, organization, and whole genome duplication

Repeat contents, homologous region proportions, genome size, and heterozygosity are the major obstacles for whole genome assembly. The former three factors are natural features that cannot be artificially changed. In contrast, heterozygosity can be decreased using gynogenesis technology or inbreeding. The gynogenetic carp was selected for sequencing in the common carp genome project. The common carp genome project was initiated in 2010 when the available sequence technologies included Roche 454, Illumina, and SOLiD. Because no previous knowledge existed concerning the assembly of an allotetraploid genome, we had to sequence the genome using all three platforms and compare the result to decide which assembly strategy was optimal. Eventually, we adapted a hybrid genome sequencing and assembly strategy as follows: contig assembly using the 454 data; scaffolding contigs using mate-pair libraries from the Illumina, SOLiD, and Roche 454 platforms; and final scaffolding using the BAC-end sequences.

Our hybrid sequencing strategy made use of the advantages of various technologies to overcome the impediments. Current genome assemblers are typed into two classes: the overlap-layout-consensus and *de bruijn* graph strategies (Li et al., 2012). The *de bruijn* graph strategy is widely used for diploid genome assembly, but is rarely applied to polyploid genome assembly. However, overlap-layout-consensus assemblers have been applied to polyploid genomes. The genome from the tetraploid rainbow trout species (Berthelot et al., 2014) was assembled with Newbler, an overlap-layout-consensus assembler. We used the Celera assembler (an assembler that functions on the basis of the overlap-layout-consensus strategy (Denisov et al., 2008)) to build the common carp genome contigs with the Roche 454 reads and the Sanger BAC-end sequences.

Using this hybrid strategy, we obtained an assembly with contig and scaffold N50 lengths reaching 68.4 kb and 1.0 Mb, respectively. The total length of all scaffolds was 1.69 Gb, covering at least 92.3% of the genome. Five indices were used to assess the quality of the assembly. First, a total of 83% of raw reads could be realigned to the assembly. Second, the alignment ratio of four full-length BAC sequences to the final assembly all reached 90%. Third, the mapping ratio of the 454 transcriptome read and Sanger ESTs were approximately 95% and 90%, respectively. Fourth, the coverages of the core eukaryotic genes (CEGs) in the common carp genome were between 82% and 95%. Fifth, we estimated the missing rate of known duplicated genes with 19 pairs of paralogs (Wang et al., 2012a) and found that 92% of homologous genes could be differentiated. Taken together, these five results provided evidence that our assembly was of high quality.

We performed gene prediction following a widely applied strategy that included a combination of de novo gene prediction, sequence homology-based predictions, and RNA-Seq data. Fgenesh+ (Salamov and Solovyev, 2000) and AUGUSTUS (Stanke et al., 2004) were used for ab initio gene prediction. The performance of AUGUSTUS is usually the best among the currently known gene prediction programs (Coghlan et al., 2008). Raw and precise alignments are required to perform sequence homology-based prediction. For the raw alignments, proteins were aligned to assembled genome sequences using Blastx to identify homologous genomic fragments. Then, proteins were realigned to these genome fragments using Genewise (Birney et al., 2004) to identify accurate splicing sites. Genewise can provide accurate and complete gene structures. To build RNA-seq gene models, gene models were constructed using Cufflinks (Trapnell et al., 2012) based on the mapping of transcriptome sequencing data by TopHat (Trapnell et al., 2009).

Finally, we combined the de novo gene prediction, sequence homology-based predictions, and RNA-seq data with EVM (Haas et al., 2008). The assembly consisted of a total of 52,610 gene models, 91.4% of which had RNA-seq evidence. Two indices were used to evaluate the quality of the predicted genes. First, sequence structures including exon number, exon length, and protein length were conserved across the teleost. The average exon number and exon length were similar across zebrafish medaka, stickleback, fugu, and tetraodon (7–10 and 140–230 bp, respectively) (Xu et al., 2014a). The exon number and length of the common carp were similar to those of other fish (Xu et al., 2014a), indicating the quality of gene prediction. The second index used to assess the quality of the predicted genes was sequence homology. Approximately 91% of common carp proteins were homologous to other species. Taken together, these two results suggested the high quality of the common carp gene annotations.

Genetic diversity studies

Genetic diversity is the variation of heritable characteristics present in a population of one species and serves as a way for populations to adapt to changing environments. It is applicable to wild populations as well as domesticated strains, which generally have lower levels of diversity. The common carp is a genetically diverse and successful species that has adapted to various environments across a broad ecological spectrum (mainly in Eurasia) with a history of domestication that extends for more than 2000 years. This species has been bred into numerous strains and local populations, producing distinct phenotypic changes in its growth rate, temperature and hypoxia tolerance, body color, scale pattern, and body shape that are partially attributable to genome diversity. It is likely that individuals with more variations in a population will possess variations in alleles that are suited for the environment, and thus are more likely to survive to produce offspring bearing that allele.

Genetic diversity has been intensively studied in the common carp by employing various types of genetic markers, from early stage markers such as allozymes, RAPD (random amplification polymorphic DNA), and AFLP (amplified fragment length polymorphism) to advanced markers such as microsatellites and SNPs (single nucleotide polymorphism). Microsatellites and SNPs represent the most popularly used genetic markers over the past decade in the field of common carp genetics. To date, thousands of microsatellite markers and millions of SNP markers have been developed from the common carp genome, thereby providing abundant genetic tools to survey and characterize the genetic diversity of various populations and strains worldwide. Microsatellite markers were developed for the common carp as early as 1997 (Crooijmans et al., 1997). However, until the early 2000s only dozens of microsatellite markers had been developed for genetics studies in the common carp (David et al., 2001; Tanck et al., 2001; Wei et al., 2001). Thus, population genetics and the genetic diversity of wild populations and domesticated strains were preliminarily studied based on a limited number of microsatellite markers. For instance, Chang et al. characterized the genetic diversity of five representative populations in China with 14 microsatellite loci and revealed high genetic similarity of three domesticated red carp populations (Xingguo red carp, Hebao carp, and Wanan red carp) and close relationships between two wild populations of Yellow River carp and Heilongjiang carp (Chang et al., 2004). Later, the authors performed an additional analysis on six wild common carp populations based on 30 microsatellite loci in a follow-up study (Li et al., 2007). The genetic diversity and heterozygosity of five mirror carp populations were analyzed using 30 microsatellite loci, which provide useful information for genetic breeding (Quan et al., 2006). Wang et al. focused their genetic diversity study on the Oujiang color carp (*C. carpio var. color*) and suggested a close relationship between the Oujiang color carp and Xingguo red carp (Wang et al., 2007). Liao et al. surveyed the genetic diversity and population structure of samples from the Yangtze River basin using six microsatellite loci and reported significant genetic diversity of the common carp in this region (Liao et al., 2006). David et al. evaluated the genetic diversity and population heterozygosity of eight aquacultured and ornamental carp populations in Israel using 12 microsatellites and 505 AFLP markers (David et al., 2001, 2007). Thai et al. surveyed the diversity and genealogical relationships of common carp from Chinese, Indonesian, and Vietnamese populations and demonstrated that Chinese and Indonesian carp strains were the most divergent; additionally, a close relationship between the Vietnamese, Koi, and Chinese color carp strains was reported (Thai et al., 2005). Similarly, the genetic diversity of European populations has also been evaluated based on various genetic markers. Strains from 2 Hungarian common carp farms were analyzed using 10 RAPD markers and 4 microsatellite markers; the results showed very similar heterozygosity values and allele frequencies (Richárd et al., 2003). A large set of domesticated/captive stocks and wild/feral populations of common carp from Europe, Central Asia, and East/South-East Asia were examined based on allozyme (23 populations), microsatellite (11 populations), and mitochondrial DNA (21 populations) variations (Kohlmann et al., 2003). All three types of genetic markers clustered the populations into two highly divergent groups (Europe/Central Asia and East/South-East Asia), demonstrating two distinct subspecies of common carp. The genetic diversity of Russian common carp breeds and populations were also investigated using similar approaches (Ludannyi et al., 2006).

In summary, population genetics studies have been intensively performed on common carp populations worldwide, thereby providing valuable information for understanding the genetic diversity and population structure of various populations. However, genetic loci remain insufficient to provide a comprehensive understanding of genetic diversity at the genome scale, especially of the tetraploidized genome of the common carp. Therefore, genetic markers or variation loci spanning the complete genome of the common carp are desired in the genomic era. To supply sufficient microsatellite loci for genetic diversity surveys and high-density linkage mapping, whole genome shotgun (WGS) sequences of the common carp were assembled and surveyed for microsatellite identification. A total of 79,014 microsatellites were identified and characterized. A MySQL database and a web-based database interface (http://genomics.cafs.ac.cn/ssrdb) were built for public access and download (Ji et al., 2012b). Genome-wide identification and genotyping of SNP markers also progressed rapidly. High throughput transcriptome sequences have been assembled, and a total of 712,042 SNPs have been discovered from four common carp strains in China (Ji et al., 2012a;

Xu et al., 2012). The more than 0.7 million SNP loci derived from the expressed sequences and potentially associated with functional genes or regulatory factors are an important resource for various genetic studies. Furthermore, the complete genome sequence of the common carp was assembled and annotated in 2012 and published in 2014, thereby enabling genetic analysis on the whole-genome scale (Xu et al., 2014a). The whole genome resequencing strategy was employed to further characterize the genetic diversity and variation in representative strains and populations. A total of 33 representative individuals were selected for genome resequencing, including four wild populations (from the Danube River, Yellow River, Heilongjiang River, and Chattahoochee River) and six domesticated strains from Asia and Europe (including Songpu, Xingguo red, Oujiang color, Hebao, Szarvas 22, and koi). A total of 18,949,596 SNPs and 1,694,102 small insertion–deletions (indels) were identified, revealing the high genetic diversity of various populations. The SNPs and small INDELs were mapped on the common carp genome to provide the first generation genetic variation map. Distinctive genetic diversities among different populations or strains would provide useful hints for dissecting the genetic basis underlying their phenotypic differences. For instance, a genetic diversity scan was performed to compare the Hebao and Songpu genomes to identify highly different genomic regions. The values of π (nucleotide diversity) were calculated using a sliding window approach, and a total of 205 genome regions were identified covering 12.67 Mb and containing 326 candidate genes. Gene Ontology (GO) and Kyoto Encyclopedia of Genes and Genomes (KEGG) analysis indicated that a significant portion of these candidate genes were associated with epithelial morphogenesis, pigmentation, the immune response, and other functional pathways. Among the selected regions, the *fgfr1a1* gene (encoding FGF receptor 1 a1) on chromosome 34 contained a 306-bp specific deletion in intron 10 (228-bp deletion) and exon 11 (78-bp deletion). Extensive investigation on large samples from four strains confirmed that the deletion was only found in Songpu. Although this INDEL has been previously reported in mirror carp, the whole genome genetic diversity scanning approach still demonstrates its power in distinguishing genetic loci associated with important traits and phenotypes. The genetic diversity of diverse populations and strains provides the genetic basis for common carp adaptation and survivability in various habitats and environments, thereby enabling its successful spread all over the world. High levels of genetic diversity and abundant germplasms also provide us with the basis for efficient selection and breeding for carp with improved performance in aquaculture. Genome resources and genetic tools will surely magnify the effects and boost breeding progress.

Genome evolution and gene fate of tetraploidization

WGD results in the duplication of genome sequences and doubles the genome size and gene content. The subsequent diploidization process deleted redundant gene copies or functionalized the paralogs, reverting the genome back to a diploid-like condition (Wolfe, 2001). Eventually, a small proportion of duplicated genes was retained, whereas the majority of the redundant copies were inactivated (Langham et al., 2004). Although it is well known that vertebrates had two rounds of WGD and that teleosts underwent three rounds of WGD (Jaillon et al., 2004), their genomes returned to the diploid status (Kasahara et al., 2007). In comparison with other teleosts, the common carp underwent an additional round of species-specific genome duplication called common carp-specific genome duplication (CcaGD). Many studies have provided evidence to support this event. The chromosome number of the common carp is twice that of most other teleosts (Ohno et al., 1967). The copies of several genes and microsatellites in the common carp were found to be increased compared to other fish (David et al., 2003; Zhang et al., 1995). These studies also demonstrated that the genome was still tetraploid and had not been completely diploidized.

Based on these observations, the question concerning the time frame of the WGD event was raised. Zhang et al. estimated that the tetraploidization event occurred 58 million years ago (MYA) based on duplicated *c-myc* genes in the common carp (Zhang et al., 1995). Larhammar et al. estimated the time to be less than 16 MYA (Larhammar and Risinger, 1994),whereas David et al. considered that the 4R genome duplication occurred approximately 12 MYA (David et al., 2003). Wang et al. generated an order-of-magnitude more transcriptome contigs than previously reported to estimate the duplication time, which they deduced to be between 5.6 and 11.3 MYA (Wang et al., 2012a). The estimated duplication time based on the whole genome sequence also supported the occurrence of an event 5.6–11.3 MYA. Although there is a large discrepancy in the estimates of the CcaGD time (from 5.6 to 58 MYA), all of these studies demonstrate that the CcaGD is the most recent genome duplication event in vertebrates.

On the basis of the whole genome sequences of the common carp, we introduced a second question concerning whether chromosome rearrangements occurred on the carp chromosomes. A total of 2114 best-matched reciprocal paralogous gene pairs were identified using blastp, and then ohnologous blocks were constructed using MCScanX (Wang et al., 2012b). We observed few chromosome rearrangements between sister chromosomes on the carp genome, demonstrating their one-to-one syntenic relationship.

The very recent duplication and one-to-one syntenic relationship on 25 paired chromosomes makes the common carp a rare and unique model to study the early evolution of duplicated genes. Although a great deal of attention has been paid to the effect of WGD on gene evolution, the majority of our knowledge regarding genome evolution following WGD events comes from plants (Roulin et al., 2013) and yeast (Wolfe and Shields, 1997). The early evolution of these processes has been poorly investigated in vertebrates to date because almost all of the well-characterized genomes have been diploidized.

At least three commonly accepted and different divergence functions of duplicated genes during the diploidization process were summarized: (1) nonfunctionalization: one copy is deleted by genomic mutation (Langham et al., 2004); (2) neo-functionalization: one of the duplicates retains the original function, whereas the other duplicate obtains a novel function different from the ancestral gene (Force et al., 1999; Rastogi and Liberles, 2005); and (3) subfunctionalization: different functions of the ancestor were retained in two duplicated genes (Semon and Wolfe, 2008). Different from functional divergence, expression divergence is another divergence driving force on the retained copies (Huminiecki and Wolfe, 2004). It has been hypothesized that massive and rapid gene deletions and intense expression differentiation follow WGDs (Wolfe and Shields, 1997). Recently, we used the common carp as a model to examine this hypothesis and studied the early effect of WGD on the evolution of duplicated genes (Li et al., 2015a). We found that most ancestral genes had two copies in the common carp genome and that most copies were conserved. Gene expression analysis revealed that most copy pairs were coexpressed in at least one tissue, while the expression of nearly half of the pairs was uncorrelated. A functional comparison revealed that 25% of the pairs exhibited functional divergence. The proportion of genes with functional divergence was less than genes with expression divergence, indicating slow functional divergence but rapid expression dissociation.

Functional genomics research and applications in the common carp

Due to the rapid advances in sequencing technologies, the genomes of many aquaculture species have been sequenced in recent years. The well-developed genomic resources promote interest in studies into genome function. Generally, functional genomics refers to the development and application of global experimental approaches to assess gene functions and is characterized by high throughput or large scale experimental methodologies combined with statistical and computational analysis of the results. Functional genomics mines genome sequence data for particularly valuable information, and therefore expands the scope of biological investigations from studying single genes or proteins to studying all genes or proteins simultaneously in a systematic fashion. This approach rapidly narrows the gap between sequence and function and yields new insights into the behavior of biological systems (Hieter and Boguski, 1997). As we mentioned earlier in this chapter, a large number of genomic resources (e.g., ESTs, BAC-end sequences, whole genome sequences, and transcriptome data) have been developed for the common carp and serve as fundamental resources to investigate the genetic basis underlying valuable traits. Here, we briefly review the functional genomics applications for skin color, immunity and disease resistance, and responses to environmental stress and toxicology of the common carp.

Skin color

As an important aquaculture species, the common carp was domesticated more than 2000 years ago. Over its long history of domestication, the common carp has been introduced into various environments worldwide, resulting in hundreds of strains or varieties that display diverse skin colors. Skin color is an important economic trait for the common carp and serves as an important criterion for visually determining quality and market value. Understanding the molecular mechanisms underlying skin pigmentation in the common carp will advance our knowledge of skin color genetics in fish and accelerate the molecular selection of fish species with consumer-favored skin colors. Interest in the genetic basis of skin coloration in the common carp is increasing. Bar et al. reported that the *MC1R* gene was associated with the development of black pigmentation in the ornamental Koi common carp (Bar et al., 2013). With the rapid development of new generation sequencing technologies, sequencing the whole transcriptome became a powerful approach. Jiang et al. sequenced the transcriptome from the skin of two common carp strains with different skin colors: the brownish-black Yellow river carp and Xingguo red carp (Jiang et al., 2014). By comparing the skin transcript profiles of the two carp strains, the authors detected 2012 unique genes that were differentially expressed and obtained candidate genes that might be involved in the skin pigmentation process, including genes encoding tyrosinase-related protein 1 (*TYRP1*), premelanosome protein (*SILV/PMEL*), cysteine/glutamate transporter (*xCT/SLC7a11*), and agouti signaling protein (*ASIP*). Further GO enrichment and pathway analysis indicated that key pigmentation-related genes were involved in at least three pathways that regulated pigment synthesis in the two strains, including the melanin biogenesis pathway, Wnt signaling pathway, and MAPK signaling pathway (Jiang et al., 2014). Similarly, comparative analysis of skin transcriptomes had been applied to another two common carp strains (the Hebao and Songpu carp) (Xu et al., 2014a). Out of the thousands of differentially expressed genes between the two skin transcriptome profiles, one interesting gene was observed with a significantly different expression level: the *slc7a11* gene that encodes solute carrier family 7 member 11, which is the plasma membrane cysteine/glutamate exchanger (xCT) that transports cysteine into malanocytes in the melanogenesis pathway. *Slc7a11* was expressed at a higher level in Hebao compared to Songpu. This result suggested that more cysteine was transported into the pigment cells in Hebao, resulting in the preponderant synthesis of pheomelanin in the skin of Hebao while eumelanin synthesis was suppressed. Higher pheomelanin accumulation in the pigment cells gives Hebao its red skin appearance (Xu et al., 2014a). However, the genetic basis for differences in *slc7a11* expression remains unclear. Some scientists have suggested that there is no pheomelanin in the teleost based on previous biochemical analyses in several fish species (Kottler et al., 2015). Therefore, further investigation is necessary to clarify and understand the roles of *slc7a11* in teleost color variation. A recent study reported that 80% of identified zebrafish pigmentation genes were also present in the Oujiang color common carp strain (Wang et al., 2014a). In addition to the direct effects of genes, other regulators such as microRNAs play crucial roles in common carp skin coloration by regulating the expression of downstream pigmentation genes (Yan et al., 2013). These results provide us with a valuable basis for understanding the molecular mechanisms of skin pigmentation in teleost fish. This understanding will facilitate genetic selection and breeding of common carp with market-favored colors. However, some details of the genetic mechanisms underlying common carp skin pigmentation are not well understood, such as the interactions among pigmentation genes and the genetic regulation underlying the synthesis of different pigments.

Immunity and disease resistance

The common carp is one of the major aquaculture fish species worldwide; however, its production is threatened by various pathogens and diseases (Li et al., 2014a). Fish kept under intensive culture conditions are constantly exposed to a wide range of stressors that could induce infection by opportunistic pathogens in the water. The immune response is the primary route by which fish eliminate pathogen infections. Therefore, culturing strains of fish with enhanced disease resistance through genetic selection and breeding is an important approach for disease prevention in aquaculture (Singh et al., 2012). Immunostimulants such as peptidoglycan, lactoferrin, glucans, and chitin are widely used to increase immunocompetency and disease resistance in fish by enhancing nonspecific defense mechanisms. Hundreds of immune-related ESTs from the head kidney of the common carp were identified after stimulation with peptidoglycan (Kawano et al., 2003). When common carp was infected with the Cyprinid herpesvirus-3 (CyHV-3) virus or *Aeromonas hydrophila* bacterium, the expression of immune-related genes was significantly regulated (Adamek et al., 2013; Chi et al., 2014; Liu et al., 2014a; Syakuri et al., 2013).

In recent years, well-developed genetic and genomic resources/tools for the common carp have been widely used to investigate the mechanisms underlying immunity and disease resistance (Zhang et al., 2011a). ESTs obtained from cDNA libraries represent the expressed portion of a genome; they have proved to be powerful tools to globally identify genes expressed in response to a given set of environmental conditions (Adams et al., 1991). For instance, Gonzalez et al. constructed two cDNA libraries from carp skin sampled 3 and 72 h after *Ichthyophthirius multifiliis* infection to detect new immune-related genes in common carp challenged with an ectoparasitic infection (Gonzalez et al., 2007). In this study, a total of 3500 ESTs were identified, of which 61 were newly described genes in the common carp, including full-length molecules of prostaglandin D2 synthase (PGDS), the CC chemokine molecule SCYA103, and a second gene for the carp β_2- microglobulin (β_2m), β_2m-2 (Gonzalez et al., 2007). Similarly, in another study two cDNA libraries from the carp immune organs—gill and intestine—were constructed in an attempt to identify immune-related genes. The authors reported that a total of 2184 EST clones were generated; the gill EST library had a higher frequency of innate immune molecules compared to the intestine EST library, including MHC class I, MHC class II, cytokines, and chemokines (Abdelkhalek et al., 2009).

Compared to EST screening, microarray is more efficient and can provide an integrated overview of the global response at the level of expression. Rakus et al. conducted microarray analysis using common carp slides containing 10,822 60-mer probes to determine the genetic basis underlying the common carp immune response to the CyHV-3 virus, which is the etiological agent of a virulent and lethal disease in the common carp (Rakus et al., 2012). Two common carp lines with different susceptibility to the virus were infected with CyHV-3 by immersion. The authors identified several immune-related genes with significantly higher expression, including a number of genes involved in pathogen recognition, complement activation, MHC class I-restricted antigen presentation, and the development of adaptive mucosal immunity (Rakus et al., 2012). Their study uncovered a wide array of immune-related genes involved in the antiviral response of the common carp toward CyHV-3, and also demonstrated that the outcome of this severe disease could to a large extent be controlled by host genetic factors (Rakus et al., 2012).

The rapid development of high-throughput deep sequencing technologies has provided a powerful platform for the characterization of the transcriptome in various species especially nonmodel species, making transcriptome profiling analysis an efficient approach for functional genomics studies. A recent published study sequenced and compared transcriptomes from two developmental stages of carp spleen and identified several thousand differently expressed unigenes between the two stages, thereby providing a valuable resource to better understand the common carp immune system and defense mechanisms (Li et al., 2015b). Deep sequencing facilitates the identification of microRNAs that are globally critical for proper immune system development and play an important role in a variety of other aspects of the immune system. A profile of microRNAs expressed in a carp immune organ (the spleen) was obtained via three major steps: construction of a small-RNA cDNA library from the spleen of the common carp; high throughput sequencing of the small RNA library; and subsequent bioinformatics analysis. This approach should facilitate our understanding of the involvement of microRNAs in immune system functions (Li et al., 2014a).

Response to environmental stress

Environmental stress, such as temperature, hypoxia and salinity, deleteriously affects the biological function of an organism by disrupting homeostasis. Understanding the response to environmental stress and the underlying mechanisms are essential. The identification of differentially expressed genes in response to environmental stress offers insights into the roles of the transcriptome in the regulation of physiological responses (Gracey, 2007). However, less attention is paid to aquaculture species, especially nonmodel species such as the common carp. The carp is hard and tolerant of a wide range of environmental changes. Gracey et al. used a microarray composed of 13,440 cDNA probes to identify the transcriptional responses of common carp subjected to a progressive cooling regimen (Gracey et al., 2004). The authors identified a very large number of genes across seven tissues in response to cold that were involved in RNA processing, translation initiation, mitochondrial metabolism, proteasomal function, and modification of higher-order structures of lipid membranes and chromosomes (Gracey et al., 2004).

Although dissolved oxygen is necessary for aquatic animals to sustain life, hypoxic and anoxic habitats appear frequently. In an effort to identify candidate genes involved in promoting hypoxia tolerance, carp were acclimated to low levels of dissolved oxygen (0.3 mg/L) at either 30°C or 17°C; then, changes in their transcriptome were profiled by hybridization to a carp array (Fraser et al., 2006). The results showed that microarray spots corresponding to different myoglobin cDNA clones were ranked consistently high as transcripts that were strongly upregulated by hypoxia treatment at both acclimation temperatures (Fraser et al., 2006).

Toxicology

With intensive agriculture, industry development, and increasing urbanization, the global environments have been challenged by water pollution, synthetic chemicals, and waste discharges that subsequently lead to increased contamination of aquatic environments. Increasing attention has been paid to exploring the response of fish to toxic environments. The common carp has been widely employed for toxicology studies. For instance, Moens et al. developed a carp cDNA microarray consisting of endocrine-related genes to determine the molecular effects induced by exposure to a variety of endocrine-disrupting compounds in carp and discriminate the specific transcriptional profiles associated with these compounds (Moens et al., 2006). Later, the authors used a custom cDNA microarray to determine the transcriptional profile of the carp liver after exposure to estrogenic compounds, for which the toxicological and physiological impact on aquatic organisms has been addressed (Moens et al., 2007b). After aqueously exposing fish to three concentrations of each estrogenic compound for 24 or 96 h, the liver was sampled and microarray analysis was conducted. The results revealed that a total of 185 different gene transcripts were differentially expressed following exposure to at least one of the estrogen(-like) compounds (Moens et al., 2007b). Custom microarrays are powerful tools that are widely used in toxicology under different toxic challenge circumstances, such as exposure to effluents that are a major source of aquatic pollutants and can cause ecotoxicological effects (Moens et al., 2007a), exposure to different concentrations of perfluorooctane sulfonate (Hagenaars et al., 2008), and exposure to a mixture of waterborne and dietary cadmium (Reynders et al., 2006). Microarray analysis provides important information that is necessary to unravel the molecular events and responses related to toxic substance exposure. Chemicals in the aquatic environment not only directly regulate the expression of transcripts but also affect DNA methylation. The effect of pesticides on DNA methylation in the common carp has been examined by investigating the mRNA levels of DNA methyltransferases and the methyl-CpG-binding protein DNA-binding domain protein2, as well as DNA methylation levels in the liver, kidney, and gill of the common carp after exposure to 40-dexposuretoatrazine (ATR) and chlorpyrifos alone or in combination after a 40-d recovery period (Wang et al., 2014b). Additionally, the effects of exposure have also been examined in the brain and gonad (Xing et al., 2015).

Genetic tool development

A number of important genetic tools have been developed over the past few decades. These efforts have provided constant progress and refinement of tools for various purposes in common carp genetic and genomic analysis, including cDNA microarrays, genetic markers, linkage maps, and SNP genotyping arrays.

In the early 2000s, a large set of ESTs were generated using traditional cloning and Sanger sequencing methods (Christoffels et al., 2006; Williams et al., 2008). Subsequently, cDNA microarrays were designed and constructed based on the EST dataset and applied for gene expression profiling and analysis of samples under various experimental treatments (Fraser et al., 2006; Gracey et al., 2004; Moens et al., 2007b). The expression signals are detected based on fluorescent-labeled cDNA probes and their hybridization onto cDNA microarrays. Therefore, the expression analysis is solely based on those "known" genes or sequences from ESTs. The expression signals are "analog" signals that are derived from fluorescent signals on the microarray. With the rapid development of next generation sequencing technologies, microarray platforms are gradually being replaced by "digital" gene expression technologies where gene expression analysis is purely based on the number of sequenced mRNA copies. The most popularly used technologies are RNA-Seq and differential gene expression (DGE) based on the Illumina platform. We will further review the application of RNA-Seq in common carp in the "Functional genomics research and applications in the common carp" section.

With the rapid development of high throughput sequencing technologies, an increasing number of genetic markers have been developed (as discussed in the "Genetic diversity studies" section), thereby providing us with the basis for genetic linkage mapping for various common carp populations. The first generation of common carp genetic mapping was constructed by Sun et al. based on 262 RAPD and SSR markers in 2000 (Sun and Liang, 2000). Thereafter, a number of genetic linkage maps were constructed based on increasing numbers of SSR and SNP markers in different mapping families (Cheng et al., 2010; Wong et al., 2004; Zhang et al., 2013; Zhao et al., 2013). As one of the most important genetic tools, genetic maps have several major applications for genetic and genomic studies. First, quantitative trait loci (QTL) associated with economically important traits can be mapped onto chromosomes or linkage groups, thereby providing the potential for positional cloning of these important traits. In past few decades, the QTLs of important traits, including growth rate, body shape, and meat quality, have been successfully mapped (Liu et al., 2013; Zhang et al., 2011b; Zheng et al., 2013). Second, the genetic maps provide us with a framework for comparative genomics analysis between the common carp and closely related teleost genomes. For instance, a dense genetic map based on 1209 genetic markers was constructed for comparative genomics studies. A total of 620 BAC-anchored markers allowed the inclusion of approximately 30% of the common carp genome onto the genetic map. Comparative genomics analysis with the zebrafish genome demonstrated that 50 common carp chromosomes were homologous with 25 zebrafish chromosomes, thereby confirming the "2 to 1" relationship in the chromosomes of these two species (Zhao et al., 2013). This study also revealed the tetraploid nature of the common carp genome. Third, genetic maps provide the ultimate chromosome framework for genome sequence assembly and scaffolding. The latest version of the common carp genetic map is a high density map based on 3470 high-quality SNPs and 773 microsatellite markers. The sole purpose of this linkage map is genome integration, and 875 Mb of common carp sequence scaffolds were successfully mapped onto the chromosomes (Xu et al., 2014a). For this purpose, an ultrahigh density genetic map is still needed to generate completely integrated genome sequences for the common carp.

To fulfil the purpose of high throughput SNP genotyping and ultrahigh density genetic mapping, the development of a high-throughput SNP genotyping platform is urgently required. This type of platform is also essential for whole genome association studies (GWAS) of important traits, as well as for genome-assisted selection in breeding programs. SNP genotyping arrays have been widely used in human genetics and genetic studies of many important model organisms and agriculture species based on either the Illumina or Affymetrix platforms. Popular SNP genotyping arrays include the human 500K array, the Genome-Wide Human SNP Array 5.0 and 6.0, the porcine 60K SNP array (Ramos et al., 2009), the bovine 50K SNP array

(Matukumalli et al., 2009), the chicken 60K (Groenen et al., 2011) and 600K SNP arrays (Kranis et al., 2013), the canine 22K SNP array (Meurs et al., 2010), and the equine 50K SNP array (McCue et al., 2012). These arrays had been used widely for research on selective sweeps, phylogeny, population structure, copy number variations, GWAS, and other aspects (Boitard and Rocha, 2013; Bourret et al., 2013; McCue et al., 2012; Utsunomiya et al., 2013; Wade et al., 2009), thereby boosting genome and genetic studies and breeding programs of these species. A few such SNP genotyping arrays for aquaculture species have also been recently developed, including arrays for Atlantic salmon (Houston et al., 2014), catfish (Liu et al., 2014b), and rainbow trout (Palti et al., 2015) in addition to our Carp 250K SNP array based on the Affymetrix Axiom platform (Xu et al., 2014b). The SNPs used on the array were selected from two resources: the transcribed sequences from RNA-seq data of four common carp strains and the whole genome resequencing data of five strains. The average interval between two loci of the final 250,000 SNPs was 6.6 kb, and the intervals between most SNPs ranged from 3 to 8 kb, thereby providing a dense "molecular ruler." The Carp 250K SNP array had been validated with 1,072 samples from various common carp populations and strains and showed that 185,150 (74.06%) were polymorphic sites. Genotyping accuracy was validated using genotyping data from a group of full siblings and their parents, and over 99.8% of the qualified SNPs were found to be reliable. The array was also validated with 80 samples from closely related cyprinid species, including *Carassius carassius*, *Ctenopharyngodon idella*, *Mylopharyngodon piceus*, *Hypophthalmichthys molitrix*, *Hypophthalmichthys nobilis*, *Megalobrama amblycephala*, *Danio rerio*, and *Leuciscus waleckii*. This validation identified 54,116 (21.65%) polymorphic sites, suggesting its universal applicability for cyprinid species. With the development of powerful genotyping tools for common carp genetics, a number of research projects, including ultrahigh genetic map construction, GWAS of important traits and a population genetics study, have been designed and performed based on the Carp 250K SNP genotyping array. Most of these projects are still ongoing, and we expect to report the results in the near future.

The high density SNP genotyping array provides high throughput genotyping tools for common carp genetics. High expense and low flexibility are two unavoidable limitations in some applications. To provide the necessary supplement to the Carp 250K SNP array, we developed a flexible low density SNP genotyping platform based on the Fluidigm SNPtype technology. Multiplexing of 48 or 96 SNPs in a single multiplex enables cost-competitive genotyping with a large number of samples. We selected 48 high quality SNPs from the 250K SNP genotyping array that were polymorphic and proved to have high genotyping accuracy. The 48 SNPs were located on 48 chromosomes, which effectively avoided linkage disequilibrium. We applied the flexible Fluidigm array for pedigree assignment in a genetic breeding program of the Yellow River carp and could reconstruct the pedigree of the mixed breeding population in a fast and efficient manner (unpublished results). This type of low throughput genotyping assay is also suitable for various breeding programs. For instance, breeding arrays could be developed based on a set of trait-associated SNP loci after GWAS analysis, which would facilitate effective genetic selection of desired alleles in a large breeding population.

Future directions

Toward the perfect genome: updating the genome and annotation

A reference genome with minimum assembly and annotation mistakes is desired by scientists in the common carp research community. There are at least three aspects that improve genome quality. First, the accuracy at the single base level needs further improvement. The accuracy of reference genomes is important for downstream analysis. Errors in CDS regions might lead to a frame shift or premature stop codons, resulting in protein errors or short proteins, respectively. The aspiration set by the Human Genome Project contained a maximum of one error per 10 kb of finished sequence (Lander et al., 2001), for a maximum error rate of 0.01%. However, due to the sequencing error of second-generation sequencing reads or third-generation sequencing reads for genomes assembled based on Illumina sequencing or other new generation sequencing platforms, the error rate could be greater than 0.01%. Moreover, the genomes assembled based on early 454 sequencing reads would have higher error rates (up to 1.07%) (Gilles et al., 2011). The published common carp genome was sequenced and assembled by largely relying on 454 sequencing reads generated in 2010. Therefore, it is reasonable to predict that the common carp genome may have a higher rate of error bases compared to genomes based on Illumina sequencing reads. A new set of genome sequencing reads based on newly developed sequencing technologies with a higher accuracy rate are desired to improve the genome assembly in the future. Second, the genome integration rate needs to be improved. Previously, we anchored approximately 875 Mb of the common carp genome onto the genetic map, accounting for only half of the whole genome. The low integration rate was largely due to the relatively low marker density of the genetic map, which retained only 4243 genetic markers. The tetraploid genome of the common carp also causes some problems for genome integration with the genetic map. The common carp underwent one more round of WGD compared to most other teleosts. Therefore, the genetic marker might have more than one best-aligned scaffolds or contig, thereby generating controversial mapping regions during the integration that are usually discarded. An ultrahigh density genetic map with more genetic markers would be the best solution to improve the integration rate for the common carp genome. Currently, an ultrahigh density genetic map based on Carp 250K SNP array genotyping is under construction. We expect to produce a map with over 10,000 markers in the near future. Third, updated gene annotation is necessary for the common carp genome. A high-quality dataset of protein-coding genes is essential for functional genomics and evolutionary analyses. Two indicators could be used to estimate the quality of the annotated genes: (1) homologous gene proportions and (2) the gene length distribution among closely related species. It is estimated that approximately 70% of human genes have at least one obvious zebrafish ortholog. Therefore, it is reasonable that the homologous gene proportion between two teleosts should be over 70% (Howe et al., 2013). Indeed, the proportions of homologous genes to other model fish in many published fish genomes was over 80% (Tine et al., 2014; Wu et al., 2014). A homologous gene proportion threshold of 80% is suitable to estimate the quality of annotated genes. Up to June 2015, ten fish genomes were available in the Ensembl genome database (Flicek et al., 2014). A comparison with the protein lengths among these fish revealed that the median protein length ranged from 350 to 430 aa and that the length distributions were similar. Hence, the protein length distribution is a second reliable indicator to measure quality. Although the homologous gene proportion of the V1.0 gene prediction in common carp is as high as 90.84% (47,795 out of 52,610), the median length of the common carp protein is just 303 aa, which is significantly shorter than other teleosts. Hence, it is necessary to improve the completeness of gene prediction of the common carp genome and release the V2.0 genes.

Construction of comprehensive genome databases

Giga data of the common carp genome have been developed, including genome sequences, resequencing, transcriptome sequencing of multiple strains (Xu et al., 2012), small RNA-sequencing (Zhu et al., 2012), genetic maps (Zhao et al., 2013), QTL analysis (Laghari et al., 2015), SNP arrays (Xu et al., 2014b), and BAC-end sequences (Xu et al., 2011b). The genome database of the common carp (named Carpbase) has been developed and used to store all of the aforementioned datasets. The web interface of the database has been constructed to allow data retrieval, blast searches, and genome viewing (http://www.carpbase.org). However, collecting more data and adding new features are necessary to facilitate better data usage in the common carp research community worldwide. First, we will add more genome data from closely related teleosts into the databases, which will enable comparative genomic analysis among the common carp and close species. Second, we will add more commonly used bioinformatics tools to facilitate online analysis of common carp data. Third, the newly generated genome data and updated genome assembly, gene prediction, and annotation will be updated in the databases.

Further develop genome resources from diverse populations

The common carp is one of the most successful species and exhibits great genetic diversity. There are hundreds of geographic populations and domesticated strains worldwide that have adapted to various wild environments with broad ecological spectra, as well as distinct aquaculture patterns in ponds, lakes, and cages. These populations have developed numerous distinct phenotypes in terms of growth rate, temperature and hypoxia tolerance, body color, scale pattern, and body shape, which are in part attributable to genetic diversity due to multiple rounds of WGD events. Although we have collected relatively abundant genetic and genome resources over the past decades for various genomic and genetic studies (especially the complete reference genome and millions of SNP loci of the common carp) (Xu et al., 2012, 2014a,b), we still just see "the tip of the iceberg." Numerous diverse loci in the genome remain undiscovered, many of which are essential resources for genetic breeding and are associated with important traits. Therefore, it is necessary to focus on population genomics and genetic diversity identification. We need to collect sufficient samples representing major wild populations of common carp that harbor rich genetic diversity and collect domesticated strains harboring distinctive traits. Whole genome resequencing with

10- to 20-fold genome coverage of these selected samples will produce millions of markers of genetic diversity, including SNPs and short INDELs. Therefore, the genetic variation map will be constructed for certain populations and strains. However, diversity in terms of single base or several bases would not represent the majority of the genetic diversity. There are still a large number of diverse loci across relatively long genome regions [i.e., large INDELs, copy number variations (CNVs), and chromosome rearrangements] that usually serve as the basis of phenotypic variations and are important for genetic breeding (Li et al., 2014b). However, resequencing strategies based on short reads cannot identify these loci. Therefore, it is necessary to perform pan-genome sequencing and assembly strategies to construct multiple sequencing libraries with various jumping lengths, sequence the genome to achieve 50 to 100-fold coverage, and then assemble the draft genome for certain strains; ultimately, this approach would generate another reference genome for certain strains. Comparative studies between multiple assembled genomes will more comprehensively unveil genetic diversity. The abundant genome resources and tools from diverse populations will facilitate uncovering the genetic basis of important traits, developing new strains with better performance, and understanding the genome and evolution history.

Unveiling the genetic basis of important traits for breeding applications

Dissection of the genetic basis of economically important traits of the common carp is one the major goals of carp geneticists. Genetic mapping and QTL localization followed by positional cloning are the most common genetic approaches used to unveil the genetic basis of target traits and have been successfully applied to genetic studies of crops and farm animals over the past few decades; however, there are few successful instances in aquaculture species, mainly due to the shortage of genetic markers and genome resources. We have developed sufficient genetic markers and a reference genome for the common carp to overcome these limitations over the past few years. The high density SNP array and high throughput genome resequencing will easily generate genome-wide genotyping data for whole genome association and localization. In the next 5–10 years, the genetic bases of important traits of the common carp, including but not limited to traits involved in growth, meat quality, disease resistance, extreme habitat tolerance, reproduction, sex determination, color pattern, and scalation, are expected to be dissected by employing comprehensive approaches, including GWAS, transcriptome analysis, gene ontology and pathway analysis, epigenomics, and gene editing with CRISPR/Cas9 technology. Gradually, the results will facilitate future highly efficient genetic breeding applications.

Concluding remarks

In recent years, the availability of genome resources and genetic tools for the common carp has increased dramatically. The complete genome assembly and gene annotation of the allotetraploidized common carp genome are the milestones. As Winston

Churchill said, "Now this is not the end. It is not even the beginning of the end. However, it is perhaps, the end of the beginning." The abundant genome resources and tools open the doors for further genetic research and breeding on the whole genome scale, thereby providing us with the fundamental resources for genome-based genetic studies and breeding. However, there is still a long journey to the destination of the genetic basis of important phenotypes and economic traits. Ultimately, we will identify the genetic loci, clarify their regulation mechanisms, and characterize the genetic basis behind those important traits and phenotypes. These well-characterized genes will gradually benefit genetic breeding programs, enabling the development of genetically improved breeds for various aquaculture patterns and environments and profiting common carp aquaculture worldwide. This is our ultimate goal for aquaculture genomics studies.

Acknowledgments

We thank Dr. Simon MacKenzie for kindly inviting us to contribute to this book. We acknowledge grant support from the National Natural Science Foundation of China (No. 31422057), National High-Technology Research and Development Program of China (863 program; 2011AA100401), and the Special Scientific Research Funds for Central Non-profit Institutes of Chinese Academy of Fishery Sciences (2014A03YQ01).

References

Abdelkhalek, N.K., Komiya, A., Kato-Unoki, Y., Somamoto, T., Nakao, M., 2009. Molecular evidence for the existence of two distinct IL-8 lineages of teleost CXC-chemokines. Fish Shellfish Immunol. 27, 763–767.

Adamek, M., Syakuri, H., Harris, S., Rakus, K.L., Brogden, G., Matras, M., Irnazarow, I., Steinhagen, D., 2013. Cyprinid herpesvirus 3 infection disrupts the skin barrier of common carp (*Cyprinus carpio L.*). Vet. Microbiol. 162, 456–470.

Adams, M.D., Kelley, J.M., Gocayne, J.D., Dubnick, M., Polymeropoulos, M.H., Xiao, H., Merril, C.R., Wu, A., Olde, B., Moreno, R.F., et al., 1991. Complementary DNA sequencing: expressed sequence tags and human genome project. Science 252, 1651–1656.

Allendorf, F.W., Thorgaard, G.H., 1984. Tetraploidy and the Evolution of Salmonid Fishes, Evolutionary Genetics of Fishes. Springer, US, pp. 1–53.

Balon, E.K., 1995. Origin and domestication of the wild carp, *Cyprinus carpio*: from Roman gourmets to the swimming flowers. Aquaculture 129, 3–48.

Bar, I., Kaddar, E., Velan, A., David, L., 2013. Melanocortin receptor 1 and black pigmentation in the Japanese ornamental carp (*Cyprinus carpio* var. Koi). Front. Genet. 4, 6.

Berthelot, C., Brunet, F., Chalopin, D., Juanchich, A., Bernard, M., Noel, B., Bento, P., Da Silva, C., Labadie, K., Alberti, A., Aury, J.M., Louis, A., Dehais, P., Bardou, P., Montfort, J., Klopp, C., Cabau, C., Gaspin, C., Thorgaard, G.H., Boussaha, M., Quillet, E., Guyomard, R., Galiana, D., Bobe, J., Volff, J.N., Genet, C., Wincker, P., Jaillon, O., Roest Crollius, H., Guiguen, Y., 2014. The rainbow trout genome provides novel insights into evolution after whole-genome duplication in vertebrates. Nat. Commun. 5, 3657.

Birney, E., Clamp, M., Durbin, R., 2004. GeneWise and Genomewise. Genome Res. 14, 988–995.

Boitard, S., Rocha, D., 2013. Detection of signatures of selective sweeps in the Blonde d'Aquitaine cattle breed. Anim. Genet 44, 579–583.

Bostock, J., McAndrew, B., Richards, R., Jauncey, K., Telfer, T., Lorenzen, K., Little, D., Ross, L., Handisyde, N., Gatward, I., 2010. Aquaculture: global status and trends. Phil. Trans. R. Soc. B 365, 2897–2912.

Bourret, V., Kent, M.P., Primmer, C.R., Vasemägi, A., Karlsson, S., Hindar, K., McGinnity, P., Verspoor, E., Bernatchez, L., Lien, S., 2013. SNP-array reveals genome-wide patterns of geographical and potential adaptive divergence across the natural range of Atlantic salmon (Salmo salar). Mol. Ecol. 22, 532–551.

Chang, Y., Sun, X., Liang, L., 2004. Genetic diversity analysis of genomic DNAs of several representative populations of common carp in China. J. Fisheries China 28, 481–486.

Cheng, L., Liu, L., Yu, X., Wang, D., Tong, J., 2010. A linkage map of common carp (*Cyprinus carpio*) based on AFLP and microsatellite markers. Anim. Genet. 41, 191–198.

Chi, C., Jiang, B., Yu, X.B., Liu, T.Q., Xia, L., Wang, G.X., 2014. Effects of three strains of intestinal autochthonous bacteria and their extracellular products on the immune response and disease resistance of common carp, *Cyprinus carpio*. Fish Shellfish Immunol. 36, 9–18.

Christoffels, A., Bartfai, R., Srinivasan, H., Komen, H., Orban, L., 2006. Comparative genomics in cyprinids: common carp ESTs help the annotation of the zebrafish genome. BMC Bioinformatics 7, S2.

Coghlan, A., Fiedler, T.J., McKay, S.J., Flicek, P., Harris, T.W., Blasiar, D., Stein, L.D., 2008. nGASP–the nematode genome annotation assessment project. BMC Bioinformatics 9, 549.

Crooijmans, R.P.M.A., Bierbooms, V.A.F., Komen, J., Van der Poel, J.J., Groenen, M.A.M., 1997. Microsatellite markers in common carp (*Cyprinus carpio L.*). Anim. Genet. 28, 129–134.

David, L., Blum, S., Feldman, M.W., Lavi, U., Hillel, J., 2003. Recent duplication of the common carp (*Cyprinus carpio L.*) genome as revealed by analyses of microsatellite loci. Mol. Biol. Evol. 20, 1425–1434.

David, L., Rajasekaran, P., Fang, J., Hillel, J., Lavi, U., 2001. Polymorphism in ornamental and common carp strains (*Cyprinus carpio L.*) as revealed by AFLP analysis and a new set of microsatellite markers. Mol. Genet. Genomics 266, 353–362.

David, L., Rosenberg, N.A., Lavi, U., Feldman, M.W., Hillel, J., 2007. Genetic diversity and population structure inferred from the partially duplicated genome of domesticated carp, *Cyprinus carpio L.* Genet. Sel. Evol. 39, 319–340.

Davidson, W.S., Koop, B.F., Jones, S.J., Iturra, P., Vidal, R., Maass, A., Jonassen, I., Lien, S., Omholt, S.W., 2010. Sequencing the genome of the Atlantic salmon (Salmo salar). Genome Biol. 11, 403.

Denisov, G., Walenz, B., Halpern, A.L., Miller, J., Axelrod, N., Levy, S., Sutton, G., 2008. Consensus generation and variant detection by Celera Assembler. Bioinformatics 24, 1035–1040.

Dong, C., Xu, J., Wang, B., Feng, J., Jeney, Z., Sun, X., Xu, P., 2015. Phylogeny and evolution of multiple common carp (*Cyprinus carpio L.*) populations clarified by phylogenetic analysis based on complete mitochondrial genomes. Mar. Biotechnol. (NY), 1–11.

FIGIS, 2013. Fisheries Global Information System (FAO-FIGIS)—Web site. Fisheries Global Information System (FIGIS). FI Institutional Websites. In: FAO Fisheries and Aquaculture Department [online]. Rome. Available from: www.fao.org/fishery/figis/en

Flicek, P., Amode, M.R., Barrell, D., Beal, K., Billis, K., Brent, S., Carvalho-Silva, D., Clapham, P., Coates, G., Fitzgerald, S., Gil, L., Giron, C.G., Gordon, L., Hourlier, T., Hunt, S., Johnson, N., Juettemann, T., Kahari, A.K., Keenan, S., Kulesha, E., Martin, F.J., Maurel,

T., McLaren, W.M., Murphy, D.N., Nag, R., Overduin, B., Pignatelli, M., Pritchard, B., Pritchard, E., Riat, H.S., Ruffier, M., Sheppard, D., Taylor, K., Thormann, A., Trevanion, S.J., Vullo, A., Wilder, S.P., Wilson, M., Zadissa, A., Aken, B.L., Birney, E., Cunningham, F., Harrow, J., Herrero, J., Hubbard, T.J., Kinsella, R., Muffato, M., Parker, A., Spudich, G., Yates, A., Zerbino, D.R., Searle, S.M., 2014. Ensembl 2014. Nucleic Acids Res. 42, D749–D755.

Force, A., Lynch, M., Pickett, F.B., Amores, A., Yan, Y.L., Postlethwait, J., 1999. Preservation of duplicate genes by complementary, degenerative mutations. Genetics 151, 1531–1545.

Fraser, J., de Mello, L.V., Ward, D., Rees, H.H., Williams, D.R., Fang, Y., Brass, A., Gracey, A.Y., Cossins, A.R., 2006. Hypoxia-inducible myoglobin expression in nonmuscle tissues. Proc. Natl. Acad. Sci. USA 103, 2977–2981.

Gilles, A., Meglecz, E., Pech, N., Ferreira, S., Malausa, T., Martin, J.F., 2011. Accuracy and quality assessment of 454 GS-FLX Titanium pyrosequencing. BMC Genomics 12, 245.

Gonzalez, S.F., Chatziandreou, N., Nielsen, M.E., Li, W., Rogers, J., Taylor, R., Santos, Y., Cossins, A., 2007. Cutaneous immune responses in the common carp detected using transcript analysis. Mol. Immunol. 44, 1664–1679.

Gracey, A.Y., 2007. Interpreting physiological responses to environmental change through gene expression profiling. J. Exp. Biol. 210, 1584–1592.

Gracey, A.Y., Fraser, E.J., Li, W., Fang, Y., Taylor, R.R., Rogers, J., Brass, A., Cossins, A.R., 2004. Coping with cold: an integrative, multitissue analysis of the transcriptome of a poikilothermic vertebrate. Proc. Natl. Acad. Sci. USA 101, 16970–16975.

Groenen, M., Megens, H.-J., Zare, Y., Warren, W., Hillier, L., Crooijmans, R., Vereijken, A., Okimoto, R., Muir, W., Cheng, H., 2011. The development and characterization of a 60K SNP chip for chicken. BMC Genomics 12, 274.

Haas, B.J., Salzberg, S.L., Zhu, W., Pertea, M., Allen, J.E., Orvis, J., White, O., Buell, C.R., Wortman, J.R., 2008. Automated eukaryotic gene structure annotation using EVidenceModeler and the Program to Assemble Spliced Alignments. Genome Biol. 9, R7.

Hagenaars, A., Knapen, D., Meyer, I.J., van der Ven, K., Hoff, P., De Coen, W., 2008. Toxicity evaluation of perfluorooctane sulfonate (PFOS) in the liver of common carp (*Cyprinus carpio*). Aquat. Toxicol. 88, 155–163.

Hasan, M.R., Hecht, T., De Silva, S., Tacon, A., 2007. Study and analysis of feeds and fertilizers for sustainable aquaculture development. Food Agric. Org. UN, 497–509.

Hieter, P., Boguski, M., 1997. Functional genomics: it's all how you read it. Science 278, 601–602.

Houston, R.D., Taggart, J.B., Cezard, T., Bekaert, M., Lowe, N.R., Downing, A., Talbot, R., Bishop, S.C., Archibald, A.L., Bron, J.E., Penman, D.J., Davassi, A., Brew, F., Tinch, A.E., Gharbi, K., Hamilton, A., 2014. Development and validation of a high density SNP genotyping array for Atlantic salmon (Salmo salar). BMC Genomics 15, 90.

Howe, K., Clark, M.D., Torroja, C.F., Torrance, J., Berthelot, C., Muffato, M., Collins, J.E., Humphray, S., McLaren, K., Matthews, L., McLaren, S., Sealy, I., Caccamo, M., Churcher, C., Scott, C., Barrett, J.C., Koch, R., Rauch, G.J., White, S., Chow, W., Kilian, B., Quintais, L.T., Guerra-Assuncao, J.A., Zhou, Y., Gu, Y., Yen, J., Vogel, J.H., Eyre, T., Redmond, S., Banerjee, R., Chi, J., Fu, B., Langley, E., Maguire, S.F., Laird, G.K., Lloyd, D., Kenyon, E., Donaldson, S., Sehra, H., Almeida-King, J., Loveland, J., Trevanion, S., Jones, M., Quail, M., Willey, D., Hunt, A., Burton, J., Sims, S., McLay, K., Plumb, B., Davis, J., Clee, C., Oliver, K., Clark, R., Riddle, C., Elliot, D., Threadgold, G., Harden, G., Ware, D., Begum, S., Mortimore, B., Kerry, G., Heath, P., Phillimore, B., Tracey, A., Corby, N., Dunn, M., Johnson, C., Wood, J., Clark, S., Pelan, S., Griffiths, G., Smith, M., Glithero, R., Howden, P., Barker, N., Lloyd, C., Stevens, C., Harley, J., Holt, K., Panagiotidis, G., Lovell,

J., Beasley, H., Henderson, C., Gordon, D., Auger, K., Wright, D., Collins, J., Raisen, C., Dyer, L., Leung, K., Robertson, L., Ambridge, K., Leongamornlert, D., McGuire, S., Gilderthorp, R., Griffiths, C., Manthravadi, D., Nichol, S., Barker, G., Whitehead, S., Kay, M., Brown, J., Murnane, C., Gray, E., Humphries, M., Sycamore, N., Barker, D., Saunders, D., Wallis, J., Babbage, A., Hammond, S., Mashreghi-Mohammadi, M., Barr, L., Martin, S., Wray, P., Ellington, A., Matthews, N., Ellwood, M., Woodmansey, R., Clark, G., Cooper, J., Tromans, A., Grafham, D., Skuce, C., Pandian, R., Andrews, R., Harrison, E., Kimberley, A., Garnett, J., Fosker, N., Hall, R., Garner, P., Kelly, D., Bird, C., Palmer, S., Gehring, I., Berger, A., Dooley, C.M., Ersan-Urun, Z., Eser, C., Geiger, H., Geisler, M., Karotki, L., Kirn, A., Konantz, J., Konantz, M., Oberlander, M., Rudolph-Geiger, S., Teucke, M., Lanz, C., Raddatz, G., Osoegawa, K., Zhu, B., Rapp, A., Widaa, S., Langford, C., Yang, F., Schuster, S.C., Carter, N.P., Harrow, J., Ning, Z., Herrero, J., Searle, S.M., Enright, A., Geisler, R., Plasterk, R.H., Lee, C., Westerfield, M., de Jong, P.J., Zon, L.I., Postlethwait, J.H., Nusslein-Volhard, C., Hubbard, T.J., Roest Crollius, H., Rogers, J., Stemple, D.L., 2013. The zebrafish reference genome sequence and its relationship to the human genome. Nature 496, 498–503.

Huminiecki, L., Wolfe, K.H., 2004. Divergence of spatial gene expression profiles following species-specific gene duplications in human and mouse. Genome Res. 14, 1870–1879.

Jaillon, O., Aury, J.M., Brunet, F., Petit, J.L., Stange-Thomann, N., Mauceli, E., Bouneau, L., Fischer, C., Ozouf-Costaz, C., Bernot, A., Nicaud, S., Jaffe, D., Fisher, S., Lutfalla, G., Dossat, C., Segurens, B., Dasilva, C., Salanoubat, M., Levy, M., Boudet, N., Castellano, S., Anthouard, V., Jubin, C., Castelli, V., Katinka, M., Vacherie, B., Biemont, C., Skalli, Z., Cattolico, L., Poulain, J., De Berardinis, V., Cruaud, C., Duprat, S., Brottier, P., Coutanceau, J.P., Gouzy, J., Parra, G., Lardier, G., Chapple, C., McKernan, K.J., McEwan, P., Bosak, S., Kellis, M., Volff, J.N., Guigo, R., Zody, M.C., Mesirov, J., Lindblad-Toh, K., Birren, B., Nusbaum, C., Kahn, D., Robinson-Rechavi, M., Laudet, V., Schachter, V., Quetier, F., Saurin, W., Scarpelli, C., Wincker, P., Lander, E.S., Weissenbach, J., Roest Crollius, H., 2004. Genome duplication in the teleost fish *Tetraodon nigroviridis* reveals the early vertebrate proto-karyotype. Nature 431, 946–957.

Jhingran, V., Pullin, R.S., 1985. A Hatchery Manual for the Common, Chinese, and Indian Major Carps. Asian Development Bank, Manila, Philippines.

Ji, P., Liu, G., Xu, J., Wang, X., Li, J., Zhao, Z., Zhang, X., Zhang, Y., Xu, P., Sun, X., 2012a. Characterization of common carp transcriptome: sequencing, de novo assembly, annotation and comparative genomics. PLoS One 7, e35152.

Ji, P., Zhang, Y., Li, C., Zhao, Z., Wang, J., Li, J., Xu, P., Sun, X., 2012b. High throughput mining and characterization of microsatellites from common carp genome. Int. J. Mol. Sci. 13, 9798–9807.

Jiang, Y., Zhang, S., Xu, J., Feng, J., Mahboob, S., Al-Ghanim, K.A., Sun, X., Xu, P., 2014. Comparative transcriptome analysis reveals the genetic basis of skin color variation in common carp. PLoS One 9, e108200.

Kasahara, M., Naruse, K., Sasaki, S., Nakatani, Y., Qu, W., Ahsan, B., Yamada, T., Nagayasu, Y., Doi, K., Kasai, Y., Jindo, T., Kobayashi, D., Shimada, A., Toyoda, A., Kuroki, Y., Fujiyama, A., Sasaki, T., Shimizu, A., Asakawa, S., Shimizu, N., Hashimoto, S., Yang, J., Lee, Y., Matsushima, K., Sugano, S., Sakaizumi, M., Narita, T., Ohishi, K., Haga, S., Ohta, F., Nomoto, H., Nogata, K., Morishita, T., Endo, T., Shin, I.T., Takeda, H., Morishita, S., Kohara, Y., 2007. The medaka draft genome and insights into vertebrate genome evolution. Nature 447, 714–719.

Kawano, H., Kono, T., Watanuki, H., Savan, R., Sakai, M., 2003. Analysis of genes expressed in head kidney of common carp *Cyprinus carpio L* treated with cortisol. Comp. Biochem. Physiol. B 136, 875–886.

Kohlmann, K., Gross, R., Murakaeva, A., Kersten, P., 2003. Genetic variability and structure of common carp (*Cyprinus carpio*) populations throughout the distribution range inferred from allozyme, microsatellite and mitochondrial DNA markers. Aquat. Living Resour. 16, 421–431.

Kottler, V.A., Künstner, A., Schartl, M., 2015. Pheomelanin in fish? Pigment Cell Melanoma Res. 28, 355–356.

Kranis, A., Gheyas, A.A., Boschiero, C., Turner, F., Yu, L., Smith, S., Talbot, R., Pirani, A., Brew, F., Kaiser, P., Hocking, P.M., Fife, M., Salmon, N., Fulton, J., Strom, T.M., Haberer, G., Weigend, S., Preisinger, R., Gholami, M., Qanbari, S., Simianer, H., Watson, K.A., Woolliams, J.A., Burt, D.W., 2013. Development of a high density 600K SNP genotyping array for chicken. BMC Genomics 14, 59.

Laghari, M.Y., Lashari, P., Zhang, X., Xu, P., Narejo, N.T., Xin, B., Zhang, Y., Sun, X., 2015. QTL mapping for economically important traits of common carp (*Cyprinus carpio L.*). J. Appl. Genet. 56, 65–75.

Lander, E.S., Linton, L.M., Birren, B., Nusbaum, C., Zody, M.C., Baldwin, J., Devon, K., Dewar, K., Doyle, M., FitzHugh, W., Funke, R., Gage, D., Harris, K., Heaford, A., Howland, J., Kann, L., Lehoczky, J., LeVine, R., McEwan, P., McKernan, K., Meldrim, J., Mesirov, J.P., Miranda, C., Morris, W., Naylor, J., Raymond, C., Rosetti, M., Santos, R., Sheridan, A., Sougnez, C., Stange-Thomann, N., Stojanovic, N., Subramanian, A., Wyman, D., Rogers, J., Sulston, J., Ainscough, R., Beck, S., Bentley, D., Burton, J., Clee, C., Carter, N., Coulson, A., Deadman, R., Deloukas, P., Dunham, A., Dunham, I., Durbin, R., French, L., Grafham, D., Gregory, S., Hubbard, T., Humphray, S., Hunt, A., Jones, M., Lloyd, C., McMurray, A., Matthews, L., Mercer, S., Milne, S., Mullikin, J.C., Mungall, A., Plumb, R., Ross, M., Shownkeen, R., Sims, S., Waterston, R.H., Wilson, R.K., Hillier, L.W., McPherson, J.D., Marra, M.A., Mardis, E.R., Fulton, L.A., Chinwalla, A.T., Pepin, K.H., Gish, W.R., Chissoe, S.L., Wendl, M.C., Delehaunty, K.D., Miner, T.L., Delehaunty, A., Kramer, J.B., Cook, L.L., Fulton, R.S., Johnson, D.L., Minx, P.J., Clifton, S.W., Hawkins, T., Branscomb, E., Predki, P., Richardson, P., Wenning, S., Slezak, T., Doggett, N., Cheng, J.F., Olsen, A., Lucas, S., Elkin, C., Uberbacher, E., Frazier, M., Gibbs, R.A., Muzny, D.M., Scherer, S.E., Bouck, J.B., Sodergren, E.J., Worley, K.C., Rives, C.M., Gorrell, J.H., Metzker, M.L., Naylor, S.L., Kucherlapati, R.S., Nelson, D.L., Weinstock, G.M., Sakaki, Y., Fujiyama, A., Hattori, M., Yada, T., Toyoda, A., Itoh, T., Kawagoe, C., Watanabe, H., Totoki, Y., Taylor, T., Weissenbach, J., Heilig, R., Saurin, W., Artiguenave, F., Brottier, P., Bruls, T., Pelletier, E., Robert, C., Wincker, P., Smith, D.R., Doucette-Stamm, L., Rubenfield, M., Weinstock, K., Lee, H.M., Dubois, J., Rosenthal, A., Platzer, M., Nyakatura, G., Taudien, S., Rump, A., Yang, H., Yu, J., Wang, J., Huang, G., Gu, J., Hood, L., Rowen, L., Madan, A., Qin, S., Davis, R.W., Federspiel, N.A., Abola, A.P., Proctor, M.J., Myers, R.M., Schmutz, J., Dickson, M., Grimwood, J., Cox, D.R., Olson, M.V., Kaul, R., Shimizu, N., Kawasaki, K., Minoshima, S., Evans, G.A., Athanasiou, M., Schultz, R., Roe, B.A., Chen, F., Pan, H., Ramser, J., Lehrach, H., Reinhardt, R., McCombie, W.R., de la Bastide, M., Dedhia, N., Blocker, H., Hornischer, K., Nordsiek, G., Agarwala, R., Aravind, L., Bailey, J.A., Bateman, A., Batzoglou, S., Birney, E., Bork, P., Brown, D.G., Burge, C.B., Cerutti, L., Chen, H.C., Church, D., Clamp, M., Copley, R.R., Doerks, T., Eddy, S.R., Eichler, E.E., Furey, T.S., Galagan, J., Gilbert, J.G., Harmon, C., Hayashizaki, Y., Haussler, D., Hermjakob, H., Hokamp, K., Jang, W., Johnson, L.S., Jones, T.A., Kasif, S., Kaspryzk, A., Kennedy, S., Kent, W.J., Kitts, P., Koonin, E.V., Korf, I., Kulp, D., Lancet, D., Lowe, T.M., McLysaght, A., Mikkelsen, T., Moran, J.V., Mulder, N., Pollara, V.J., Ponting, C.P., Schuler, G., Schultz, J., Slater, G., Smit, A.F., Stupka, E., Szustakowski, J., Thierry-Mieg, D., Thierry-Mieg, J., Wagner, L., Wallis, J., Wheeler, R., Williams, A., Wolf, Y.I., Wolfe, K.H., Yang, S.P., Yeh, R.F., Collins, F., Guyer, M.S.,

Peterson, J., Felsenfeld, A., Wetterstrand, K.A., Patrinos, A., Morgan, M.J., de Jong, P., Catanese, J.J., Osoegawa, K., Shizuya, H., Choi, S., Chen, Y.J., 2001. Initial sequencing and analysis of the human genome. Nature 409, 860–921.

Langham, R.J., Walsh, J., Dunn, M., Ko, C., Goff, S.A., Freeling, M., 2004. Genomic duplication, fractionation and the origin of regulatory novelty. Genetics 166, 935–945.

Larhammar, D., Risinger, C., 1994. Molecular genetic aspects of tetraploidy in the common carp *Cyprinus carpio*. Mol. Phylogenet. Evol. 3, 59–68.

Li, Z., Chen, Y., Mu, D., Yuan, J., Shi, Y., Zhang, H., Gan, J., Li, N., Hu, X., Liu, B., Yang, B., Fan, W., 2012. Comparison of the two major classes of assembly algorithms: overlap-layout-consensus and de-bruijn-graph. Brief Funct. Genomics 11, 25–37.

Li, J.T., Hou, G.Y., Kong, X.F., Li, C.Y., Zeng, J.M., Li, H.D., Xiao, G.B., Li, X.M., Sun, X.W., 2015a. The fate of recent duplicated genes following a fourth-round whole genome duplication in a tetraploid fish, common carp (*Cyprinus carpio*). Sci. Rep. 5, 8199.

Li, D., Kang, D., Yin, Q., Sun, X., Liang, L., 2007. Microsatellite DNA marker analysis of genetic diversity in wild common carp (*Cyprinus carpio L.*) populations. J. Genet. Genomics 34, 984–993.

Li, Y., Xu, P., Zhao, Z., Wang, J., Zhang, Y., Sun, X.W., 2011. Construction and characterization of the BAC library for common carp *Cyprinus carpio L.* and establishment of microsynteny with zebrafish Danio rerio. Mar. Biotechnol. (NY) 13, 706–712.

Li, G., Zhao, Y., Liu, Z., Gao, C., Yan, F., Liu, B., Feng, J., 2015b. De novo assembly and characterization of the spleen transcriptome of common carp (*Cyprinus carpio*) using Illumina paired-end sequencing. Fish Shellfish Immunol. 44, 420–429.

Li, G., Zhao, Y., Wen, L., Liu, Z., Yan, F., Gao, C., 2014a. Identification and characterization of microRNAs in the spleen of common carp immune organ. J. Cell Biochem. 115, 1768–1778.

Li, Y.-h., Zhou, G., Ma, J., Jiang, W., Jin, L.-g., Zhang, Z., Guo, Y., Zhang, J., Sui, Y., Zheng, L., 2014b. De novo assembly of soybean wild relatives for pan-genome analysis of diversity and agronomic traits. Nat. Biotechnol 32, 1045–1052.

Liao, X., Yu, X., Tong, J., 2006. Genetic diversity of common carp from two largest Chinese lakes and the Yangtze River revealed by microsatellite markers. Hydrobiologia 568, 445–453.

Liu, J., Liu, Z.Z., Zhao, X.J., Wang, C.H., 2014a. MHC class IIalpha alleles associated with resistance to Aeromonas hydrophila in purse red common carp, *Cyprinus carpio Linnaeus*. J. Fish Dis. 37, 571–575.

Liu, S., Sun, L., Li, Y., Sun, F., Jiang, Y., Zhang, Y., Zhang, J., Feng, J., Kaltenboeck, L., Kucuktas, H., Liu, Z., 2014b. Development of the catfish 250K SNP array for genome-wide association studies. BMC Res. Notes 7, 135.

Liu, J., Zhang, L., Xu, L., Ren, H., Lu, J., Zhang, X., Zhang, S., Zhou, X., Wei, C., Zhao, F., Du, L., 2013. Analysis of copy number variations in the sheep genome using 50K SNP BeadChip array. BMC Genomics 14, 229.

Ludannyi, R.I., Khrisanfova, G.G., Vasi'ev, V.A., Prizenko, V.K., Bogeruk, A.K., Ryskov, A.P., Semenov, S.K., 2006. [Genetic diversity and differentiation of Russian common carp (*Cyprinus carpio L.*) breeds inferred from RAPD markers]. Genetika 42, 1121–1129.

Matukumalli, L.K., Lawley, C.T., Schnabel, R.D., Taylor, J.F., Allan, M.F., Heaton, M.P., O'Connell, J., Moore, S.S., Smith, T.P., Sonstegard, T.S., Van Tassell, C.P., 2009. Development and characterization of a high density SNP genotyping assay for cattle. PLoS One 4, e5350.

McCue, M.E., Bannasch, D.L., Petersen, J.L., Gurr, J., Bailey, E., Binns, M.M., Distl, O., Guerin, G., Hasegawa, T., Hill, E.W., Leeb, T., Lindgren, G., Penedo, M.C., Roed, K.H.,

Ryder, O.A., Swinburne, J.E., Tozaki, T., Valberg, S.J., Vaudin, M., Lindblad-Toh, K., Wade, C.M., Mickelson, J.R., 2012. A high density SNP array for the domestic horse and extant Perissodactyla: utility for association mapping, genetic diversity, and phylogeny studies. PLoS Genet. 8, e1002451.

Meurs, K.M., Mauceli, E., Lahmers, S., Acland, G.M., White, S.N., Lindblad-Toh, K., 2010. Genome-wide association identifies a deletion in the 3' untranslated region of striatin in a canine model of arrhythmogenic right ventricular cardiomyopathy. Hum. Genet. 128, 315–324.

Moens, L.N., van der Ven, K., Van Remortel, P., Del-Favero, J., De Coen, W.M., 2006. Expression profiling of endocrine-disrupting compounds using a customized *Cyprinus carpio* cDNA microarray. Toxicol. Sci. 93, 298–310.

Moens, L.N., Smolders, R., van der Ven, K., van Remortel, P., Del-Favero, J., De Coen, W.M., 2007a. Effluent impact assessment using microarray-based analysis in common carp: a systems toxicology approach. Chemosphere 67, 2293–2304.

Moens, L.N., van der Ven, K., Van Remortel, P., Del-Favero, J., De Coen, W.M., 2007b. Gene expression analysis of estrogenic compounds in the liver of common carp (*Cyprinus carpio*) using a custom cDNA microarray. J. Biochem. Mol. Toxicol. 21, 299–311.

Ohno, S., 1970. Evolution by Gene Duplication. Springer-Verlag, Berlin, New York.

Ohno, S., Muramoto, J., Christian, L., Atkin, N.B., 1967. Diploid-tetraploid relationship among old-world members of the fish family Cyprinidae. Chromosoma 23, 1–9.

Palti, Y., Gao, G., Liu, S., Kent, M.P., Lien, S., Miller, M.R., Rexroad, 3rd, C.E., Moen, T., 2015. The development and characterization of a 57K single nucleotide polymorphism array for rainbow trout. Mol. Ecol. Resour. 15, 662–672.

Phillips, R., Rab, P., 2001. Chromosome evolution in the *Salmonidae* (Pisces): an update. Biol. Rev. Cambr. Phil. Soc. 76, 1–25.

Quan, Y., Li, D., Cao, D., Sun, X., Liang, L., 2006. [Population genetic variation and structure analysis on five populations of mirror carp *Cyprinus carpio L.* using microsatellites]. Yi Chuan 28, 1541–1548.

Rakus, K.L., Irnazarow, I., Adamek, M., Palmeira, L., Kawana, Y., Hirono, I., Kondo, H., Matras, M., Steinhagen, D., Flasz, B., Brogden, G., Vanderplasschen, A., Aoki, T., 2012. Gene expression analysis of common carp (*Cyprinus carpio L.*) lines during Cyprinid herpesvirus 3 infection yields insights into differential immune responses. Dev. Comp. Immunol. 37, 65–76.

Ramos, A.M., Crooijmans, R.P., Affara, N.A., Amaral, A.J., Archibald, A.L., Beever, J.E., Bendixen, C., Churcher, C., Clark, R., Dehais, P., Hansen, M.S., Hedegaard, J., Hu, Z.L., Kerstens, H.H., Law, A.S., Megens, H.J., Milan, D., Nonneman, D.J., Rohrer, G.A., Rothschild, M.F., Smith, T.P., Schnabel, R.D., Van Tassell, C.P., Taylor, J.F., Wiedmann, R.T., Schook, L.B., Groenen, M.A., 2009. Design of a high density SNP genotyping assay in the pig using SNPs identified and characterized by next generation sequencing technology. PLoS One 4, e6524.

Rastogi, S., Liberles, D.A., 2005. Subfunctionalization of duplicated genes as a transition state to neofunctionalization. BMC Evol. Biol. 5, 28.

Reynders, H., van der Ven, K., Moens, L.N., van Remortel, P., De Coen, W.M., Blust, R., 2006. Patterns of gene expression in carp liver after exposure to a mixture of waterborne and dietary cadmium using a custom-made microarray. Aquat. Toxicol. 80, 180–193.

Richárd, B., Sándor, E., Yue, G.H., Kovács, B., Urbányic, B., Tamásd, G., Horváthc, L., Orbán, L., 2003. Genetic analysis of two common carp broodstocks by RAPD and microsatellite markers. Aquaculture 219, 157–167.

Roulin, A., Auer, P.L., Libault, M., Schlueter, J., Farmer, A., May, G., Stacey, G., Doerge, R.W., Jackson, S.A., 2013. The fate of duplicated genes in a polyploid plant genome. Plant J. 73, 143–153.

Salamov, A.A., Solovyev, V.V., 2000. Ab initio gene finding in Drosophila genomic DNA. Genome Res. 10, 516–522.

Semon, M., Wolfe, K.H., 2008. Preferential subfunctionalization of slow-evolving genes after allopolyploidization in Xenopus laevis. Proc. Natl. Acad. Sci. USA 105, 8333–8338.

Singh, A., Sood, N., Chauhan, U.K., Mohindra, V., 2012. EST-based identification of immune-relevant genes from spleen of Indian catfish, Clarias batrachus (Linnaeus, 1758). Gene 502, 53–59.

Stanke, M., Steinkamp, R., Waack, S., Morgenstern, B., 2004. AUGUSTUS: a web server for gene finding in eukaryotes. Nucleic Acids Res. 32, W309–W312.

Sun, X., Liang, L., 2000. A genetic linkage map of common carp. J. Fishery Sci. China 7, 1–5.

Syakuri, H., Adamek, M., Brogden, G., Rakus, K.L., Matras, M., Irnazarow, I., Steinhagen, D., 2013. Intestinal barrier of carp (*Cyprinus carpio L.*) during a cyprinid herpesvirus 3-infection: molecular identification and regulation of the mRNA expression of claudin encoding genes. Fish Shellfish Immunol. 34, 305–314.

Tanck, M.W., Palstra, A.P., van der Weerd, M., Leffering, C.P., van der Poel, J.J., Bovenhuis, H., Komen, J., 2001. Segregation of microsatellite alleles and residual heterozygosity at single loci in homozygous androgenetic common carp (*Cyprinus carpio L.*). Genome 44, 743–751.

Thai, B.T., Burridge, C.P., Pham, T.A., Austin, C.M., 2005. Using mitochondrial nucleotide sequences to investigate diversity and genealogical relationships within common carp (*Cyprinus carpio L.*). Anim. Genet. 36, 23–28.

Tine, M., Kuhl, H., Gagnaire, P.A., Louro, B., Desmarais, E., Martins, R.S., Hecht, J., Knaust, F., Belkhir, K., Klages, S., Dieterich, R., Stueber, K., Piferrer, F., Guinand, B., Bierne, N., Volckaert, F.A., Bargelloni, L., Power, D.M., Bonhomme, F., Canario, A.V., Reinhardt, R., 2014. European sea bass genome and its variation provide insights into adaptation to euryhalinity and speciation. Nat. Commun. 5, 5770.

Trapnell, C., Pachter, L., Salzberg, S.L., 2009. TopHat: discovering splice junctions with RNA-Seq. Bioinformatics 25, 1105–1111.

Trapnell, C., Roberts, A., Goff, L., Pertea, G., Kim, D., Kelley, D.R., Pimentel, H., Salzberg, S.L., Rinn, J.L., Pachter, L., 2012. Differential gene and transcript expression analysis of RNA-seq experiments with TopHat and Cufflinks. Nat. Protoc. 7, 562–578.

Utsunomiya, Y.T., Perez O'Brien, A.M., Sonstegard, T.S., Van Tassell, C.P., do Carmo, A.S., Meszaros, G., Solkner, J., Garcia, J.F., 2013. Detecting loci under recent positive selection in dairy and beef cattle by combining different genome-wide scan methods. PLoS One 8, e64280.

Wade, C.M., Giulotto, E., Sigurdsson, S., Zoli, M., Gnerre, S., Imsland, F., Lear, T.L., Adelson, D.L., Bailey, E., Bellone, R.R., Blocker, H., Distl, O., Edgar, R.C., Garber, M., Leeb, T., Mauceli, E., MacLeod, J.N., Penedo, M.C., Raison, J.M., Sharpe, T., Vogel, J., Andersson, L., Antczak, D.F., Biagi, T., Binns, M.M., Chowdhary, B.P., Coleman, S.J., Della Valle, G., Fryc, S., Guerin, G., Hasegawa, T., Hill, E.W., Jurka, J., Kiialainen, A., Lindgren, G., Liu, J., Magnani, E., Mickelson, J.R., Murray, J., Nergadze, S.G., Onofrio, R., Pedroni, S., Piras, M.F., Raudsepp, T., Rocchi, M., Roed, K.H., Ryder, O.A., Searle, S., Skow, L., Swinburne, J.E., Syvanen, A.C., Tozaki, T., Valberg, S.J., Vaudin, M., White, J.R., Zody, M.C., Lander, E.S., Lindblad-Toh, K., 2009. Genome sequence, comparative analysis, and population genetics of the domestic horse. Science 326, 865–867.

Wang, C.-h., Li, S.-f., 2004. Phylogenetic relationships of ornamental (koi) carp, Oujiang color carp and Long-fin carp revealed by mitochondrial DNA COII gene sequences and RAPD analysis. Aquaculture 231, 83–91.

Wang, J.T., Li, J.T., Zhang, X.F., Sun, X.W., 2012a. Transcriptome analysis reveals the time of the fourth round of genome duplication in common carp (*Cyprinus carpio*). BMC Genomics 13, 96.

Wang, D., Liao, X., Cheng, L., Yu, X., Tong, J., 2007. Development of novel EST-SSR markers in common carp by data mining from public EST sequences. Aquaculture 271, 558–574.

Wang, Y., Tang, H., Debarry, J.D., Tan, X., Li, J., Wang, X., Lee, T.H., Jin, H., Marler, B., Guo, H., Kissinger, J.C., Paterson, A.H., 2012b. MCScanX: a toolkit for detection and evolutionary analysis of gene synteny and collinearity. Nucleic Acids Res. 40, e49.

Wang, C., Wachholtz, M., Wang, J., Liao, X., Lu, G., 2014a. Analysis of the skin transcriptome in two oujiang color varieties of common carp. PLoS One 9, e90074.

Wang, C., Zhang, Z., Yao, H., Zhao, F., Wang, L., Wang, X., Xing, H., Xu, S., 2014b. Effects of atrazine and chlorpyrifos on DNA methylation in the liver, kidney and gill of the common carp (*Cyprinus carpio L.*). Ecotoxicol. Environ. Saf. 108, 142–151.

Wei, D., Lou, Y., Sun, X., Shen, J., 2001. Isolation of microsatellite markers in the common carp (*Cyprinus carpio*). Zool. Res. 22, 238–241.

Welcomme, R.L., 1988. International Introductions of Inland Aquatic Species. Food & Agriculture Organization, Rome, Italy.

Williams, D.R., Li, W., Hughes, M.A., Gonzalez, S.F., Vernon, C., Vidal, M.C., Jeney, Z., Jeney, G., Dixon, P., McAndrew, B., Bartfai, R., Orban, L., Trudeau, V., Rogers, J., Matthews, L., Fraser, E.J., Gracey, A.Y., Cossins, A.R., 2008. Genomic resources and microarrays for the common carp *Cyprinus carpio L.* J. Fish Biol. 72, 2095–2117.

Wolf, U., Ritter, H., Atkin, N., Ohno, S., 1969. Polyploidization in the fish family Cyprinidae, order Cypriniformes. Humangenetik 7, 240–244.

Wolfe, K.H., 2001. Yesterday's polyploids and the mystery of diploidization. Nat. Rev. Genet. 2, 333–341.

Wolfe, K.H., Shields, D.C., 1997. Molecular evidence for an ancient duplication of the entire yeast genome. Nature 387, 708–713.

Wong, G.K., Liu, B., Wang, J., Zhang, Y., Yang, X., Zhang, Z., Meng, Q., Zhou, J., Li, D., Zhang, J., Ni, P., Li, S., Ran, L., Li, H., Li, R., Zheng, H., Lin, W., Li, G., Wang, X., Zhao, W., Li, J., Ye, C., Dai, M., Ruan, J., Zhou, Y., Li, Y., He, X., Huang, X., Tong, W., Chen, J., Ye, J., Chen, C., Wei, N., Dong, L., Lan, F., Sun, Y., Yang, Z., Yu, Y., Huang, Y., He, D., Xi, Y., Wei, D., Qi, Q., Li, W., Shi, J., Wang, M., Xie, F., Zhang, X., Wang, P., Zhao, Y., Li, N., Yang, N., Dong, W., Hu, S., Zeng, C., Zheng, W., Hao, B., Hillier, L.W., Yang, S.P., Warren, W.C., Wilson, R.K., Brandstrom, M., Ellegren, H., Crooijmans, R.P., van der Poel, J.J., Bovenhuis, H., Groenen, M.A., Ovcharenko, I., Gordon, L., Stubbs, L., Lucas, S., Glavina, T., Aerts, A., Kaiser, P., Rothwell, L., Young, J.R., Rogers, S., Walker, B.A., van Hateren, A., Kaufman, J., Bumstead, N., Lamont, S.J., Zhou, H., Hocking, P.M., Morrice, D., de Koning, D.J., Law, A., Bartley, N., Burt, D.W., Hunt, H., Cheng, H.H., Gunnarsson, U., Wahlberg, P., Andersson, L., Kindlund, E., Tammi, M.T., Andersson, B., Webber, C., Ponting, C.P., Overton, I.M., Boardman, P.E., Tang, H., Hubbard, S.J., Wilson, S.A., Yu, J., Yang, H., 2004. A genetic variation map for chicken with 2.8 million single-nucleotide polymorphisms. Nature 432, 717–722.

Wu, C., Zhang, D., Kan, M., Lv, Z., Zhu, A., Su, Y., Zhou, D., Zhang, J., Zhang, Z., Xu, M., Jiang, L., Guo, B., Wang, T., Chi, C., Mao, Y., Zhou, J., Yu, X., Wang, H., Weng, X., Jin, J.G., Ye, J., He, L., Liu, Y., 2014. The draft genome of the large yellow croaker reveals well-developed innate immunity. Nat. Commun. 5, 5227.

Xing, H., Wang, C., Wu, H., Chen, D., Li, S., Xu, S., 2015. Effects of atrazine and chlorpyrifos on DNA methylation in the brain and gonad of the common carp. Comp. Biochem. Physiol. C 168, 11–19.

Xu, J., Ji, P., Zhao, Z., Zhang, Y., Feng, J., Wang, J., Li, J., Zhang, X., Zhao, L., Liu, G., Xu, P., Sun, X., 2012. Genome-wide SNP discovery from transcriptome of four common carp strains. PLoS One 7, e48140.

Xu, P., Li, J., Li, Y., Cui, R., Wang, J., Wang, J., Zhang, Y., Zhao, Z., Sun, X., 2011a. Genomic insight into the common carp (*Cyprinus carpio*) genome by sequencing analysis of BAC-end sequences. BMC Genomics 12, 188.

Xu, P., Li, J., Li, Y., Cui, R., Wang, J., Zhang, Y., Zhao, Z., Sun, X., 2011b. Genomic insight into the common carp (*Cyprinus carpio*) genome by sequencing analysis of BAC-end sequences. BMC Genomics 12, 188.

Xu, P., Zhang, X., Wang, X., Li, J., Liu, G., Kuang, Y., Xu, J., Zheng, X., Ren, L., Wang, G., Zhang, Y., Huo, L., Zhao, Z., Cao, D., Lu, C., Li, C., Zhou, Y., Liu, Z., Fan, Z., Shan, G., Li, X., Wu, S., Song, L., Hou, G., Jiang, Y., Jeney, Z., Yu, D., Wang, L., Shao, C., Song, L., Sun, J., Ji, P., Wang, J., Li, Q., Xu, L., Sun, F., Feng, J., Wang, C., Wang, S., Wang, B., Li, Y., Zhu, Y., Xue, W., Zhao, L., Wang, J., Gu, Y., Lv, W., Wu, K., Xiao, J., Wu, J., Zhang, Z., Yu, J., Sun, X., 2014a. Genome sequence and genetic diversity of the common carp, *Cyprinus carpio*. Nat. Genet. 46, 1212–1219.

Xu, J., Zhao, Z., Zhang, X., Zheng, X., Li, J., Jiang, Y., Kuang, Y., Zhang, Y., Feng, J., Li, C., Yu, J., Li, Q., Zhu, Y., Liu, Y., Xu, P., Sun, X., 2014b. Development and evaluation of the first high-throughput SNP array for common carp (*Cyprinus carpio*). BMC Genomics 15, 307.

Yan, B., Liu, B., Zhu, C.D., Li, K.L., Yue, L.J., Zhao, J.L., Gong, X.L., Wang, C.H., 2013. microRNA regulation of skin pigmentation in fish. J. Cell Sci. 126, 3401–3408.

Zhang, H., Okamoto, N., Ikeda, Y., 1995. Two c-myc genes from a tetraploid fish, the common carp (*Cyprinus carpio*). Gene 153, 231–236.

Zhang, Y., Stupka, E., Henkel, C.V., Jansen, H.J., Spaink, H.P., Verbeek, F.J., 2011a. Identification of common carp innate immune genes with whole-genome sequencing and RNA-Seq data. J. Integr. Bioinform. 8, 169.

Zhang, Y., Xu, P., Lu, C., Kuang, Y., Zhang, X., Cao, D., Li, C., Chang, Y., Hou, N., Li, H., 2011b. Genetic linkage mapping and analysis of muscle fiber-related QTLs in common carp (*Cyprinus carpio L.*). Mar. Biotechnol. (NY) 13, 376–392.

Zhang, X., Zhang, Y., Zheng, X., Kuang, Y., Zhao, Z., Zhao, L., Li, C., Jiang, L., Cao, D., Lu, C., Xu, P., Sun, X., 2013. A consensus linkage map provides insights on genome character and evolution in common carp (*Cyprinus carpio L.*). Mar. Biotechnol. (NY) 15, 275–312.

Zhao, L., Zhang, Y., Ji, P., Zhang, X., Zhao, Z., Hou, G., Huo, L., Liu, G., Li, C., Xu, P., Sun, X., 2013. A dense genetic linkage map for common carp and its integration with a BAC-based physical map. PLoS One 8, e63928.

Zheng, X., Kuang, Y., Lv, W., Cao, D., Zhang, X., Li, C., Lu, C., Sun, X., 2013. A consensus linkage map of common carp (*Cyprinus carpio L.*) to compare the distribution and variation of QTLs associated with growth traits. Sci. China Life Sci. 56, 351–359.

Zhou, J., Wu, Q., Wang, Z., Ye, Y., 2004. Molecular phylogeny of three subspecies of common carp *Cyprinus carpio*, based on sequence analysis of cytochrome b and control region of mtDNA. J. Zool. Syst. Evol. Res. 42, 266–269.

Zhu, Y.P., Xue, W., Wang, J.T., Wan, Y.M., Wang, S.L., Xu, P., Zhang, Y., Li, J.T., Sun, X.W., 2012. Identification of common carp (*Cyprinus carpio*) microRNAs and microRNA-related SNPs. BMC Genomics 13, 413.

Future perspective

11

Simon MacKenzie, Sissel Jentoft**,†*
*Institute of Aquaculture, University of Stirling, Stirling, United Kingdom;
**Department of Biosciences, Centre for Ecological and Evolutionary Synthesis (CEES), University of Oslo, Oslo, Norway; †Department of Natural Sciences, University of Agder, Kristiansand, Norway

From the excellent contributions throughout this book it has become very apparent how genomics-driven research is impacting upon aquaculture. Moving through the chapters from the cutting edge technologies developed in the zebrafish to their translational application highlighted, for example, in the turbot exemplify how genomics is shaping fish biology and aquaculture development. Genomics research has attained a worldwide impact in aquaculture contributing to sustainability and food security across our continents.

Huge and continuing technological advances in the field of sequencing has "opened the door" to many scientists in the field of aquaculture to vast swathes of information at the nucleotide level. This has greatly facilitated the emergence and in some cases application of marker assisted selection (MAS) programs that will significantly impact aquaculture production in the coming decades. Indeed commercial fish breeding is being revolutionized by NGS technologies that have the potential to uncover increasingly more important production traits such as disease resistance and tolerance to environmental variations. An excellent example is the infectious pancreatic necrosis (IPN) QTL in Atlantic salmon that has been successfully exploited in the aquaculture industry. Furthermore the importance of these increasingly accurate analyses can effectively reduce mortalities under classical selection regimes thus incorporating a 3R's approach to selection regimes.

However as commented throughout this volume there remains much work to do in order to unlock the potential of genomics in aquaculture. A major issue commented on throughout is the pressing need to increase the quality of annotations in target species genomes. Whole genome resources are central to developments in animal breeding programs and provide the foundations to build informed MAS programs. Critical advances described by Jentoft and colleagues addressing the assembly of the latest cod genome describe the current state of the art and provide a framework for future efforts in assembly and annotation. Here the integrated use of different sequencing technologies plays a key role in this strategy and is critical in order to advance our understanding of fish genomes and their relationships both within and between species.

Translational developments directed to breeding programs can be expected in the future as economic resequencing of the whole genomes becomes widespread. The development of high-throughput genotyping platforms devoted to understanding variation across entire genomes will fuel such developments and become key for commercial success in intensive systems. The effective explosion of available SNP markers

Genomics in Aquaculture. http://dx.doi.org/10.1016/B978-0-12-801418-9.00011-1

for many species and their respective reference genomes will drive optimization of genotyping-by-sequencing (GBS) technologies for whole-genome profiling. In this case genotyping and SNP discovery are simultaneous facilitating genome-wide association studies and QTL development by marker discovery. Such information will lead to the further development of other cost-effective genotyping platforms, possibly in array-type formats, for routine applications in the industry.

A much commented on theme throughout the chapters of the book and highlighted by Yann Guiguen and colleagues in Chapter 3 is the necessity to integrate rapidly expanding resources in their respective species repositories and across species in order to fully exploit the resources generated by our community. Initiatives such as the Functional Annotation of Animal Genomes resource (http://www.faang.org) which is the equivalent of the ENCODE Human genome should be supported. Related programs such as DANIO-CODE will also further enhance these multilayered resources. Increasingly a major constraint to individual team and smaller networks is to maintain skilled staff and construct adequate infrastructures to carry out the required bioinformatics analyses. Indeed this has been flagged up as a major issue in view of training staff, accessing facilities, pooling efforts, and being able to maintain an organized structure around genomics research in aquaculture and indeed in most fields of research. As sequencing and sequencing service providers provide increasing outputs in line with decreasing costs the cost of maintaining a skilled technical staff will increase. The costs involved for the industrial sector to develop analytical pipelines and dedicated workflows to streamline MAS programs and manage data will be of utmost importance in order to exploit the potential of genomics research in the aquaculture industry. Such costs are likely prohibitive and therefore reemphasize the importance of academia-industry consortia working together to produce a sustainable industry enhancing food security in the aquaculture industry.

The future contribution of the model organism, *Danio rerio*, to improving selective breeding among other critical areas in aquaculture has been extensively described by Alestrom and Winther-Larsen. The addition of a SWOT analysis here is an excellent exercise for those wishing to further examine the potential of the zebrafish system and understand the cost benefits of developing research in this organism. The availability of the CASPR/Cas9 system and its relative ease of application will undoubtedly change how we approach genome biology over the next few decades as functional analyses come to the forefront. Human biomedical initiatives such as The Zebrafish Mutation Project (http://www.sanger.ac.uk/resources/zebrafish/zmp) that aims to develop knockout for each allele in the zebrafish, one of the 26,000 genes carrying a knock-out mutation, looks at disease models and will provide a robust source of information directly relevant to fish biology and aquaculture. Bridging the genotype–phenotype gap in aquaculture to fuel sustainable aquaculture requires crossover and discussion between these often separated sets of researchers. From a technological viewpoint advances and novel methodologies are developed in the laboratory animal models that "trickle-down" to aquaculture researchers. Current advances in epigenomics, transcriptomics, and functional analyses in laboratory models will provide the basic resources required for comparative epigenomic studies, for example. How these studies translate into an industrial aquaculture setting, however, remains a significant challenge to the field. However it would appear that zebrafish research will likely play an increasingly important role in genomic research aimed at aquaculture.

Aquaculture production systems are very diverse and are composed of very different markets and outcomes dependent upon geographical location and extensive versus intensive culture. As highlighted in the Tilapia chapter by McAndrew and colleagues it is highly likely that for the moment genomics-driven research in aquaculture will benefit the intensive production systems such as the Atlantic salmon. In such production systems genetic gain can have a significant impact on growth potential or disease resistance. However as highlighted throughout this book rapid advances are being made with many aquacultured species. At the current time, there are more than 100 species being cultured across our planet including finfish, crustaceans, and shellfish. The contribution to global food security of aquaculture and associated products is expected to significantly increase in the future and we believe genomics-based research to be a major contributor to driving this blue revolution.

Index

D

H

I

J

K

L

N

O

P

Kolibri 360 GmbH
Ferdinand-Jühlke-Straße 7
99095 Erfurt, Deutschland
produktsicherheit@kolibri360.de